Computer Solution
of Large Sparse
Positive Definite Systems

Computer Solution
of Large Sparse
Positive Definite Systems

Alan George
University of Waterloo
Waterloo, Ontario

Joseph W-H Liu

Prentice-Hall, Inc.
Englewood Cliffs, New Jersey 07632

Library of Congress Cataloging in Publication Data

George, Alan.
    Computer solution of large sparse positive definite systems.

    (Prentice-Hall series in computational mathematics)
    Bibliography: p. 314
    Includes index.
    1. Sparse matrices–Data processing. 2. Fortran
(Computer program language) I. Liu, Joseph W. H.,
joint author. II. Title. III. Series.

QA188.G46   512.9'434   80–10357
ISBN 0–13–165274–5

PRENTICE-HALL SERIES IN COMPUTATIONAL MATHEMATICS
Cleve Moler, advisor

Printed in the United States of America

10 9 8 7 6 5 4 3 2 1

Editorial/production supervision: Nancy Milnamow
Cover design: Edsal Enterprises
Manufacturing buyer: Joyce Levatino and Gordon Osbourne

PRENTICE-HALL INTERNATIONAL, INC., *London*
PRENTICE-HALL OF AUSTRALIA PTY. LIMITED, *Sydney*
PRENTICE-HALL OF CANADA, LTD., *Toronto*
PRENTICE-HALL OF INDIA PRIVATE LIMITED, *New Delhi*
PRENTICE-HALL OF JAPAN, INC., *Tokyo*
PRENTICE-HALL OF SOUTHEAST ASIA PTE. LTD., *Singapore*
WHITEHALL BOOKS LIMITED, *Wellington, New Zealand*

Contents

**Chapter 6/Quotient Tree Methods for Finite Element
and Finite Difference Problems**

Chapter 7/One-Way Dissection Methods for Finite Element Problems

Preface

This book is intended to introduce the reader to the important practical problem of solving large systems of sparse linear equations on a computer. The problem has many facets, from fundamental questions about the inherent complexity of certain problems, to less precisely specified questions about the design of efficient data structures and computer programs. In order to limit the size of the book, and yet consider the problems in detail, we have restricted our attention to symmetric positive definite systems of equations. Such problems are very common, arising in numerous fields of science and engineering. For similar reasons, we have limited our treatment to one specific method for each general approach to solving large sparse positive definite systems. For example, among the numerous methods for approximately minimizing the bandwidth of a matrix, we have selected only one, which through our experience has appeared to perform well. Our objective is to expose the reader to the important ideas, rather than the method which is necessarily best for his particular problem. Our hope is that someone familiar with the contents of the book could make sound judgements about the applicability and appropriateness of proposed ideas and methods for solving sparse systems.

The quality of the computer implementation of sparse matrix algorithms can have a profound effect on their performance, and the difficulty of implementation varies a great deal from one algorithm to another. Thus, while "paper and pencil" analysis of sparse matrix algorithms are useful and important, they are not enough. Our view is that studying and using subroutines which implement these algorithms is an essential component in a good introduction to this important area of scientific computation. To this end, we provide listings of Fortran subroutines, and discuss them in detail. The procedure for obtaining machine readable copies of these is provided in Appendix A.

We are grateful to Mary Wang for doing a superb job of typing the original manuscript, and to Anne Trip de Roche and Heather Pente for coping with our numerous revisions. We are also grateful to the many students who debugged early versions of the manuscript, and in particular to Mr. Hamza Rachwan and Mr. Esmond Ng for a careful reading of the final manuscript.

Writing a book consumes time that might otherwise be spent with ones wife and children. We are grateful to our wives for their patience and understanding, and we dedicate this book to them.

Alan George
Joseph W-H Liu

1/ Introduction

1.0 About the Book

This book deals with efficient computer methods for solving large sparse systems of linear algebraic equations. The reader is assumed to have a basic knowledge of linear algebra, and should be familiar with standard matrix notation and manipulation. Some basic knowledge of graph theory notation would be helpful, but it is not required since all the relevant notions and notations are introduced as they are needed.

This is a book about *computing*, and it contains numerous Fortran subroutines which are to be studied and used. Thus, the reader should have at least a basic understanding of Fortran, and ideally one should have access to a computer to execute programs using the subroutines in the book. The success of algorithms for sparse matrix computations, perhaps more than in any other area of numerical computation, depends on the quality of their *computer implementation*; i.e., the computer program which executes the algorithm. Implementations of these algorithms characteristically involve fairly complicated storage schemes, and the degree of complication varies substantially for different algorithms. Some algorithms which appear extremely attractive "on paper" may be much less so in practice because their implementation is complicated and inefficient. Other less theoretically attractive algorithms may be more desirable in practical terms because their implementation is simple and incurs very little "overhead."

For these and other reasons which will be apparent later, we have included Fortran subroutines which implement many of the important algorithms discussed in the book. We have also included some numerical experiments which illustrate the implementation issues noted above, and which provide the reader with some information about the absolute time and storage that sparse matrix computations require on a typical computer. The subroutines have been carefully tested, and are written in a portable subset of Fortran (Ryder 1974). Thus, they should execute correctly on most computer systems without any changes. They would be a useful addition to the library of any computer center which does scientific computing. Machine readable

1

copies of the subroutines, along with the test problems described and used in Chapter 9, are available from the authors.

Our hope is that this book will be valuable in at least two capacities. First, it can serve as a text for senior or graduate students in computer science or engineering. The exercises at the end of each chapter are designed to test the reader's understanding of the material, to provide avenues for further investigation, and to suggest some important research problems. Some of the exercises involve using and/or changing the programs we provide, so it is desirable to have access to a computer which supports the Fortran language, and to have the programs available in a computer library.

This book should also serve as a useful reference for all scientists and engineers involved in solving large sparse positive definite matrix problems. Although this class of problems is special, a substantial fraction (perhaps the majority) of linear equation problems arising in science and engineering have this property. It is a large enough class to warrant separate treatment. In addition, as we shall see later, the solution of sparse problems with this property is fundamentally different from that for the general case.

1.1   Cholesky's Method and the Ordering Problem

All the methods we discuss in this book are based on a single numerical algorithm known as *Cholesky's method*, a symmetric variant of Gaussian elimination tailored to symmetric positive definite matrices. We shall define this class of matrices and describe the method in detail in Section 2.1. Suppose the given system of equations to be solved is

$$Ax = b, \qquad (1.1.1)$$

where A is an N by N, symmetric, positive definite *coefficient matrix*, b is a vector of length N, called the *right hand side*, and x is the *solution vector* of length N, whose components are to be computed. Applying Cholesky's method to A yields the *triangular factorization*

$$A = LL^T, \qquad (1.1.2)$$

where L is *lower triangular* with positive diagonal elements. A matrix M is lower {upper} triangular if $m_{ij} = 0$ for $i < j$ $\{i > j\}$. The superscript T indicates the *transpose* operation. In Section 2.1 we show that such a factorization always exists when A is symmetric and positive definite.

Using (1.1.2) in (1.1.1) we have

$$LL^Tx = b, \qquad (1.1.3)$$

and by substituting $y = L^T x$, it is clear we can obtain x by solving the triangular systems

$$Ly = b, \tag{1.1.4}$$

and

$$L^T x = y. \tag{1.1.5}$$

As an example, consider the problem

$$
\begin{pmatrix}
4 & 1 & 2 & \frac{1}{2} & 2 \\
1 & \frac{1}{2} & 0 & 0 & 0 \\
2 & 0 & 3 & 0 & 0 \\
\frac{1}{2} & 0 & 0 & \frac{5}{8} & 0 \\
2 & 0 & 0 & 0 & 16
\end{pmatrix}
\begin{pmatrix}
x_1 \\ x_2 \\ x_3 \\ x_4 \\ x_5
\end{pmatrix}
=
\begin{pmatrix}
7 \\ 3 \\ 7 \\ -4 \\ -4
\end{pmatrix}. \tag{1.1.6}
$$

The Cholesky factor of the coefficient matrix of (1.1.6) is given by

$$
L =
\begin{pmatrix}
2 & & & & 0 \\
0.50 & 0.50 & & & \\
1 & -1 & 1 & & \\
0.25 & -0.25 & -0.50 & 0.50 & \\
1 & -1 & -2 & -3 & 1
\end{pmatrix}. \tag{1.1.7}
$$

Solving $Ly = b$, we obtain

$$
y =
\begin{pmatrix}
3.5 \\ 2.5 \\ 6 \\ -2.5 \\ -0.50
\end{pmatrix},
$$

and then solving $L^T x = y$ yields

$$
x =
\begin{pmatrix}
2 \\ 2 \\ 1 \\ -8 \\ -0.50
\end{pmatrix}.
$$

The example above illustrates the most important fact about applying Cholesky's method to a sparse matrix A: the matrix usually suffers *fill-in*. That is, L has nonzeros in positions which are zero in the lower triangular part of A.

Now suppose we relabel the variables x_i according to the recipe $x_i \rightarrow \tilde{x}_{5-i+1}, i = 1, 2, \ldots, 5$, and rearrange the equations so that the last one becomes the first, the second last becomes the second, and so on, with the first equation finally becoming the last one. We then obtain the equivalent system of equations (1.1.8).

$$
\begin{pmatrix}
16 & 0 & 0 & 0 & 2 \\
0 & \frac{5}{8} & 0 & 0 & \frac{1}{2} \\
0 & 0 & 3 & 0 & 2 \\
0 & 0 & 0 & \frac{1}{2} & 1 \\
2 & \frac{1}{2} & 2 & 1 & 4
\end{pmatrix}
\begin{pmatrix}
\tilde{x}_1 \\
\tilde{x}_2 \\
\tilde{x}_3 \\
\tilde{x}_4 \\
\tilde{x}_5
\end{pmatrix}
=
\begin{pmatrix}
-4 \\
-4 \\
7 \\
3 \\
7
\end{pmatrix}
\qquad (1.1.8)
$$

It should be clear that this relabelling of the variables and reordering of the equations amounts to a symmetric permutation of the rows and columns of A, with the same permutation applied to b. We refer to this new system as $\tilde{A}\tilde{x} = \tilde{b}$. Using Cholesky's method on this new system as before, we factor $\tilde{A}$ into $\tilde{L}\tilde{L}^T$, obtaining (to three significant figures)

$$
\tilde{L} =
\begin{pmatrix}
4 & & & & \mathbf{0} \\
0 & 0.791 & & & \\
0 & 0 & 1.73 & & \\
0 & 0 & 0 & 0.707 & \\
0.500 & 0.632 & 1.15 & 1.41 & 0.129
\end{pmatrix}.
$$

Solving $\tilde{L}\tilde{y} = \tilde{b}$ and $\tilde{L}^T\tilde{x} = \tilde{y}$ yields the solution $\tilde{x}$, which is simply a rearranged form of x. The crucial point is that our reordering of the equations and variables provided a triangular factor $\tilde{L}$ which is now *just as sparse as the lower triangle of A*. Although it is rarely possible in practice to achieve this, for most sparse matrix problems a judicious reordering of the rows and columns of the coefficient matrix can lead to enormous reductions in fill-in, and hence savings in computer execution time and storage (assuming of course that sparsity is exploited.) The study of algorithms which automatically perform this reordering process is one of the major topics of this book, along with a study of effective computational and storage schemes for the sparse factors $\tilde{L}$ that these reorderings provide.

The 5 by 5 matrix example above illustrates the basic characteristics of sparse elimination and the effect of reordering. To emphasize these points, we consider a somewhat larger example, the zero-nonzero pattern of which is given in Figure 1.1.1. On factoring this matrix into LL^T, we obtain the structure shown in Figure 1.1.2. Evidently the matrix in its present ordering is not good for sparse elimination, since it has suffered a lot of fill.

Figure 1.1.3 and 1.1.5 display the structure of two symmetric permutations A' and A'' of the matrix A whose structure is shown in Figure 1.1.1. The structure of their Cholesky factors L' and L'' is shown in Figures 1.1.4 and 1.1.6 respectively. The matrix A' has been permuted into so-called *band form*, to be discussed in Chapter 4. The matrix A'' has been ordered to reduce fill-in; a method for obtaining this type of ordering is the topic of Chapter 5. The number of nonzeros in L, L' and L'' is 369, 189, and 177 respectively.

As our example shows, some orderings can lead to dramatic reductions in the amount of fill, or confine it to certain specific parts of L which can be easily stored. This task of finding a "good" ordering, which we refer to as the "ordering problem," is central to the study of the solution of sparse positive definite systems.

Figure 1.1.1   Nonzero pattern of a 35 by 35 matrix *A*.

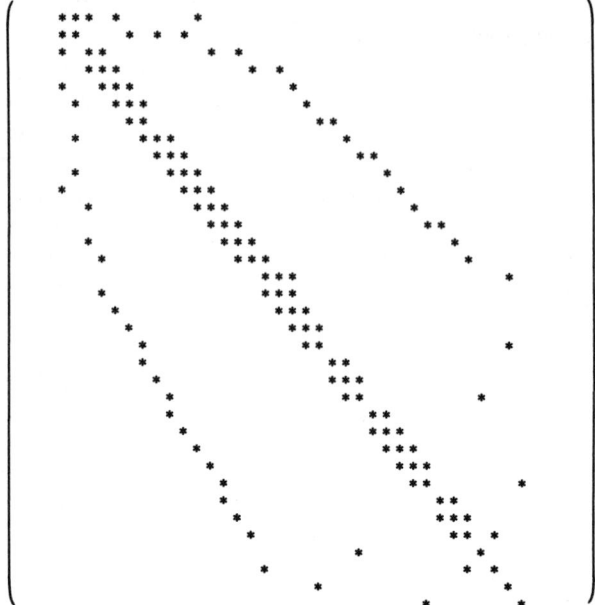

Figure 1.1.2   Nonzero pattern of the Cholesky factor *L* for the matrix whose structure is shown in Figure 1.1.1.

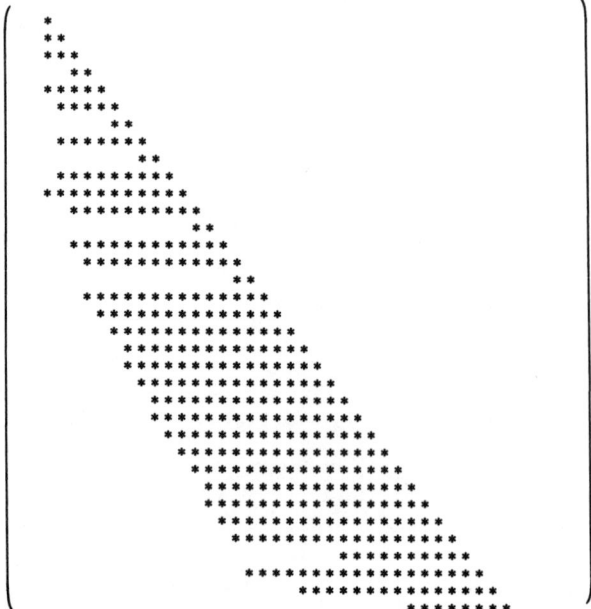

Figure 1.1.3   The structure of A', a symmetric permutation of the matrix A whose structure is shown in Figure 1.1.1.

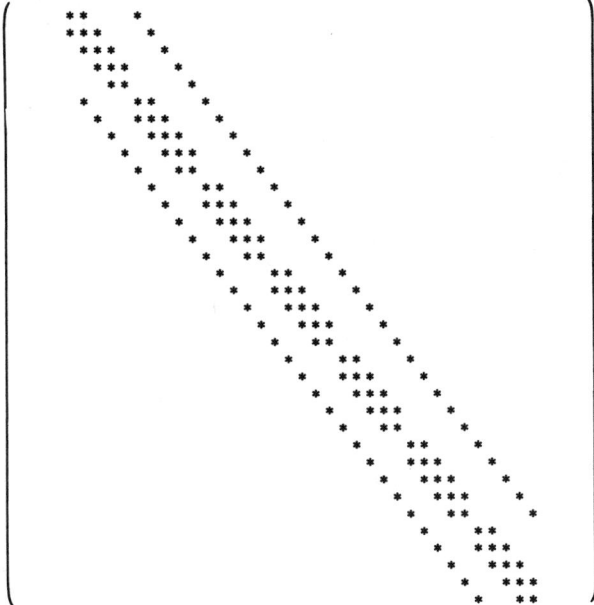

Figure 1.1.4   The structure of L', the Cholesky factor of A', whose structure is shown in Figure 1.1.3.

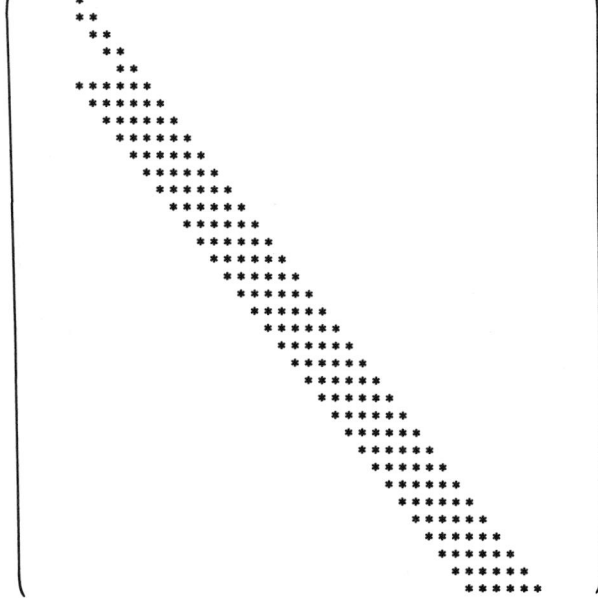

Figure 1.1.5   The structure of A'', a symmetric permutation of the matrix A whose structure is shown in Figure 1.1.1.

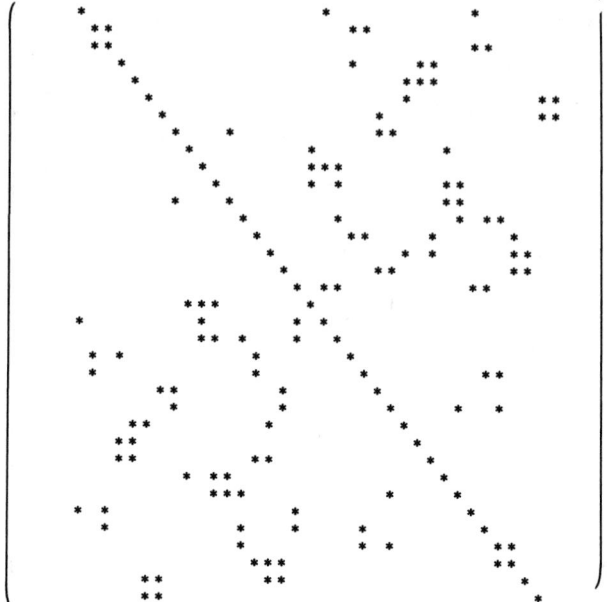

Figure 1.1.6   The structure of L'', the Cholesky factor of A'' whose structure is shown in Figure 1.1.5.

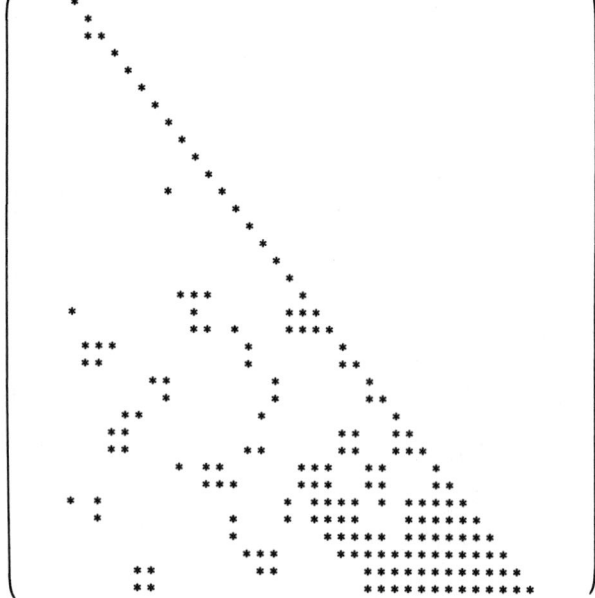

1.2   Positive Definite Versus Indefinite Matrix Problems

In this book we deal exclusively with the case when A is symmetric and positive definite. As we noted earlier, a substantial portion of linear equation problems arising in science and engineering have this property, and the ordering problem is both different from and easier than that for general sparse A. For a general indefinite sparse matrix A, some form of *pivoting* (row and/or column interchanges) is necessary to ensure numerical stability (Forsythe 1967). Thus given A, one normally obtains a factorization of PA or PAQ, where P and Q are permutation matrices of the appropriate size. (Note that the application of P on the left permutes the rows of A, and the application of Q on the right permutes the columns of A.) These permutations are determined *during the factorization* by a combination of (usually competing) numerical stability and sparsity requirements (Duff 1974). Different matrices, even though they may have the same zero/nonzero pattern, will normally yield different P and Q, and therefore have factors with different sparsity patterns. In other words, it is in general not possible to predict where fill-in will occur for general sparse matrices before the computation begins. Thus, we are obliged to use some form of *dynamic* storage scheme which allocates storage for fill-in as the computation proceeds.

On the other hand, symmetric Gaussian elimination (Cholesky's method, or one of its variants, described in Chapter 2) applied to a symmetric positive definite matrix does not require interchanges (pivoting) to maintain numerical stability. Since PAP^T is also symmetric and positive definite for any permutation matrix P, this means we can choose to reorder A symmetrically i) without regard to numerical stability and ii) before the actual numerical factorization begins.

These options, *which are normally not available to us when A is a general indefinite matrix*, have enormous practical implications. Since the ordering can be determined before the factorization begins, the locations of the fill-in suffered during the factorization can also be determined. Thus, the data structure used to store L can be constructed before the actual numerical factorization, and spaces for fill components can be reserved. The computation then proceeds with the storage structure remaining *static* (unaltered). Thus, the three problems of i) finding a suitable ordering, ii) setting up the appropriate storage scheme, and iii) the actual numerical computation, can be isolated as separate objects of study, as well as separate computer software modules, as depicted below.

This independence of tasks has a number of distinct advantages. It encourages software modularity, and, in particular, allows us to tailor storage methods to the given task at hand. For example, the use of lists to store matrix subscripts may be quite appropriate for an implementation of an ordering algorithm, but decidedly inappropriate in connection with actually storing the matrix or its factors. In the same vein, knowing that we can use a storage scheme in a static manner during the factorization sometimes allows us to select a method which is very efficient in terms of storage requirements, but would be a disaster in terms of bookkeeping overhead if it had to be altered during the factorization. Finally, in many engineering design applications, numerous *different* positive definite matrix problems having the *same* structure must be solved. Obviously, the ordering and storage scheme set-up only needs to be performed once, so it is desirable to have these tasks isolated from the actual numerical computation.

In numerous practical situations matrix problems arise which are unsymmetric but have symmetric structure, and for which it can be shown that pivoting for numerical stability is not required when Gaussian elimination is applied. Almost all the ideas and algorithms described in this book extend immediately to this class of problems. Some hints on how this can be done are provided in Exercise 4.5.1, Chapter 4.

1.3   Iterative Versus Direct Methods

Numerical methods for solving systems of linear equations fall into two general classes, *iterative* and *direct*. A typical iterative method involves the initial selection of an approximation $x^{(1)}$ to x, and the determination of a sequence $x^{(2)}$, $x^{(3)}$, . . . such that $\lim_{i \to \infty} x^{(i)} = x$. Usually the calculation of $x^{(i+1)}$ involves only A, b, and one or two of the previous iterates. In theory, when we use an iterative method we must perform an infinite number of arithmetic operations in order to obtain x, but in practice we stop the iteration when we believe our current approximation is acceptably close to x. On the other hand, in the absence of rounding errors, direct methods provide the solution after a finite number of arithmetic operations have been performed.

Which class of method is better? The question cannot be answered in general since it depends upon how we define "better," and also upon the particular problem or class of problems to be solved. Iterative methods are attractive in terms of computer storage requirements since their implementations typically require only A, b, $x^{(i)}$ and perhaps one or two other vectors to be stored. On the other hand, when A is factored, it typically suffers some fill-in, so that the *filled matrix* $F = L + L^T$ has nonzeros in positions which are zero in A. Thus, it is often true that direct methods for sparse systems require more storage than implementations of iterative methods. The actual ratio depends very much on the problem being solved, and also on the ordering used.

A comparison of iterative and direct methods in terms of computational requirements is even more complicated. As we have seen, the ordering used can dramatically affect the amount of arithmetic performed using Gaussian elimination. The number of iterations performed by an iterative scheme depends very much on the characteristics of A, and on the sometimes delicate problem of determining, on the basis of computable quantities, when $x^{(i)}$ is "close enough" to x.

In some situations, such as in the design of some mechanical devices, or the simulation of some time-dependent phenomena, many systems of equations having the same coefficient matrix must be solved. In this case, the cost of the direct scheme may be essentially that of solving the triangular system *given the factorization*, since the factorization cost amortized over all solutions may be negligible. In these situations it is also often the case that the number of iterations required by an iterative scheme is quite small, since a good starting vector $x^{(1)}$ is often available.

The above remarks should make it clear that unless the question of which class of method should be used is posed in a quite narrow and well defined context, it is either very complicated or impossible to answer. Our justification for considering only direct methods in this book is that several excellent references dealing with iterative methods are already available (Varga 1962, Young 1971), while there is no such comparable reference known to the authors for direct methods for large sparse systems. In addition, there are situations where it can be shown quite convincingly that direct methods are far more desirable than any conceivable iterative scheme.

2/ Fundamentals

2.0 Introduction

In this chapter we examine the basic numerical algorithm used throughout the book to solve symmetric positive definite matrix problems. The method, known as Cholesky's method, was discussed briefly in Section 1.1. In what follows we prove that the factorization always exists for positive definite matrices, and examine several ways in which the computation can be performed. Although these are mathematically and (usually) numerically equivalent, they differ in the order in which the numbers are computed and used. These differences are important with respect to computer implementation of the method. We also derive expressions for the amount of arithmetic performed by the method.

As we indicated in Section 1.1, when Cholesky's method is applied to a sparse matrix A, it generally suffers some fill-in, so that its Cholesky factor L has nonzeros in positions which are zero in A. For some permutation matrix P, we can instead factor PAP^T into $\tilde{L}\tilde{L}^T$, and $\tilde{L}$ may be much more attractive than L, according to some criterion. In Section 2.3 we discuss some of these criteria, and indicate how practical implementation factors complicate the comparison of different methods.

2.0.1 Notations

The reader is assumed to be familiar with the elementary theory and properties of matrices as presented in Stewart (1973). In this section, we shall describe the matrix notations used throughout the book.

We shall use bold face capital italic letters for matrices. The entries of a matrix will be represented by lower case italic letters with two subscripts. For example, let A be an N by N matrix. Its (i,j)-th element is denoted by a_{ij}. The number N is called the *order* of the matrix A.

A vector will be denoted by a lower case bold italic letter and its elements by lower case letters with a single subscript. Thus, we have

$$v = \begin{pmatrix} v_1 \\ v_2 \\ \cdot \\ \cdot \\ \cdot \\ v_N \end{pmatrix},$$

a vector of length N.

For a given matrix A, its i-th row and i-th column are denoted by A_{i*} and A_{*i} respectively. When A is symmetric, we have $A_{i*} = A_{*i}$ for $i = 1, \ldots, N$.

We shall use I_N to represent the *identity matrix* of order N; that is, the matrix with all entries zero except for ones on the diagonal.

In sparse matrix analysis, we often need to count the number of nonzeros in a vector or matrix. We use $\eta(\square)$ to denote the number of nonzero components in $\square$, where $\square$ stands for a vector or a matrix. Obviously,

$$\eta(I_N) = N .$$

We also often need to refer to the number of members in a *set S*; we denote this number by $|S|$.

Let $f(n)$ and $g(n)$ be functions of the independent variable n. We use the notation

$$f(n) = O(g(n))$$

if for some constant K and all sufficiently large n,

$$\left| \frac{f(n)}{g(n)} \right| \le K .$$

We say that $f(n)$ has at most the *order of magnitude* of $g(n)$. This is a useful notation in the analysis of sparse matrix algorithms, since often we are only interested in the dominant term in arithmetic and nonzero counts.

For example, if $f(n) = \frac{1}{6}n^3 + \frac{1}{2}n^2 - \frac{2}{3}n$, we can write

$$f(n) = O(n^3) .$$

For large enough n, the relative contribution from the terms $\frac{1}{2}n^2$ and $-\frac{2}{3}n$ is negligible.

Expressions such as $f(n)$ above arise in counting arithmetic operations or numbers of nonzeros, and are often the result of some fairly complicated summations. Since we are usually only concerned with the dominant term, a very common device used to simplify the computation is to replace the summation by an integral sign. For example, for large n,

$$\sum_{k=1}^{n} (2n + k)(n - k) \simeq \int_{0}^{n} (2n + k)(n - k)dk .$$

Exercises

2.0.1)   Compute directly the sum $\sum_{i=1}^{N} i^2(N - i)$, and also approximate it using an integral, as described at the end of this section.

2.0.2)   Compute directly the sum $\sum_{i=1}^{N} \sum_{j=1}^{N-i+1} (i + j)$, and also approximate it using a double integral.

2.0.3)   Let A and B be two N by N sparse matrices. Show that the number of multiplications required to compute $C = AB$ is given by

$$\sum_{i=1}^{N} \eta(A_{*i})\eta(B_{i*}).$$

2.0.4)   Let B be a given M by N sparse matrix. Show that the product $B^T B$ can be computed using

$$\frac{1}{2} \sum_{i=1}^{M} \eta(B_{i*})(\eta(B_{i*}) + 1)$$

multiplications.

2.0.5)   A common scheme to store a sparse vector has a main storage array which contains all the nonzero entries in the vector, and an accompanying vector which gives the subscripts of the nonzeros. Let u and v be two sparse vectors of size N stored in this format. Consider the computation of the inner product $w = u^T v$.

a)   If the subscript vectors are in ascending (or descending) order, show that the inner product can be done using only $O(\eta(u) + \eta(v))$ comparisons.

b)   What if the subscripts are in random order?

c)   How would you perform the computation if the subscripts are in random order and a temporary real array of size N with all zero entries is provided?

2.1 The Factorization Algorithm

2.1.1 Existence and Uniqueness of the Factorization

A symmetric matrix A is *positive definite* if $x^TAx > 0$ for all nonzero vectors x. Such matrices arise in many applications; typically x^TAx represents the energy of some physical system which is positive for any configuration x. In a positive definite matrix A the diagonal entries are always positive since

$$e^TAe = a_{ij},$$

where e is the i-th *characteristic vector*, the components of which are all zeros except for a one in the i-th position. This observation will be used in proving the following factorization theorem due to Cholesky (Stewart 1973).

Theorem 2.1.1

If A is an N by N symmetric positive definite matrix, it has a unique triangular factorization LL^T, where L is a lower triangular matrix with positive diagonal entries.

Proof   The proof is by induction on the order of the matrix A.   The result is certainly true for one by one matrices since a_{11} is positive.

Suppose the assertion is true for matrices of order $N - 1$. Let A be a symmetric positive definite matrix of order N. It can be partitioned into the form

$$A = \begin{pmatrix} d & v^T \\ v & \overline{H} \end{pmatrix},$$

where d is a positive scalar and $\overline{H}$ is an $N - 1$ by $N - 1$ submatrix. The partitioned matrix can be written as the product

$$\begin{pmatrix} \sqrt{d} & 0 \\ \dfrac{v}{\sqrt{d}} & I_{N-1} \end{pmatrix} \begin{pmatrix} 1 & 0 \\ 0 & H \end{pmatrix} \begin{pmatrix} \sqrt{d} & v^T/\sqrt{d} \\ 0 & I_{N-1} \end{pmatrix},$$

where $H = \overline{H} - \dfrac{vv^T}{d}$. Clearly the matrix H is symmetric. It is also positive definite since for any nonzero vector x of length $N - 1$,

$$\left(-\dfrac{x^Tv}{d}, \; x^T \right) \begin{pmatrix} d & v^T \\ v & \overline{H} \end{pmatrix} \begin{pmatrix} -\dfrac{x^Tv}{d} \\ x \end{pmatrix} = x^T \left(\overline{H} - \dfrac{vv^T}{d} \right) x$$

$$= x^T H x \ ,$$

which implies $x^T H x > 0$. By the induction assumption, H has a triangular factorization $L_H L_H^T$ with positive diagonals. Thus, A can be expressed as

$$\begin{pmatrix} \sqrt{d} & 0 \\[2mm] \dfrac{v}{\sqrt{d}} & I_{N-1} \end{pmatrix} \begin{pmatrix} 1 & 0 \\ 0 & L_H \end{pmatrix} \begin{pmatrix} 1 & 0 \\ 0 & L_H^T \end{pmatrix} \begin{pmatrix} \sqrt{d} & v^T/\sqrt{d} \\ 0 & I_{N-1} \end{pmatrix}$$

$$= \begin{pmatrix} \sqrt{d} & 0 \\[2mm] \dfrac{v}{\sqrt{d}} & L_H \end{pmatrix} \begin{pmatrix} \sqrt{d} & v^T/\sqrt{d} \\ 0 & L_H^T \end{pmatrix}$$

$$= L L^T.$$

It is left to the reader to show that the factor L is unique.   □

If we apply the result to the matrix example

$$\begin{pmatrix} 4 & 8 \\ 8 & 25 \end{pmatrix},$$

we obtain the factors

$$\begin{pmatrix} 2 & 0 \\ 4 & 3 \end{pmatrix} \begin{pmatrix} 2 & 4 \\ 0 & 3 \end{pmatrix}.$$

It is appropriate here to point out that there is a closely related factorization of a symmetric positive definite matrix (Martin 1971). Since the Cholesky factor L has positive diagonal elements, one can factor out a diagonal matrix $D^{1/2}$ from L, yielding $L = \tilde{L} D^{1/2}$ whence we have

$$A = \tilde{L} D \tilde{L}^T. \tag{2.1.1}$$

In the above matrix example, this alternative factorization is

$$\begin{pmatrix} 1 & 0 \\ 2 & 1 \end{pmatrix} \begin{pmatrix} 4 & 0 \\ 0 & 9 \end{pmatrix} \begin{pmatrix} 1 & 2 \\ 0 & 1 \end{pmatrix}.$$

This factorization is as easy to compute as the original, and can be obtained without square root calculation (see Exercise 2.1.4). We do not use it in our book because in some circumstances it leads to certain disagreeable asymmetries in calculations involving partitioned matrices.

2.1.2 Computing the Factorization

Theorem 2.1.1 guarantees the existence and uniqueness of the Cholesky factor for a symmetric positive definite matrix, but the order and the way in which the components of the factor L are actually computed can vary. In this section, we examine some different ways in which L can be computed; these options are important because they provide us with flexibility in the design of storage schemes for the sparse matrix factor L.

The constructive proof of Theorem 2.1.1 suggests a computational scheme to determine the factor L. It is the so-called *outer product form* of the algorithm. The scheme can be described step by step in matrix terms as follows.

$$A = A_0 = H_0 = \begin{pmatrix} d_1 & v_1^T \\ v_1 & \overline{H}_1 \end{pmatrix} \tag{2.1.2}$$

$$= \begin{pmatrix} \sqrt{d_1} & 0 \\ \dfrac{v_1}{\sqrt{d_1}} & I_{N-1} \end{pmatrix} \begin{pmatrix} 1 & 0 \\ 0 & \overline{H}_1 - \dfrac{v_1 v_1^T}{d_1} \end{pmatrix} \begin{pmatrix} \sqrt{d_1} & v_1^T/\sqrt{d_1} \\ 0 & I_{N-1} \end{pmatrix}$$

$$= L_1 \begin{pmatrix} 1 & 0 \\ 0 & H_1 \end{pmatrix} L_1^T$$

$$= L_1 A_1 L_1^T,$$

$$A_1 = \begin{pmatrix} 1 & 0 \\ 0 & H_1 \end{pmatrix} = \begin{pmatrix} 1 & 0 & 0 \\ 0 & d_2 & v_2^T \\ 0 & v_2 & \overline{H}_2 \end{pmatrix}$$

$$= \begin{pmatrix} 1 & 0 \\ 0 & \sqrt{d_2} \\ 0 & \dfrac{v_2}{\sqrt{d_2}} & I_{N-2} \end{pmatrix} \begin{pmatrix} 1 & 0 & 0 \\ 0 & 1 & 0 \\ 0 & 0 & \overline{H}_2 - \dfrac{v_2 v_2^T}{d_2} \end{pmatrix} \begin{pmatrix} 1 & 0 & 0 \\ & \sqrt{d_2} & v_2^T/\sqrt{d_2} \\ 0 & & I_{N-2} \end{pmatrix}$$

$$= L_2 A_2 L_2^T,$$

$$\cdot$$
$$\cdot$$
$$\cdot$$

$$A_{N-1} = L_N I_N L_N^T .$$

Here, for $1 \le i \le N$, d_i is a positive scalar, v_i is a vector of length $N - i$, and H_i is an $N - i$ by $N - i$ positive definite symmetric matrix. After N steps of the algorithm, we have

$$A = L_1 L_2 \cdots L_N L_N^T \cdots L_2^T L_1^T = LL^T,$$

where it can be shown (see Exercise 2.1.6) that

$$L = L_1 + L_2 + \cdots + L_N - (N-1)I_N. \qquad (2.1.3)$$

Thus, the i-th column of L is precisely the i-th column of L_i.

In this scheme, the columns of L are computed one by one. At the same time, each step involves the modification of the submatrix $\overline{H}_i$ by the outer product $v_i v_i^T / d_i$ to give H_i, which is simply the submatrix remaining to be factored. The access to the components of A during the factorization is depicted as follows.

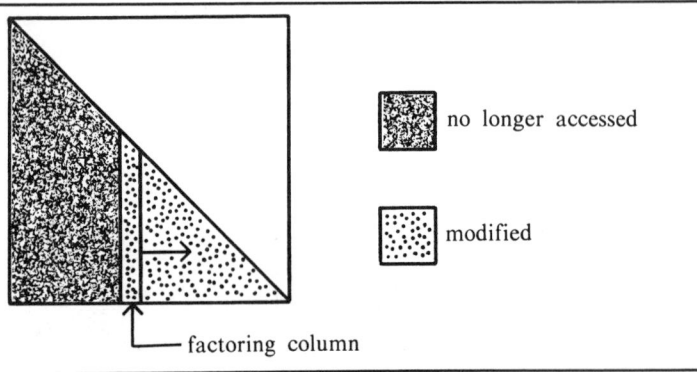

no longer accessed

modified

factoring column

An alternative formulation of the factorization process is the *bordering method*. Suppose the matrix A is partitioned as

$$A = \begin{pmatrix} M & u \\ u^T & s \end{pmatrix},$$

where the symmetric factorization $L_M L_M^T$ of the $N - 1$ by $N - 1$ leading principal submatrix M has already been obtained. (Why is M positive definite?) Then the factorization of A is given by

$$A = \begin{pmatrix} L_M & 0 \\ w^T & t \end{pmatrix} \begin{pmatrix} L_M^T & w \\ 0 & t \end{pmatrix}, \qquad (2.1.4)$$

where

$$w = L_M^{-1} u \qquad (2.1.5)$$

and

$$t = (s - w^T w)^{1/2}.$$

(Why is $s - w^T w$ positive?)

Note that the factorization $L_M L_M^T$ of the submatrix M is also obtained by the bordering technique. So, the scheme can be described as follows.

For $i = 1, 2, ..., N$,

Solve
$$\begin{pmatrix} \ell_{1,1} & & & \mathbf{0} \\ & \cdot & \cdot & \\ & \cdot & & \cdot \\ & \cdot & & \cdot \\ \ell_{i-1,1} & \cdot & \cdot & \ell_{i-1,i-1} \end{pmatrix} \begin{pmatrix} \ell_{i,1} \\ \cdot \\ \cdot \\ \cdot \\ \ell_{i,i-1} \end{pmatrix} = \begin{pmatrix} a_{i,1} \\ \cdot \\ \cdot \\ \cdot \\ a_{i,i-1} \end{pmatrix}.$$

Compute $\ell_{i,i} = \left(a_{i,i} - \sum_{k=1}^{i-1} \ell_{i,k}^2 \right)^{1/2}.$

In this scheme, the rows of L are computed one at a time; the part of the matrix remaining to be factored is not accessed until the corresponding part of L is to be computed. The sequence of computations can be depicted as follows.

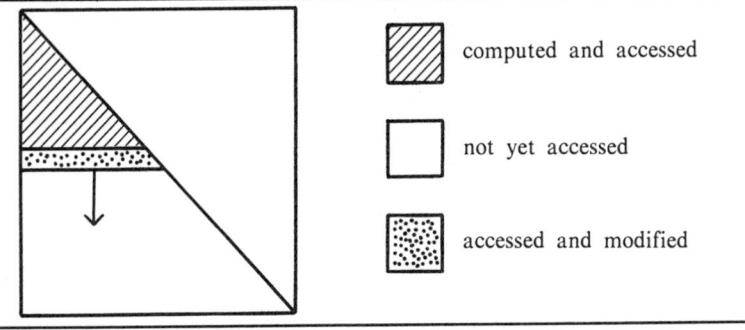

computed and accessed

not yet accessed

accessed and modified

The final scheme for computing the components of L is the *inner product form* of the algorithm. It can be described as follows.

For $j = 1, 2, \ldots, N$

$$\text{Compute } \ell_{j,j} = \left(a_{j,j} - \sum_{k=1}^{j-1} \ell_{j,k}^2 \right)^{\frac{1}{2}}.$$

For $i = j+1, j+2, \ldots, N$

$$\text{Compute } \ell_{i,j} = \left(a_{i,j} - \sum_{k=1}^{j-1} \ell_{i,k} \ell_{j,k} \right) / \ell_{j,j}.$$

These formulae can be derived directly by equating the elements of A to the corresponding elements of the product LL^T.

Like the outer product version of the algorithm, the columns of L are computed one by one, but the part of the matrix remaining to be factored is not accessed during the scheme. The sequence of computations and the relevant access to the components of A (or L) is depicted as follows.

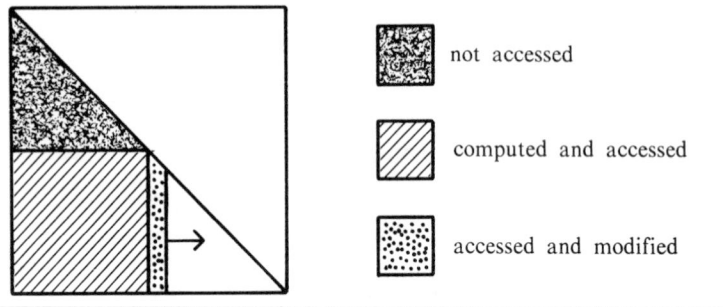

not accessed

computed and accessed

accessed and modified

The latter two formulations can be organized so that only inner products are involved. This can be used to improve the accuracy of the numerical factorization by accumulating the inner products in double precision. On some computers, this can be done at little extra cost.

2.1.3 Sparse Matrix Factorization

As we have seen in Chapter 1, when a sparse matrix is factored, it usually suffers some fill-in; that is, the lower triangular factor L has nonzero components in positions which are zero in the original matrix. Recall in Section 1.1 the factorization of the matrix example

$$A = \begin{pmatrix} 4 & 1 & 2 & \frac{1}{2} & 2 \\ 1 & \frac{1}{2} & 0 & 0 & 0 \\ 2 & 0 & 3 & 0 & 0 \\ \frac{1}{2} & 0 & 0 & \frac{5}{8} & 0 \\ 2 & 0 & 0 & 0 & 16 \end{pmatrix}.$$

Its triangular factor L is given by

$$L = \begin{pmatrix} 2 & 0 & 0 & 0 & 0 \\ 0.5 & 0.5 & 0 & 0 & 0 \\ 1 & -1 & 1 & 0 & 0 \\ 0.25 & -0.25 & -0.5 & 0.5 & 0 \\ 1 & -1 & -2 & -3 & 1 \end{pmatrix}$$

so that the matrix A suffers fill at a_{32}, a_{42}, a_{43}, a_{52}, a_{53} and a_{54}. This phenomenon of *fill-in*, which is usually ignored in solving dense systems, plays a crucial role in sparse elimination.

The creation of nonzero entries can be best understood using the outer-product formulation of the factorization process. At the i-th step, the submatrix $\overline{H}_i$ is modified by the matrix $v_i v_i^T / d_i$ to give H_i. As a result, the submatrix H_i may have nonzeros in locations which are zero in $\overline{H}_i$. In the example above,

$$\overline{H}_1 = \begin{pmatrix} \frac{1}{2} & 0 & 0 & 0 \\ 0 & 3 & 0 & 0 \\ 0 & 0 & \frac{5}{8} & 0 \\ 0 & 0 & 0 & 16 \end{pmatrix}$$

and it is modified at step 1 to give (to three significant figures)

$$H_1 = \overline{H}_1 - \frac{1}{4} \begin{pmatrix} 1 \\ 2 \\ \frac{1}{2} \\ 2 \end{pmatrix} \begin{pmatrix} 1 & 2 & \frac{1}{2} & 2 \end{pmatrix}$$

$$= \begin{pmatrix} .25 & -.5 & -.125 & -.5 \\ -.5 & 2 & -.25 & -1 \\ -.125 & -.25 & .563 & -.25 \\ -.5 & -1 & -.25 & 15 \end{pmatrix}.$$

If zeros are exploited in solving a sparse system, fill-in affects both the storage and computation requirements. Recall that $\eta(\square)$ is the number of nonzero components in $\square$, where $\square$ stands for a vector or a matrix. Clearly, from (2.1.2) and (2.1.3), the number of nonzeros in L is given by

$$\eta(L) = N + \sum_{i=1}^{N-1} \eta(v_i). \qquad (2.1.6)$$

In the following theorem, and throughout the book, we measure arithmetic requirements by the number of multiplicative operations (multiplications and divisions), which we simply refer to as "operations." The majority of the arithmetic performed in matrix operations involves sequences of arithmetic operations which occur in multiply-add pairs, so the number of additive operations is about equal to the number of multiplicative operations.

Theorem 2.1.2

The number of operations required to compute the triangular factor L of the matrix A is given by

$$\frac{1}{2} \sum_{i=1}^{N-1} \eta(v_i)[\eta(v_i)+3] = \frac{1}{2} \sum_{i=1}^{N-1} [\eta(L_{*i})-1][\eta(L_{*i})+2].$$

$$(2.1.7)$$

Proof   The three formulations of the factorization differ only in the order in which operations are performed. For the purpose of counting operations, the outer-product formulation (2.1.2) is used. At the i-th step, $\eta(v_i)$ operations are required to compute $v_i/\sqrt{d_i}$, and $\frac{1}{2}\eta(v_i)[\eta(v_i)+1]$ operations are needed to form the symmetric matrix

$$\frac{v_i v_i^T}{d_i} = \left(\frac{v_i}{\sqrt{d_i}}\right) \left(\frac{v_i}{\sqrt{d_i}}\right)^T.$$

The result follows from summing over all the steps. $\square$

For the dense case, the number of nonzeros in L is

$$\frac{1}{2} N(N+1) \qquad (2.1.8)$$

and the arithmetic cost is

$$\frac{1}{2}\sum_{i=1}^{N-1} i(i+3) = \frac{1}{6}N^3 + \frac{1}{2}N^2 - \frac{2}{3}N. \qquad (2.1.9)$$

Consider also the Cholesky factorization of a symmetric positive definite tridiagonal matrix , an example of a sparse matrix. It can be shown (see Chapter 5) that if L is its factor,

$$\eta(L_{*i}) = 2, \quad \text{for } i = 1, \ldots, N-1.$$

In this case, the number of nonzeros in L is

$$\eta(L) = 2N - 1,$$

and the arithmetic cost of computing L is

$$\frac{1}{2}\sum_{i=1}^{N-1} 1(4) = 2(N-1).$$

Comparing these results with the counts for the dense case, we see a dramatic difference in storage and computational costs.

The costs for solving equivalent sparse systems with different orderings can also be very different. As illustrated in Section 1.1, the matrix example A at the beginning of this section can be ordered so that it does not suffer any fill-in at all! The permutation matrix used is

$$P = \begin{pmatrix} 0 & 0 & 0 & 0 & 1 \\ 0 & 0 & 0 & 1 & 0 \\ 0 & 0 & 1 & 0 & 0 \\ 0 & 1 & 0 & 0 & 0 \\ 1 & 0 & 0 & 0 & 0 \end{pmatrix},$$

which reverses the ordering of A when applied. We obtain the permuted matrix

$$PAP^T = \begin{pmatrix} 16 & 0 & 0 & 0 & 2 \\ 0 & \frac{5}{8} & 0 & 0 & \frac{1}{2} \\ 0 & 0 & 3 & 0 & 2 \\ 0 & 0 & 0 & \frac{1}{2} & 1 \\ 2 & \frac{1}{2} & 2 & 1 & 4 \end{pmatrix}$$

This simple example illustrates that a judicious choice of P can result in dramatic reductions in fill-in and arithmetic requirements. Therefore, in solving a given linear equation problem

$$Ax = b,$$

the *general procedure* involves first finding a permutation or ordering *P* of the given problem. Then the system is expressed as

$$(PAP^T)(Px) = Pb$$

and Cholesky's method is applied to the symmetric positive definite matrix PAP^T yielding the triangular factorization LL^T. By solving the equivalent permuted system, we can often achieve a reduction in the computer storage and execution time requirements.

Exercises

2.1.1)   Show that the Cholesky factorization for a symmetric positive definite matrix is unique.

2.1.2)   Let *A* be an *N* by *N* symmetric positive definite matrix. Show that

   a)   any principal submatrix of *A* is positive definite,
   b)   *A* is nonsingular and A^{-1} is also positive definite,
   c) $\max\limits_{1 \leq i \leq N} a_{ii} = \max\limits_{1 \leq i,j \leq N} |a_{ij}|$.

2.1.3)   Let *A* be a symmetric positive definite matrix. Show that

   a) B^TAB is positive definite if and only if *B* is non-singular,
   b)   the augmented matrix

$$\begin{pmatrix} A & u \\ u^T & s \end{pmatrix}$$

is positive definite if and only if $s > u^TA^{-1}u$.

2.1.4)   Write out equations similar to those in (2.1.2) and (2.1.3) which yield the factorization LDL^T, where *L* is now lower triangular with ones on the diagonal, and *D* is a diagonal matrix with positive diagonal elements.

2.1.5)   Let *E* and *F* be two *N* by *N* lower triangular matrices which for some *k* $(1 \leq k \leq N)$ satisfy

$$e_{jj} = 1 \quad \text{for} \quad j > k$$

$$e_{ij} = 0 \quad \text{for} \quad i > j \quad \text{and} \quad j > k$$

$$f_{jj} = 1 \quad \text{for} \quad j \leq k$$

$$f_{ij} = 0 \quad \text{for} \quad i > j \quad \text{and} \quad j \leq k .$$

The case when $N = 6$ and $k = 3$ is depicted below.

$$E \; = \; \begin{pmatrix} * & & & & & \\ * & * & & \mathbf{0} & & \\ * & * & * & & & \\ * & * & * & 1 & & \\ * & * & * & 0 & 1 & \\ * & * & * & 0 & 0 & 1 \end{pmatrix} \qquad F \; = \; \begin{pmatrix} 1 & & & & & \\ 0 & 1 & & \mathbf{0} & & \\ 0 & 0 & 1 & & & \\ 0 & 0 & 0 & * & & \\ 0 & 0 & 0 & * & * & \\ 0 & 0 & 0 & * & * & * \end{pmatrix}$$

Show that $EF = E + F - I$, and hence prove that (2.1.3) holds.

2.1.6) Give an example of a symmetric matrix which does not have a triangular factorization LL^T and one which has more than one factorization.

2.2 Solving Triangular Systems

2.2.1 Computing the Solution

Once we have computed the factorization, we must solve the triangular systems $Ly = b$ and $L^Tx = y$. In this section, we consider the numerical solution of triangular systems.

Consider the N by N linear system

$$Tx = b$$

where T is nonsingular and triangular. Without loss of generality, we assume that T is lower triangular. There are two common ways of solving the system, which differ only in the order in which the operations are performed.

The first one involves the use of inner-products and the defining equations are given by:

For $i = 1, 2, \ldots, N$,

$$x_i = \left(b_i - \sum_{k=1}^{i-1} t_{i,k} x_k \right) / t_{i,i} \, . \tag{2.2.1}$$

The sequence of computations is depicted by the following diagram.

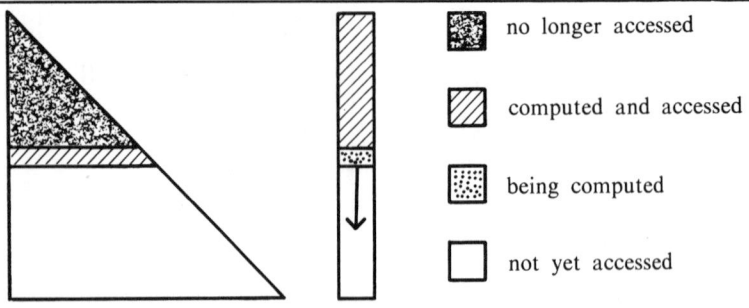

The second method uses the matrix components of T in the same way as the outer-product version of the factorization. The defining equations are as follows.

For $i = 1, 2, \ldots, N$,

$$x_i = b_i/t_{i,i}$$

$$
\begin{pmatrix} b_{i+1} \\ \cdot \\ \cdot \\ \cdot \\ b_N \end{pmatrix} \leftarrow \begin{pmatrix} b_{i+1} \\ \cdot \\ \cdot \\ \cdot \\ b_N \end{pmatrix} - x_i \begin{pmatrix} t_{i+1,i} \\ \cdot \\ \cdot \\ \cdot \\ t_{N,i} \end{pmatrix} \tag{2.2.2}
$$

Note that this scheme lends itself to exploiting sparsity in the solution x. If b_i turns out to be zero at the beginning of the i-th step, x_i is zero and the entire step can be skipped. The access to components of the system is shown as follows.

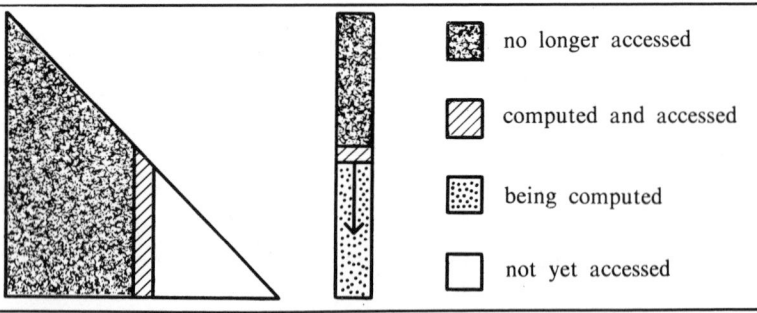

The former solution method accesses the components of the lower triangular matrix row by row and therefore lends itself to row-wise storage schemes. If the matrix is stored column by column, the latter method is more appropriate. It is interesting to note that this column oriented method is often used to solve the *upper* triangular system

$$L^T x = y \; ,$$

where L is a lower triangular matrix stored using a row-wise scheme.

2.2.2   Operation Counts

We now establish some simple results about solving triangular systems. They will be helpful later in obtaining operation counts.

Consider the solution of

$$Tx = b \; ,$$

where T is nonsingular and lower triangular.

Lemma 2.2.1

The number of operations required to solve for x is

$$\sum_i \{\eta(T_{*i}) \mid x_i \neq 0\}$$

Proof   It follows from (2.2.2) that if $x_i \neq 0$, the i-th step requires $\eta(T_{*i})$ operations.   □

Corollary 2.2.2

If the sparsity of the solution vector x is not exploited (that is, x is assumed to be full), then the number of operations required to compute x is $\eta(T)$.

Thus, it follows that the operation count for solving $Tx = b$, when T and x are full, is

$$\frac{1}{2} N(N + 1) \; . \tag{2.2.3}$$

The following results give some relationships between the structure of the right hand side b and the solution x of a lower triangular system. Lemma 2.2.3 and Corollary 2.2.5 appeal to a *no-cancellation assumption*; that is, whenever two nonzero quantities are added or subtracted, the result is nonzero. This means that in the analysis we ignore any zeros which might be created through exact cancellation. Such cancellation rarely occurs, and in order to predict such cancellation we would have to know the numerical values of T and b. Such a prediction would be difficult in general, particularly in floating point arithmetic which is subject to rounding error.

Lemma 2.2.3

With the no-cancellation assumption, if $b_i \neq 0$ then $x_i \neq 0$.

Proof   Since T is non-singular, $t_{ii} \neq 0$ for $1 \leq i \leq N$. The result then follows from the no-cancellation assumption and the defining equation (2.2.1) for x_i.   □

Lemma 2.2.4
Let x be the solution to $Tx = b$. If $b_i = 0$ for $1 \leq i \leq k$, then $x_i = 0$ for $1 \leq i \leq k$.

Corollary 2.2.5
With the no-cancellation assumption, $\eta(b) \leq \eta(x)$.

Exercises

2.2.1)   Use Lemma 2.2.1 to show that factorization by the bordering scheme requires

$$\frac{1}{2} \sum_{i=1}^{N-1} (\eta(L_{*i}) - 1)(\eta(L_{*i}) + 2)$$

operations.

2.2.2)   Show that the inverse of a nonsingular lower triangular matrix is lower triangular (use Lemma 2.2.4).

2.2.3)   Let T be a nonsingular lower triangular matrix with the *propagation property*, that is, $t_{i,i-1} \neq 0$ for $2 \leq i \leq N$.

   a)   Show that in solving $Tx = b$, if $b_i \neq 0$ then $x_j \neq 0$ for $i \leq j \leq N$.
   b)   Show that T^{-1} is a full lower triangular matrix.

2.2.4)   Does Lemma 2.2.1 depend upon the no-cancellation assumption? Explain. What about Theorem 2.1.2 and Lemma 2.2.4?

2.2.5)   Prove a result analogous to Lemma 2.2.4 for upper triangular matrices.

2.2.6)   Suppose you have numerous N by N lower triangular systems of the form $Ly = b$ to solve, where L and b are both sparse. It is known that the solution y is also sparse for these problems. You have a choice of two storage schemes for L, as illustrated by the 5 by 5 example below; one is column oriented and one is row oriented. Which one would you choose, and why would you choose it? If you wrote a Fortran program to solve such systems using your choice of data structures, would the execution time be proportional to the number of operations performed? Explain. (Assume that the number of operations performed is at least $O(N)$.)

$$L = \begin{pmatrix} 3 & & & & 0 \\ 0 & 2 & & & \\ 2 & 0 & 4 & & \\ 0 & 3 & 9 & 5 & \\ 0 & 0 & 7 & 0 & 7 \end{pmatrix}$$

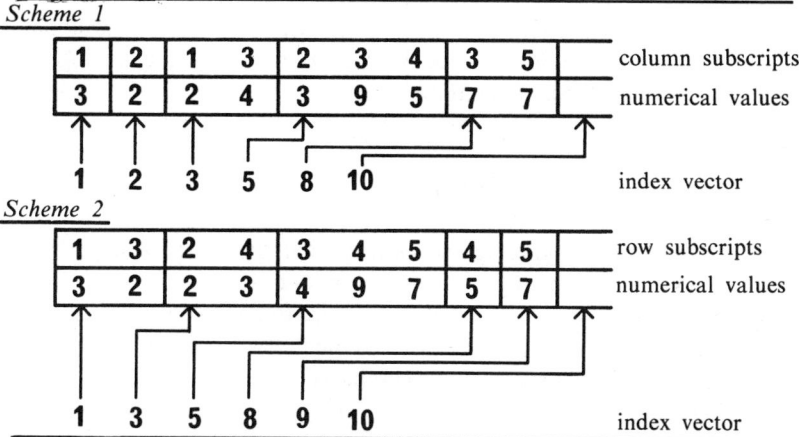

2.2.7)   Let L and W be N by N non-sparse lower triangular matrices, with nonzero diagonal elements. Approximately how many operations are required to compute $L^{-1}W$? How many operations are required to compute $W^T W$?

2.2.8)   Suppose that $n = 1 + k(p-1)$ for some positive integer k, and that W is an n by p full (pseudo) lower triangular matrix. That is, W has zeros above position $1 + (i-1)k$ in column i of W, and is nonzero otherwise. An example with $n = 7$ and $p = 4$ appears below. Roughly how many operations are required to compute $W^T W$, in terms of n and p?

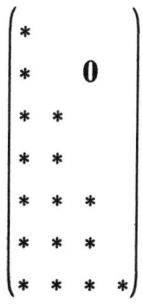

2.2.9)   Suppose L is a nonsingular n by n lower triangular matrix, and W is as described in Exercise 2.2.8. Approximately how many operations are required to compute $L^{-1}W$, as a function of n and p?

2.2.10)   a) Suppose $A = LL^T$, where L is as in Exercise 2.2.9 and $\eta(L_{*i}) \geq 2$, $1 \leq i \leq n$. Assuming the no-cancellation assumption, show that computing A^{-1} by solving $LW = I$ and $L^TZ = W$ yields a full matrix.

   b) Suppose A is unsymmetric with triangular factorization LU, where L is unit lower triangular and U is upper triangular. State the conditions and results analogous to those in a) above.

2.3   Some Practical Considerations

The objective of studying sparse matrix techniques for solving linear systems is to reduce *cost* by exploiting sparsity of the given system. We have seen in Section 2.1.3 that it is possible to achieve drastic reductions in storage and arithmetic requirements, when the solutions of dense and tridiagonal systems are compared.

There are various kinds of sparse storage schemes, which differ in the way zeros are exploited. Some might store some zeros in exchange for a simpler storage scheme; others exploit all the zeros in the system. In Chapters 4 to 8, we discuss the commonly used sparse schemes for solving linear systems.

The choice of a storage method naturally affects the storage requirement, and the use of ordering strategies (choice of permutation matrix P). Moreover, it has significant impact on the implementation of the factorization and solution, and hence on the complexity of the programs and the execution time.

However, irrespective of what sparse storage scheme is used, there are four distinct phases that can be identified in the entire computational process.

Step 1   (*Ordering*) Find a "good" ordering (permutation P) for the given matrix A, with respect to the chosen storage method.

Step 2   (*Storage allocation*) Determine the necessary information about the Cholesky factor L of PAP^T to set up the storage scheme.

Step 3   (*Factorization*) Factor the permuted matrix PAP^T into LL^T.

Step 4   (*Triangular solution*) Solve $Ly = b$ and $L^Tz = y$. Then set $x = P^Tz$.

Even with a prescribed storage method, there are many ways for finding orderings, determining the corresponding storage structure of

L, and performing the actual numerical computation. We shall refer to a sparse storage scheme and an associated ordering/allocation/factorization/solution combination collectively as a *solution method.*

The most commonly cited objectives for choosing a solution method are to a) reduce computer storage, b) reduce computer execution time or c) reduce some combination of storage and execution which reflects the way charges are assessed to the user of the computer system. Although there are other criteria which sometimes govern the choice of method, these are the main ones and serve to illustrate the complications involved in evaluating a strategy.

In order to be able to declare that one method is better than another with respect to one of the measures cited above, we must be able to evaluate precisely that measure for each method, and this evaluation is substantially more complicated than one would expect. We deal first with the computer storage criterion.

2.3.1   Storage Requirements

Computer storage used for sparse matrices typically consists of two parts, *primary storage* used to hold the numerical values, and *overhead storage*, used for pointers, subscripts and other information needed to record the structure of the matrix and to facilitate access to the numerical values.   Since we must pay for computer storage regardless of how it is used, any evaluation of storage requirements for a solution method must include a description of the way the matrix or matrices involved are to be stored, so that the storage overhead can be included along with the primary storage in the storage requirement. The comparison of two different strategies with respect to the storage criterion may involve basically different data structures, having very different storage overheads. Thus, a method which is superior in terms of reducing primary storage may be inferior when overhead storage is included in the comparison. This point is illustrated pictorially in Figure 2.3.1.

Figure 2.3.1   Primary/Overhead storage for two different methods.

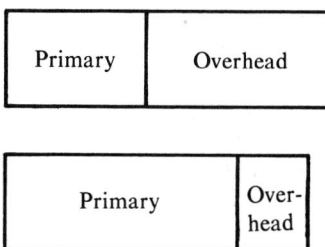

As a simple example, consider the two orderings of a matrix problem in Figure 2.3.2, along with their corresponding factors L and $\tilde{L}$. The elements of the lower triangle of L (excluding the diagonal) are stored row by row in a single array, with a parallel array holding their column subscripts. A third array indicates the position of each row, and a fourth array contains the diagonal elements of L. The matrix $\tilde{L}$ is stored using the so-called envelope storage scheme, described in Chapter 4. Nonzeros in A are denoted by ∗, with ⊛ denoting fill-in components in L or $\tilde{L}$.

Figure 2.3.2   Two different orderings for a sparse matrix A, along with the sparsity patterns of their respective triangular factors L and $\tilde{L}$.

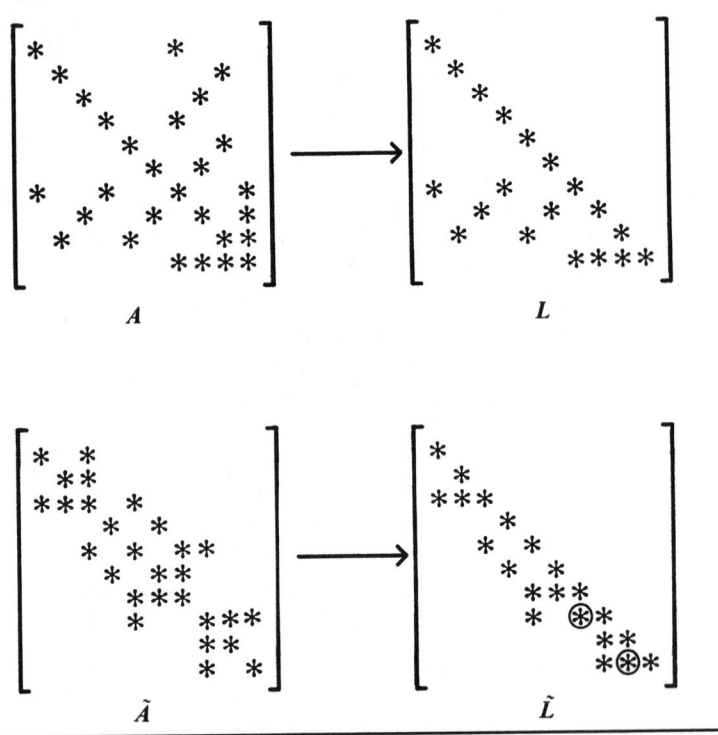

The examples in Figures 2.3.2 and 2.3.3 illustrate some important points about orderings and storage schemes. On the surface, ordering 1, corresponding to A appears to be better than ordering 2 since it yields no fill-in at all, whereas the latter ordering causes two fill components. Moreover, the storage scheme used for $\tilde{L}$ appears to be inferior to that used for L, since the latter actually ignores some sparsity, while all the sparsity in L is exploited. However, because of differences in overhead, the second ordering/storage combination yields the lower *total* storage requirement. Of course the differences here are trivial, but the point is valid. As we increase the sophistication of our storage scheme, exploiting more and more zeros, the primary storage decreases, but the overhead usually increases. There is usually a point where it pays to ignore some zeros, because the overhead storage required to exploit them is more than the decrease in primary storage.

Figure 2.3.3   Storage schemes for the matrices L and $\tilde{L}$ of Figure 2.3.2.

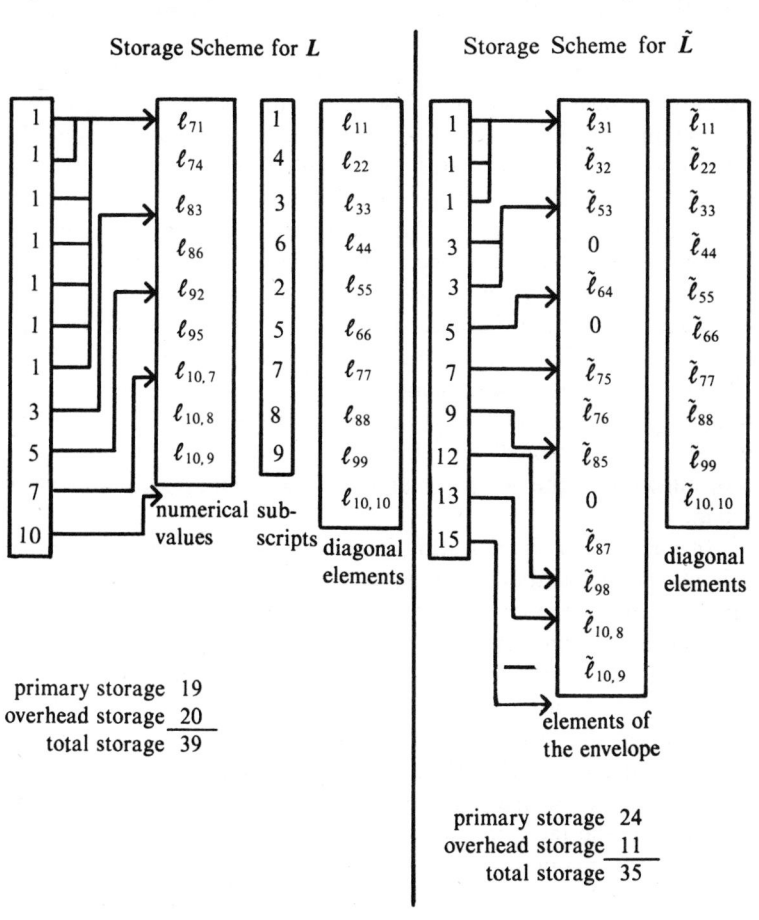

primary storage 19
overhead storage 20
total storage 39

primary storage 24
overhead storage 11
total storage 35

To summarize, the main points in this section are:

1. Storage schemes for sparse matrices involve two components: *primary* storage and *overhead* storage.
2. Comparisons of ordering strategies must take into account the storage scheme to be used, if the comparison is to be practically relevant.

2.3.2   Execution Time

We now turn to computer execution time as a criterion. It is helpful in the discussion to consider the four steps in the entire computation process: ordering, allocation, factorization and solution.

As we shall see in Chapter 9, the execution times required to find different orderings can vary dramatically. But even after we have found the ordering, there is much left to do before we can actually begin the numerical computation. We must set up the appropriate storage scheme for L, and in order to do this we must determine its structure. This allocation step also varies in cost, depending on the ordering and storage scheme used. Finally, as we shall see through the numerous experiments supplied in Chapter 9, differences in storage schemes can lead to substantial differences in the arithmetic operations-per-second output of the factorization and triangular solution subroutines. Normally, the execution of a sparse matrix program will be (or should be) roughly *proportional* to the amount of arithmetic performed. However, differences in orderings and data structures can lead to large differences in the constant of proportionality. Thus, arithmetic operation counts may not be a very reliable measure for comparing solution methods, or at best must be used with care. The constant of proportionality is affected not only by the data structure used, but also by the computer architecture, compiler, and operating system.

In addition to the variation in the respective costs of executing each of the steps above, comparisons of different strategies often depend on the particular context in which a problem is being solved. If the given matrix problem is to be solved only once, a comparison of strategies should surely include the execution time required to produce the ordering and set up the storage scheme.

However, sometimes *many different* problems having the *same structure* must be solved, and it may be reasonable to ignore this initialization cost in comparing methods, since the bulk of the execution time involves the factorization and triangular solutions. In still other circumstances, many systems differing only in their right hand sides must be solved. In this case, it may be reasonable to compare strategies simply on the basis of their respective triangular solution times.

To summarize, the main points of the section are:

1. The overall solution of $Ax = b$ involves four basic steps. Their relative execution times in general vary substantially over different orderings and storage schemes.
2. Depending on the problem context, the execution times of some of the steps mentioned above may be practically irrelevant when comparing methods.

Exercises

2.3.1)   Suppose you have a choice of two methods (method 1 and method 2) for solving a sparse system of equations $Ax = b$ and the criterion for the choice of method is execution time. The ordering and allocation steps for method 1 require a total of 20 seconds, while the corresponding time for method 2 is only 2 seconds. The factorization time for method 1 is 6 seconds and the solve time is .5 seconds, while for method 2 the corresponding execution times are 10 seconds and 1.5 seconds.

   a)   What method would you choose if the system is to be solved only once?

   b)   What method would you choose if twelve systems $Ax = b$, having the *same* sparsity structure but different numerical values in A and b are to be solved?

   c)   What is your answer to b) if only the numerical values of the right side b differ among the different systems?

2.3.2)   Suppose for a given class of sparse positive definite matrix problems you have a choice between two orderings, "turtle" and "hare." Your friend P.C.P. (Pure Complexity Pete, Esq.), shows that the turtle ordering yields triangular factors having $\eta_t(N) \simeq N^{3/2} + N - \sqrt{N}$ nonzeros, where N is the size of the problem. He also shows that the corresponding function for the hare ordering is $\eta_h(N) \simeq 7.75N \log_2(\sqrt{N} + 1) - 24N + 11.5\sqrt{N} \log_2(\sqrt{N} + 1) + 11\sqrt{N} + .75 \log_2(\sqrt{N} + 1)$. Another friend, C.H.H. (Computer Hack Harold), implements linear equation solvers which use storage schemes appropriate for each ordering. Harold finds that for the hare implementation he needs one integer data item (a subscript) for each nonzero element of L, together with 3 pointer arrays of length N. For the turtle implementation, the overhead storage is only N pointers.

   a)   Suppose your choice of methods is based strictly on the total computer storage used to hold L, and that integers and floating point numbers each require one computer word. For what values of N would you use the hare implementation?

   b)   What is your answer if Harold changes his programs so that integers are packed three to a computer word?

3/ Some Graph Theory Notation and Its Use in the Study of Sparse Symmetric Matrices

3.0 Introduction

In this chapter we introduce a few basic graph theory notions, and establish their correspondence to matrix concepts. Although rather few results from graph theory have found direct application to the analysis of sparse matrix computations, the notation and concepts are convenient and helpful in describing algorithms and identifying or characterizing matrix structure. Nevertheless, it is easy to become over-committed to the use of graph theory in such analyses, and the result is often to obscure some basically simple ideas in exchange for notational elegance. Thus, although we may sacrifice uniformity, where it is appropriate and aids the presentation, we will give definitions and results in both graph theory and matrix terms. In the same spirit, our intention is to introduce most graph theory notions only as they are required, rather than introducing them all in this section and then referring to them later.

3.1 Basic Terminology and Some Definitions

For our purposes, a *graph* $G = (X,E)$ consists of a finite set of *nodes* or *vertices* together with a set E of *edges*, which are unordered pairs of vertices. An *ordering* {*labelling*} α of $G = (X,E)$ is simply a mapping of $\{1, 2, \ldots, N\}$ onto X, where N denotes the number of nodes of G. Unless we specifically state otherwise, a graph will be unordered; the graph G labelled by α will be denoted by $G^\alpha = (X^\alpha, E)$.

Since our objective in introducing graphs is to facilitate the study of sparse matrices, we now establish the relationship between graphs and matrices. Let A be an N by N symmetric matrix. The *ordered graph of A*, denoted by $G^A = (X^A, E^A)$ is one for which the N vertices of G^A are numbered from 1 to N, and $\{x_i, x_j\} \in E^A$ if and only if $a_{ij} = a_{ji} \neq 0$, $i \neq j$. Here x_i denotes the node of X^A with label i. Figure 3.1.1 illustrates the structure of a matrix and its labelled graph. We denote the i-th diagonal element of a matrix by $\odot$ to emphasize its correspondence with node i of the corresponding graph. Off-diagonal nonzeros are depicted by "*".

Figure 3.1.1   A matrix and its labelled graph, with * denoting a nonzero entry of A.

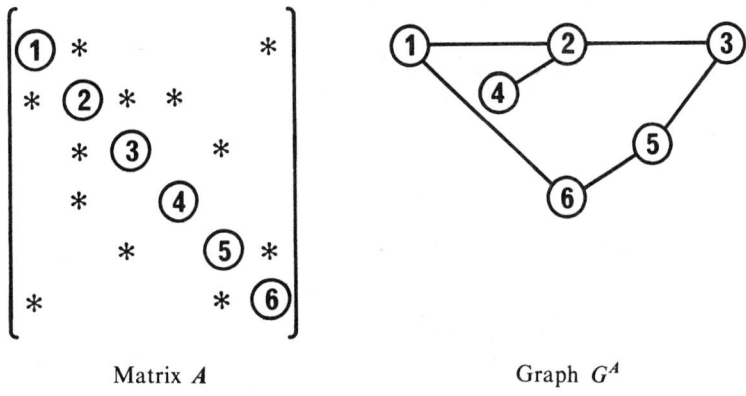

Matrix A Graph G^A

For any N by N *permutation* matrix $P \neq I$, the unlabelled graphs of A and PAP^T are the same but the associated labellings are different. Thus, the unlabelled graph of A represents the structure of A without suggesting any particular ordering. It represents the equivalence class of matrices PAP^T, where P is any N by N permutation matrix. Finding a "good" permutation for A can be regarded as finding a good labelling for its graph. Figure 3.1.2 illustrates these points.

Some graph theory definitions involve *unlabelled* graphs. In order to interpret these definitions in matrix terms, we must have a matrix to refer to, and this immediately implies an ordering on the graph. Although this should not cause confusion, the reader should be careful not to attach any significance to the particular ordering chosen in our matrix examples and interpretations. When we refer to "the matrix corresponding to G," we must either specify some ordering α of G, or understand that some arbitrary ordering is assumed.

Two nodes x and y in G are *adjacent* if $\{x,y\} \in E$. For $Y \subset X$,[1] the *adjacent set* of Y, denoted by $Adj(Y)$, is

$$Adj(Y) = \left\{ x \in X - Y \mid \{x,y\} \in E \text{ for some } y \in Y \right\} . \quad (3.1.1)$$

In words, $Adj(Y)$ is simply the set of nodes in G which are not in Y but are adjacent to at least one node in Y. Figure 3.1.3 illustrates the matrix interpretation of $Adj(Y)$. For convenience, the set Y has been labelled consecutively. When Y is the single node y, we will write $Adj(y)$ rather than the formally correct $Adj(\{y\})$.

For $Y \subset X$, the *degree* of Y, denoted by $Deg(Y)$, is simply the

---

[1] Here and elsewhere in this book, the notation $Y \subset X$ means that Y may be equal to X. When Y is intended to be a proper subset of X, we will explicitly indicate so.

number $|Adj(Y)|$, where $|S|$ denotes the number of members in the set S. Again, when Y is a single node y we write $Deg(y)$ rather than $Deg(\{y\})$. For example, in Figure 3.1.3, $Deg(x_2) = 3$.

Figure 3.1.2 Graph of Figure 3.1.1 with different labellings, and the corresponding matrix structures. Here P and Q denote permutation matrices.

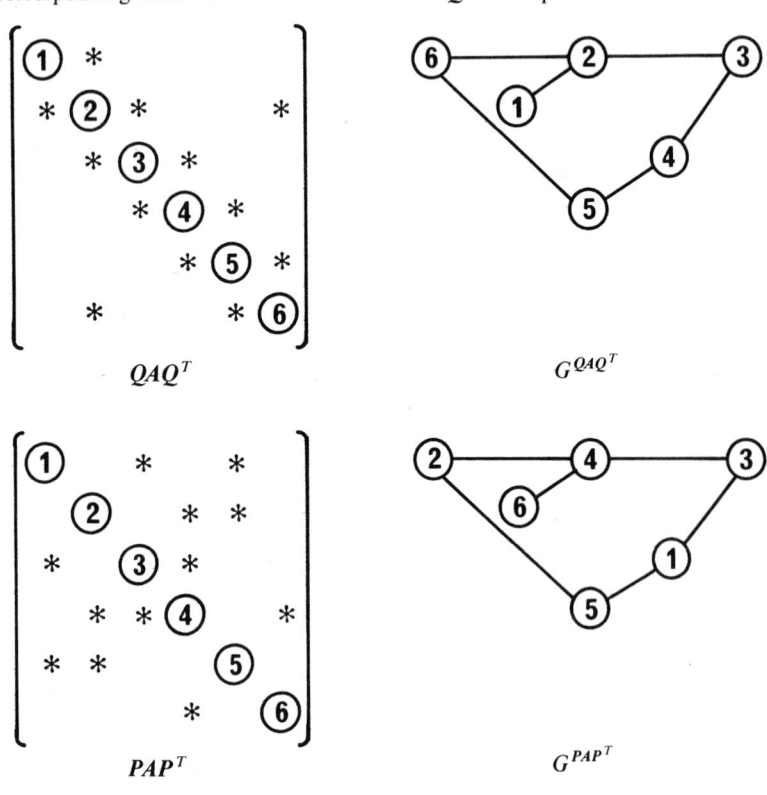

A *subgraph* $G' = (X',E')$ of G is a graph for which $X' \subset X$ and $E' \subseteq E$. For $Y \subset X$, the *section graph* $G(Y)$ is the subgraph $(Y,E(Y))$, where

$$E(Y) = \{\{x,y\} \in E \mid x \in Y,\ y \in Y\} . \tag{3.1.2}$$

In matrix terms, the section graph $G(Y)$ is the graph of the matrix obtained by deleting all rows and columns from the matrix of G except those corresponding to Y. This is illustrated in Figure 3.1.4.

A section graph is said to be a *clique* if the nodes in the subgraph are pairwise adjacent. In matrix terms, a clique corresponds to a full submatrix. For example $G(\{x_2,x_4\})$ is a clique.

Figure 3.1.3   An illustration of the adjacent set of a set $Y \subset X$.

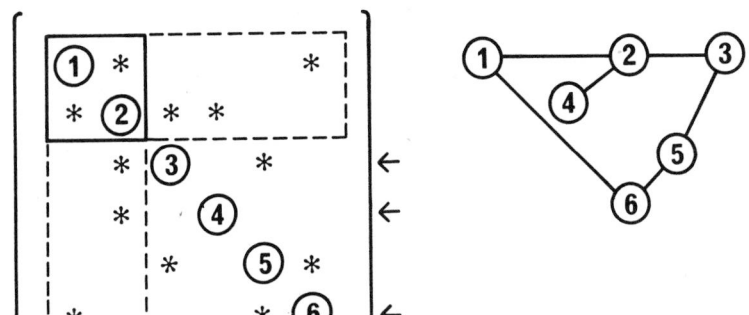

$$A \qquad\qquad\qquad G^A$$

$$Y = \{x_1, x_2\}, \qquad Adj(Y) = \{x_3, x_4, x_6\}.$$

Figure 3.1.4   Example of a section graph $G(Y)$ and the matrix correspondence.   The original graph G is that of Figure 3.1.1.

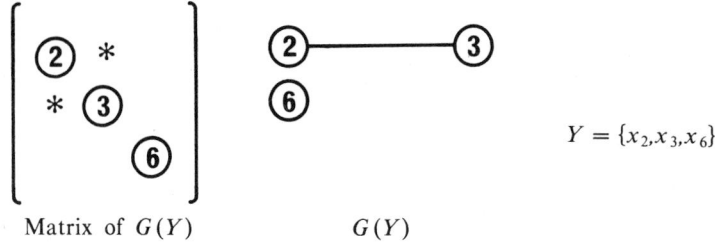

Matrix of $G(Y)$ $\qquad\qquad$ $G(Y)$

$$Y = \{x_2, x_3, x_6\}$$

The example in Figure 3.1.4 illustrates a concept we now explore, namely that of the *connectedness* of a graph. For distinct nodes x and y in G, a *path* from x to y of length $\ell \geq 1$ is an ordered set of $\ell + 1$ distinct nodes $(v_1, v_2, \ldots, v_{\ell+1})$ such that $v_{i+1} \in Adj(v_i)$, $i = 1$, $2, \ldots, \ell$ with $v_1 = x$ and $v_{\ell+1} = y$. A graph is *connected* if every pair of distinct nodes is joined by at least one path. Otherwise G is disconnected, and consists of two or more *connected components*.   In matrix terms, it should be clear that if G is disconnected and consists of k connected components and each component is labelled consecutively, the corresponding matrix will be *block diagonal*, the corresponding matrix will be *block diagonal*, with each diagonal block corresponding to a connected component. The graph $G(Y)$ in Figure 3.1.4 is so ordered, and the corresponding matrix is block diagonal.   Figure 3.1.5 shows a path in a graph and its interpretation in matrix terms.

Figure 3.1.5   A path in a graph and the corresponding matrix interpretation.

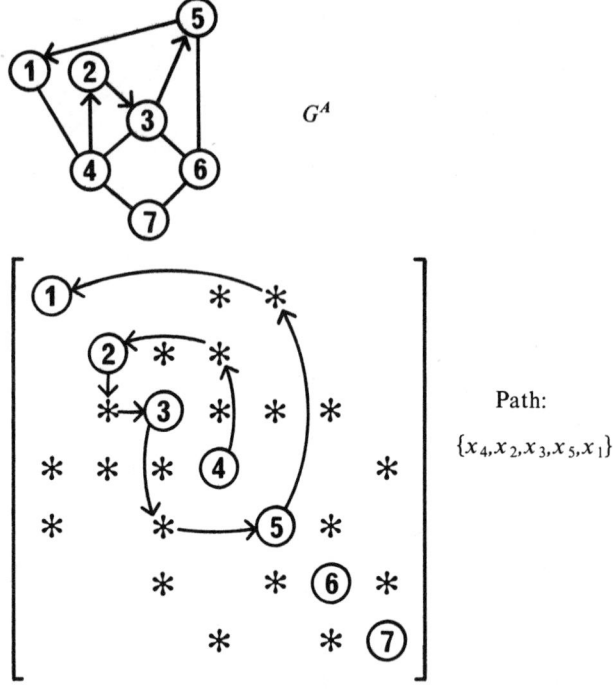

G^A

Path:

$\{x_4,x_2,x_3,x_5,x_1\}$

Finally, the set $Y \subset X$ is a *separator* of the connected graph G if the section graph $G(X-Y)$ is disconnected. Thus, for example, $Y = \{x_3,x_4,x_5\}$ is a separator of the graph of Figure 3.1.5, since $G(X-Y)$ has three components having node sets $\{x_1\}$, $\{x_2\}$, and $\{x_6,x_7\}$.

Exercises

3.1.1)   A symmetric matrix A is said to be *reducible* if there exists a permutation matrix P such that

$$P^T A P = \begin{pmatrix} A_{11} & 0 \\ 0 & A_{22} \end{pmatrix}.$$

Otherwise, A is said to be *irreducible*. Show that a symmetric matrix A is irreducible if and only if its associated graph G^A is connected.

3.1.2)   Let A be a symmetric matrix. Show that the matrix A has the *propagation property* (see Exercise 2.2.3) if and only if there exists the path $(x_1, x_2, \ldots, x_N)$ in the associated graph G^A.

3.1.3) Characterize the graphs associated with the following matrices.

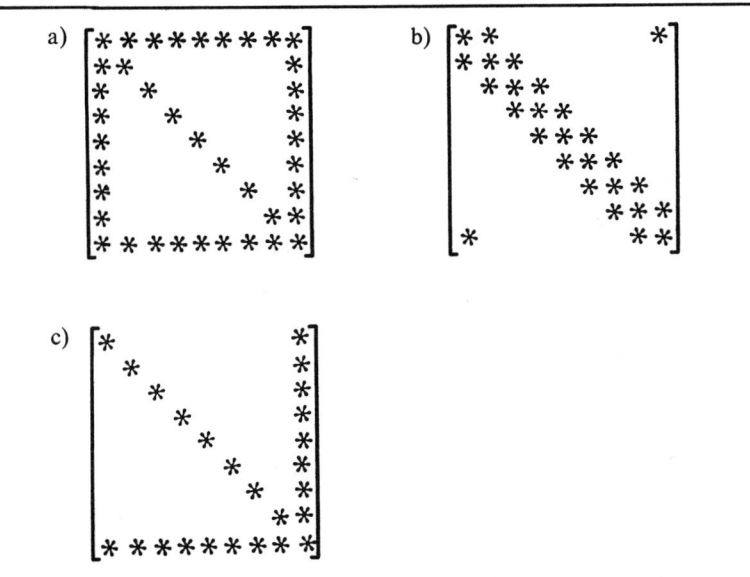

3.2 Computer Representation of Graphs

In general, the performances of graph algorithms are quite sensitive to the way the graphs are represented. For our purposes, the basic operation used is that of retrieving adjacency relations between nodes. So, we need a representation which provides the adjacency properties of the graph and which is economical in storage.

Let $G = (X, E)$ be a graph with N nodes. An *adjacency list* for $x \in X$ is a list containing all the nodes in $Adj(x)$. An *adjacency structure* for G is simply the set of adjacency lists for all $x \in X$. Such a structure can be implemented quite simply and economically by storing the adjacency lists sequentially in a one-dimensional array ADJNCY along with an index array XADJ of length $N + 1$ containing pointers to the beginning of each adjacency list in ADJNCY. An example is shown in Figure 3.2.1. It is often convenient for programming purposes to have an extra entry in XADJ such that XADJ$(N + 1)$ points to the next available storage location in ADJNCY, as shown in Figure 3.2.1. Clearly the total storage requirement for this storage scheme is then $|X| + 2|E| + 1$.

Figure 3.2.1 Example of an adjacency structure.

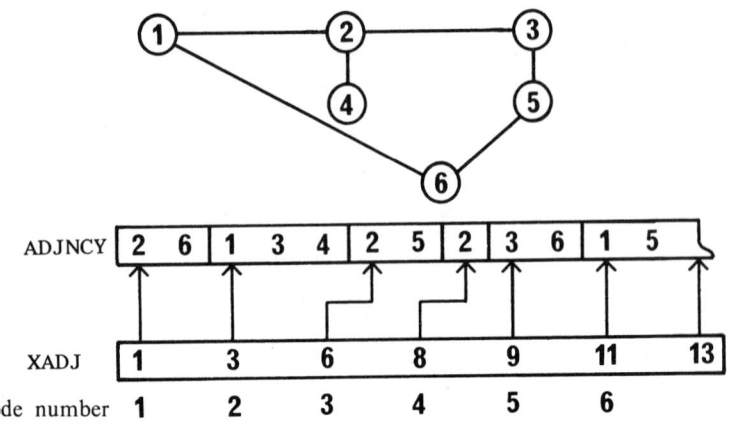

To examine all the neighbors of a node, the following program segment can be used.

```
       ⋮
NBRBEG = XADJ(NODE)
NBREND = XADJ(NODE + 1) - 1
IF (NBREND .LT. NBRBEG) GO TO 200
DO 100 I = NBRBEG, NBREND
        NABOR = ADJNCY(I)
            ⋮
100    CONTINUE
200        ⋮
```

Although all our implementations involving graphs use the storage scheme described above, several others are often used. A common storage scheme is a simple *connection table*, having N rows and m columns, where $m = \max\{Deg(x) \mid x \in X\}$. The adjacency list for node i is stored in row i. This storage scheme may be quite inefficient if a substantial number of the nodes have degrees less than m. An example of a connection table for the graph of Figure 3.2.1 is given in Figure 3.2.2.

The first two schemes described have a distinct disadvantage. Unless the degrees of the nodes are known *a priori*, it is difficult to construct the storage scheme when the graph is provided as a list of edges because we do not know the ultimate size of the adjacency lists. We can overcome this difficulty by introducing a *link field*. Figure 3.2.3 illustrates an example of such a scheme for the graph of Figure 3.2.1. The pointer HEAD(i) starts the adjacency list for node i, with NBRS containing a neighbor of node i and LINK containing the pointer to the location of the next neighbor of node i. For example, to

Figure 3.2.2 Connection table for the graph of Figure 3.2.1. Unused positions in the table are indicated by – .

Node	Neighbors		
1	2	6	–
2	1	3	4
3	2	5	–
4	2	–	–
5	3	6	–
6	1	5	–

retrieve the neighbors of node 5, we retrieve HEAD(5) which is 8. We then examine NBRS(8) which yields 3, one of the neighbors of node 5. We then retrieve LINK(8), which is 2, implying that the next neighbor of node 5 is NBRS(2), which is 6. Finally, we discover that LINK(2) $= -5$, which indicates the end of the adjacency list for node 5. (In general, a negative link of $-i$ indicates the end of the adjacency list for node i.) The storage requirement for this graph representation is $|X| + 4|E|$, which is substantially more than the adjacency list scheme we use in our programs.

Provided there is enough space in the arrays NBRS and LINK, new edges can be added with ease. For example, to add the edge $\{3, 6\}$ to the adjacency structure, we would adjust the adjacency list of node 3 by setting LINK(13) to 1, NBRS(13) to 6, and HEAD(3) to 13. The adjacency list of node 6 would be similarly changed by setting LINK(14) to 5, NBRS(14) to 3, and HEAD(6) to 14.

3.3   Some General Information on the Subroutines which Operate on Graphs

Numerous subroutines that operate on graphs are described in subsequent chapters. In all these subroutines, the graph $G = (X,E)$ is stored using the integer array pair (XADJ, ADJNCY), as described in Section 3.2. In addition, many of the subroutines share other common parameters. In order to avoid repeatedly describing these parameters in subsequent chapters, we discuss their role here, and refer to them later as required.

It should be clear that the mere fact that a graph is stored using the (XADJ, ADJNCY) array pair *implies* a particular labelling of the graph. This ordering will be referred to as the *original* numbering, and when we refer to "node i," it is this numbering we mean. When a subroutine finds a new ordering, the ordering is stored in an array PERM, where PERM(i) $= k$ means the original node number k is the i-th node in the new ordering. We often use a related permutation vector

Figure 3.2.3 Adjacency linked lists for the graph of Figure 3.2.1.

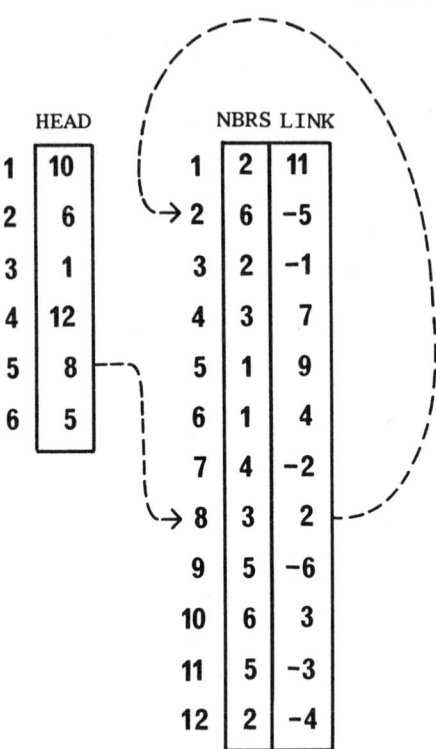

INVP of length N (the inverse permutation) which satisfies INVP(PERM(i)) = i. That is, INVP(k) gives the position in PERM where the node originally numbered k resides.

It is necessary in many of our algorithms to perform operations only on certain section subgraphs of the graph G. To implement these operations, many of our subroutines have an integer array MASK, of length N, which is used to prescribe such a subgraph. The subroutines only consider those nodes i for which MASK(i) $\neq 0$. Figure 3.3.1 contains an example illustrating the role of the integer array MASK.

Finally, some of our subroutines have a single node number, usually called ROOT, as an argument, with MASK(ROOT) $\neq 0$. These subroutines typically operate on the *connected component* of the section subgraph prescribed by MASK which contains the node ROOT. That is, the combination of ROOT and MASK determine the connected subgraph of G to be processed. We will often use the phrase "the component

Figure 3.3.1 An example showing how the array MASK can be used to prescribe a subgraph of G.

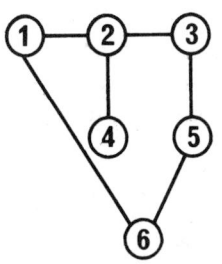

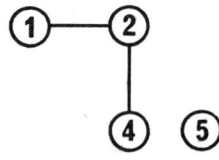

i	MASK(i)
1	**1**
2	**1**
3	**0**
4	**1**
5	**1**
6	**0**

Graph G, labelled as implied by the storage scheme, as in Figure 3.2.1.

Subgraph of G prescribed by MASK.

prescribed by ROOT and MASK" to refer to this connected subgraph.   For example, the combination of ROOT $= 2$ along with the array MASK and graph G in Figure 3.3.1 would specify the graph

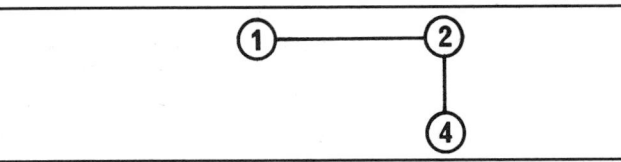

To summarize, some frequently used parameters in our subroutines, along with their contents are listed as follows:

(XADJ, ADJNCY)	the array pair which stores the graph in its original ordering. The original labels of the nodes adjacent to node i are found in ADJNCY(k), XADJ(i) $\leq k <$ XADJ($i+1$), with XADJ($N+1$) $= 2\lvert E \rvert + 1$.
PERM	an integer array of length N containing a new ordering.
INVP	an integer array of length N containing the inverse of the permutation.
MASK	an integer array of length N used to prescribe a section subgraph of G. Subroutines ignore nodes for which MASK(i) $= 0$.
ROOT	a node number of which MASK(ROOT) $\neq 0$. The subroutine usually operates on the component of the subgraph specified by MASK which contains the node ROOT.

Exercises

3.3.1)   Suppose we represent a graph $G = (X,E)$ using a *lower adjacency structure*. That is, instead of storing the entire $Adj(x)$ for each node x, we only store those nodes in $Adj(x)$ with labels larger than that of x. For example, the graph of Figure 3.2.1 could be represented as shown below, using the pair of arrays LADJ and XLADJ.

LADJ	2	6	3	4	5	6	

XLADJ	1	3	5	6	6	7	7
node number	1	2	3	4	5	6	

Design a subroutine that transforms a lower adjacency structure to the entire adjacency structure. Assume you have an array LADJ of length $2|E|$ containing the lower adjacency structure in its first $|E|$ positions, and the array XLADJ. In addition, you have a temporary array of length $|X|$. When the subroutine completes execution, the arrays XLADJ and LADJ should contain the elements of XADJ and ADJNCY as described in Section 3.2.

3.3.2)   Suppose a disconnected graph $G = (X,E)$ is stored in the pair of arrays XADJ and ADJNCY, as described in Section 3.2. Design a subroutine which accepts as input a node $x \in X$, and returns the nodes in the connected component of G which contains x. Be sure to describe the parameters of the subroutine, and any auxiliary storage you require.

3.3.3)   Suppose a (possibly disconnected) graph $G = (X,E)$ is stored in the pair of arrays XADJ and ADJNCY as described in Section 3.1. Suppose a subset $Y \subset X$ is specified by an integer array MASK of length N as described in Section 3.3. Design and implement a subroutine which accepts as input the number N, the arrays XADJ, ADJNCY, and MASK, and returns the number of connected components in the section subgraph $G(Y)$. You may need a temporary array of length N in order to make your implementation simple and easy to understand.

3.3.4)   Suppose the graph of the matrix A is stored in the array pair (XADJ, ADJNCY), as described in Section 3.2, and suppose the arrays PERM and INVP correspond to the permutation matrices P and P^T, as described in Section 3.3. Write a subroutine to list the column subscript of the first nonzero element in each row of the matrix PAP^T. Your subroutine should also print the *number* of nonzeros to the left of the diagonal in each row of PAP^T.

3.3.5) Design a subroutine as described in Exercise 3.3.4, with the additional feature that it only operates on the submatrix of PAP^T specified by the array MASK.

3.3.6) Suppose a graph is to be input as a sequence of edges (pairs of node numbers), and the size of the adjacency lists is not known beforehand. Design and implement a subroutine called INSERT which could be used to construct the linked data structure as exemplified by Figure 3.2.3. Be sure to describe the parameter list carefully, and consider how the arrays are to be initialized. You should not assume that $|X|$ and $|E|$ are known beforehand. Be sure to handle abnormal conditions, such as when the arrays are not large enough to accommodate all the edges, repeated input of the same edge, etc.

3.3.7) Suppose the graph of a matrix A is stored in the array pair (XADJ, ADJNCY), as described in Section 3.2. Design and implement a subroutine which accepts as input this array pair, along with two node numbers i and j, and determines whether there is a path joining them in the graph. If there is, then the subroutine returns the length of a shortest such path; otherwise it returns zero. Describe any temporary arrays you need.

3.3.8) Design and implement a subroutine as described in Exercise 3.3.7, with the additional feature that it only operates on the subgraph specified by the array MASK.

4/ Band and Envelope Methods

4.0 Introduction

In this chapter we consider one of the simplest methods for solving sparse systems, the band schemes and the closely related envelope or profile methods. Loosely speaking, the objective is to order the matrix so that the nonzeros in PAP^T are clustered "near" the main diagonal. Since this property is retained in the corresponding Cholesky factor L, such orderings appear to be attractive in reducing fill, and are widely used in practice (Cuthill 1972, Felippa 1975, Melosh and Bamford 1969).

Although these orderings are often far from optimal in the least-arithmetic or least-fill senses, they are often an attractive practical compromise. In general the programs and data structures needed to exploit the sparsity that these orderings provide are relatively simple; that is, the storage and computational overhead involved in using the orderings tends to be small compared to more sophisticated orderings. (Recall our remarks in Section 2.3.) The orderings themselves also tend to be much cheaper to obtain than more (theoretically) efficient orderings. For small problems, and even moderate size problems which are to be solved only a few times, the methods described in this chapter should be seriously considered.

4.1 The Band Method

Let A be an N by N symmetric positive definite matrix, with entries a_{ij}. For the i-th row of A, $i = 1, 2, \ldots, N$, let

$$f_i(A) = \min\{j \mid a_{ij} \neq 0\} \, ,$$

and

$$\beta_i(A) = i - f_i(A) \, .$$

The number $f_i(A)$ is simply the column subscript of the first nonzero component in row i of A. Since the diagonal entries a_{ii} are positive, we have

$$f_i(A) \leq i \quad \text{and} \quad \beta_i(A) \geq 0 \, .$$

Following Cuthill and McKee, we define the *bandwidth* of A by[1]

$$\beta(A) = \max\{\beta_i(A) \mid 1 \leq i \leq N\}$$

$$= \max\{|i-j| \mid a_{ij} \neq 0\} \, .$$

The number $\beta_i(A)$ is called the *i-th bandwidth* of A. We define the *band* of A as

$$Band\,(A) = \{\{i,j\} \mid 0 < i-j \leq \beta(A)\} \, , \qquad (4.1.1)$$

which is the region within $\beta(A)$ locations of the main diagonal. Unordered pairs $\{i,j\}$ are used in (4.1.1) instead of ordered pairs because A is symmetric. The matrix example in Figure 4.1.1 has a bandwidth of 3. Matrices with a bandwidth of one are called tridiagonal matrices.

Figure 4.1.1   Example showing $f_i(A)$ and $\beta_i(A)$.

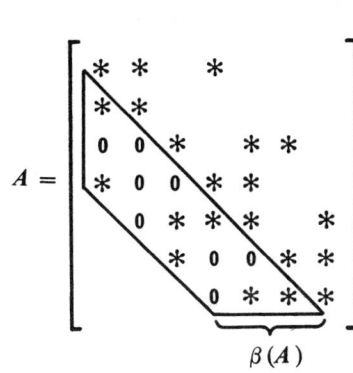

i	$f_i(A)$	$\beta_i(A)$
1	1	0
2	1	1
3	3	0
4	1	3
5	3	2
6	3	3
7	5	2

Implicit in the use of the band method is that zeros outside *Band*(A) are ignored; zeros inside the band are usually stored, although often exploited as far as the actual computation is concerned. This exploitation of zeros is possible in the direct solution because

$$Band\,(A) = Band\,(L + L^T) \, ,$$

a relation that will be proved in Section 4.2 when the envelope method is considered.

---

[1] Other authors define the bandwidth of A to be $2\beta(A)+1$.

A common method for storing a symmetric band matrix A is the so-called *diagonal storage scheme* (Martin 1971). The $\beta(A)$ sub-diagonals of the lower triangle of A which comprise *Band*(A) and the main diagonal of A are stored as the columns of an N by $(\beta(A)+1)$ rectangular array, as shown in Figure 4.1.2. This storage scheme is very simple, and is quite efficient as long as $\beta_i(A)$ does not vary too much with i.

Figure 4.1.2 The diagonal storage scheme.

$$
\begin{pmatrix}
a_{11} \\
a_{21} & a_{22} & & \text{symmetric} \\
0 & 0 & a_{33} \\
a_{41} & 0 & 0 & a_{44} \\
 & 0 & a_{53} & a_{54} & a_{55} \\
 & & a_{63} & 0 & 0 & a_{66} \\
 & & & 0 & a_{75} & a_{76} & a_{77}
\end{pmatrix}
$$

Matrix A

$$
\begin{bmatrix}
- & - & - & a_{11} \\
- & - & a_{21} & a_{22} \\
- & 0 & 0 & a_{33} \\
a_{41} & 0 & 0 & a_{44} \\
0 & a_{53} & a_{54} & a_{55} \\
a_{63} & 0 & 0 & a_{66} \\
0 & a_{75} & a_{76} & a_{77}
\end{bmatrix}
$$

Storage Array

Theorem 4.1.1

The number of operations required to factor the matrix A having bandwidth β, assuming *Band*$(L + L^T)$ is full, is

$$
\frac{1}{2}\beta(\beta+3)N - \frac{\beta^3}{3} - \beta^2 - \frac{2}{3}\beta .
$$

Proof   The result follows from Theorem 2.1.2 and the observation
that

$$\eta(L_{*i}) = \begin{cases} \beta + 1 & \text{for } 1 \le i \le N - \beta \\ N - i + 1 & \text{for } N - \beta < i \le N \end{cases}$$

$\square$

Theorem 4.1.2
Let A be as in Theorem 4.1.1. Then the number of operations required
to solve the matrix problem $Ax = b$, given the Cholesky factor L of A,
is

$$2(\beta + 1)N - \beta(\beta + 1) .$$

Proof   The result follows from Theorem 2.1.2 and the definition of
$\eta(L_{*i})$ given in the proof of Theorem 4.1.1. $\square$

As mentioned above, the attraction of this approach is its simplicity.
However, it has some potentially serious weaknesses. First, if $\beta_i(A)$
varies widely with i, the diagonal storage scheme illustrated in
Figure 4.1.2 will be inefficient. Moreover, as we shall see later, there
are some very sparse problems which can be solved very efficiently, but
which cannot be ordered to have a small bandwidth (see Figure 4.2.3).
Thus, there are problems for which band methods are simply
inappropriate. Perhaps the most persuasive reason for not being very
enthusiastic about band schemes is that the envelope schemes discussed
in the next section share all the advantages of simplicity enjoyed by
band schemes, with very few of the disadvantages.

Exercises

4.1.1)   Suppose A is an N by N symmetric positive definite matrix
with bandwidth β. You have two sets of numerical subroutines for solving $Ax = b$. One set stores A (over-written by L
during the factorization) as a full lower triangular matrix by
storing the rows of the lower triangular part row by row in a
one dimensional array, in the sequence $a_{11}, a_{21}, a_{22}, a_{31}, \ldots,$
$a_{n,n-1}, a_{n,n}$. The other set of subroutines stores A (again
over-written by L during the factorization) using the diagonal
storage scheme described in this section. For a given β and
N, which scheme would you use if you were trying to
minimize storage requirements?

4.1.2)   Consider the star graph of N nodes, as shown in Figure 4.2.3(a). Prove that *any* ordering of this graph yields a bandwidth of at least $\lceil (N-1)/2 \rceil$.

4.2   The Envelope Method

4.2.1   *Matrix Formulation*

A slightly more sophisticated scheme for exploiting sparsity is the so-called *envelope* or *profile* method, which simply takes advantage of the variation in $\beta_i(A)$ with i. The *envelope* of A, denoted by $Env(A)$, is defined by

$$Env(A) = \{\{i,j\} \mid 0 < i - j \le \beta_i(A)\} .$$

In terms of the column subscripts $f_i(A)$, we have

$$Env(A) = \{\{i,j\} \mid f_i(A) \le j < i\} .$$

The quantity $|Env(A)|$ is called the *profile* or *envelope size* of A, and is given by

$$|Env(A)| = \sum_{i=1}^{N} \beta_i(A) .$$

Figure 4.2.1   Illustration of the envelope of A. Circled elements denote fill elements of L.

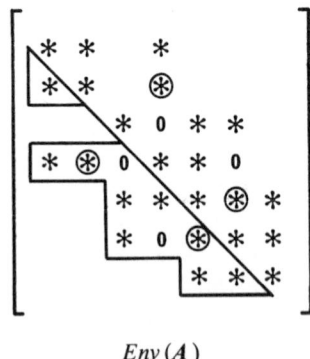

$Env(A)$

Lemma 4.2.1

$$Env(A) = Env(L + L^T) .$$

Proof   We prove the lemma by induction on the dimension N. Assume that the result holds for $N-1$ by $N-1$ matrices. Let A be an N by N symmetric matrix partitioned as

$$A = \begin{pmatrix} M & u \\ u^T & s \end{pmatrix} ,$$

where s is a scalar, u is a vector of length $N-1$, and M is an $N-1$ by $N-1$ nonsingular matrix factored as $L_M L_M^T$. By the inductive assumption, we have $Env(M) = Env(L_M + L_M^T)$. If LL^T is the symmetric factorization of A, the triangular factor L can be partitioned as

$$L = \begin{pmatrix} L_M & 0 \\ w^T & t \end{pmatrix} ,$$

where t is a scalar, and w is a vector of length $N-1$. It is then sufficient to show that $f_N(A) = f_N(L + L^T)$.

From (2.1.4), the vectors u and w are related by

$$L_M w = u .$$

But $u_i = 0$ for $1 \le i < f_N(A)$ and the entry $u_{f_N(A)}$ is nonzero. By Lemmas 2.2.3 and 2.2.4, we have $w_i = 0$ for $1 \le i < f_N(A)$ and $w_{f_N(A)} \ne 0$. Hence $f_N(A) = f_N(L + L^T)$, so that

$$Env(A) = Env(L + L^T) .$$

$\square$

Theorem 4.2.2

$$Env(A) \subset Band(A) .$$

Proof   It follows from the definitions of *Band* and *Env*. $\square$

Lemma 4.2.1 justifies the exploitation of zeros outside the envelope or the band region. Assuming that only those zeros outside $Env(A)$ are exploited, we now determine the arithmetic cost in performing the direct solution. In order to compute operation counts, it is helpful to introduce the notion of *frontwidth*.   For a matrix A, the i-th *frontwidth* of A is defined to be

$$\omega_i(A) = \left| \{ k \mid k > i \text{ and } a_{k\ell} \ne 0 \text{ for some } \ell \le i \} \right| .$$

Note that $\omega_i(A)$ is simply the number of "active" rows at the i-th step in the factorization; that is, the number of rows of the envelope of A, which intersect column i. The quantity

$$\omega(A) = \max\{\omega_i(A) \mid 1 \le i \le N\}$$

is usually referred to as the *frontwidth* or *wave front* of A (Irons 1970, Melosh 1969). Figure 4.2.2 illustrates these definitions.

Figure 4.2.2   Illustration of the i-th bandwidth and frontwidth.

$$A = \begin{bmatrix} * & * & & * & & & \\ * & * & & \circledast & & & \\ & & * & 0 & * & * & \\ * & \circledast & 0 & * & * & 0 & \\ & & * & * & * & \circledast & * \\ & & * & 0 & \circledast & * & * \\ & & & & * & * & * \end{bmatrix}$$

i	$\omega_i(A)$	$\beta_i(A)$
1	2	0
2	1	1
3	3	0
4	2	3
5	2	2
6	1	3
7	0	2

The relevance of the notion of frontwidth in the analysis of the envelope method is illustrated by the following.

Lemma 4.2.3

$$|Env(A)| = \sum_{i=1}^{N} \omega_i(A).$$

Theorem 4.2.4

If only those zeros outside the envelope are exploited, the number of operations required to factor A into LL^T is given by

$$\frac{1}{2} \sum_{i=1}^{N} \omega_i(A)(\omega_i(A)+3),$$

and the number of operations required to solve the system $Ax = b$, given the factorization LL^T is

$$2 \sum_{i=1}^{N} (\omega_i(A)+1).$$

Proof   If we treat the envelope of A as full, the number of nonzeros in L_{*i} is simply $\omega_i(A)+1$. The result then follows from Theorem 2.1.2 and Lemma 2.2.1.   □

Although profile schemes appear to represent a rather minor increase in sophistication over band schemes, they can sometimes lead to quite spectacular improvements. To see this, consider the example in Figure 4.2.3 showing two orderings of the same matrix.

Figure 4.2.3(a)   Star graph of N nodes.

Figure 4.2.3(b)   Minimum profile ordering and minimum band ordering for the star graph on N nodes with $N = 9$.

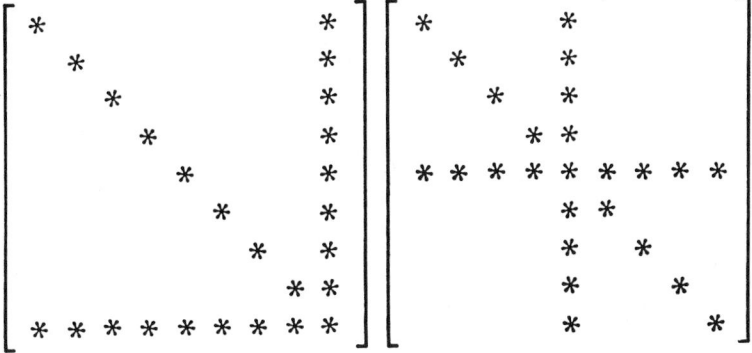

Ordering corresponding to numbering the center node last

Minimum bandwidth ordering

It is not hard to verify that the number of operations required to factor the minimum profile ordered matrix, and the number of nonzeros in the corresponding factor are both $O(N)$, as is the bandwidth. On the other hand, the minimum bandwidth ordering yields an $O(N^3)$ operation count and an L having $O(N^2)$ nonzeros.

Although this example is contrived, numerous practical examples exist where envelope schemes are much more efficient than band schemes. For some examples, see Liu and Sherman (1975).

4.2.2   Graph Interpretation

For an N by N symmetric matrix A, let its associated undirected graph be

$$G^A = (X^A, E^A) \ ,$$

where the node set is labelled as implied by A:

$$X^A = \{x_1, \ldots, x_N\} \ .$$

To provide insight into the combinatorial nature of the envelope method, it is important to give graph theoretic interpretation to the matrix definitions introduced in the previous subsection.

Theorem 4.2.5
For $i < j$, $\{i,j\} \in Env(A)$ if and only if $x_j \in Adj(\{x_1, \ldots, x_i\})$.

Proof   If $x_j \in Adj(\{x_1, \ldots, x_i\})$, then $a_{jk} \neq 0$ for some $k \leq i$ so that $f_j(A) \leq i$ and $\{i,j\} \in Env(A)$.
    Conversely, if $f_j(A) \leq i < j$, this means $x_j \in Adj(x_{f_j(A)})$ which implies $x_j \in Adj(\{x_1, \ldots, x_i\})$.   □

Corollary 4.2.6
For $i = 1, \ldots, N$, $\omega_i(A) = |Adj(\{x_1, \ldots, x_i\})|$.

Proof   From the definition of $\omega_i(A)$, we have

$$\omega_i(A) = \left| \{j > i \mid \{i,j\} \in Env(A)\} \right| \ ,$$

so that the result follows from Theorem 4.2.5.   □

Consider the matrix example and its associated labelled graph in Figure 4.2.4. The respective adjacent sets are

$$
\begin{aligned}
Adj(x_1) &= \{x_2, x_4\} \ , \\
Adj(\{x_1, x_2\}) &= \{x_4\} \ , \\
Adj(\{x_1, x_2, x_3\}) &= \{x_4, x_5, x_6\} \ , \\
Adj(\{x_1, x_2, x_3, x_4\}) &= \{x_5, x_6\} \ , \\
Adj(\{x_1, \ldots, x_5\}) &= \{x_6, x_7\} \ , \\
Adj(\{x_1, \ldots, x_6\}) &= \{x_7\} \ , \\
Adj(\{x_1, \ldots, x_7\}) &= \emptyset \ .
\end{aligned}
$$

Compare them with the row subscripts of the envelope entries in each column.

Figure 4.2.4   A matrix and its associated labelled graph.

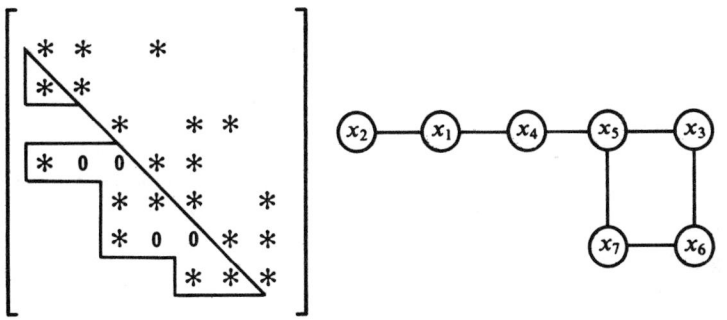

The set $Adj(\{x_1, \ldots, x_i\})$ shall be referred to as the *i-th front* of the labelled graph, and its size the *i-th frontwidth* (as before).

Exercises

4.2.1)   Prove that

$$\frac{1}{2}\sum_{i=1}^{N}\omega_i(A)(\omega_i(A)+3) \leq \frac{1}{2}\sum_{i=1}^{N}\beta_i(A)(\beta_i(A)+3) \ .$$

4.2.2)   A symmetric matrix A is said to have the *monotone profile property* if $f_j(A) \leq f_i(A)$ for $j \leq i$. Show that for monotone profile matrices,

$$\frac{1}{2}\sum_{i=1}^{N}\omega_i(A)(\omega_i(A)+3) = \frac{1}{2}\sum_{i=1}^{N}\beta_i(A)(\beta_i(A)+3) \ .$$

4.2.3)   Prove that the following conditions are equivalent.

a)   for $1 \leq i \leq N$, the section graphs $G(\{x_1, \ldots, x_i\})$ are connected

b)   for $2 \leq i \leq N$, $f_i(A) < i$.

4.2.4)   (*Full Envelope*) Prove that the matrix $L + L^T$ has a full envelope if $f_i(A) < i$ for $2 \leq i \leq N$. Show that $L + L^T$ has a full envelope for monotone profile matrix A.

4.2.5)   Let L be an N by N lower triangular matrix with bandwidth $\beta << N$, and let V be an N by P (pseudo) lower triangular matrix as defined in Exercise 2.2.8. Approximately how many operations are required to compute $L^{-1}V$?

4.2.6)   Let $\{x_1, \ldots, x_N\}$ be the nodes in the graph G^A associated
with a symmetric matrix A.   Show that the following conditions are equivalent.

    a) $Env(A)$ is full,
    b) $Adj(\{x_1, \ldots, x_i\}) \subset Adj(x_i)$ for $1 \leq i \leq N$,
    c) $Adj(\{x_1, \ldots, x_i\}) \cup \{x_i\}$ is a clique for $1 \leq i \leq N$.

4.2.7)   Show that if the graph G^A is connected, then $\omega_i(A) \neq 0$ for
$1 \leq i \leq N - 1$.

4.3   Envelope Orderings

4.3.1   The Reverse Cuthill-McKee Algorithm

Perhaps the most widely used profile reduction ordering algorithm is a
variant of the Cuthill-McKee ordering. In 1969, Cuthill and McKee
(Cuthill 1969) published their algorithm which was primarily designed
to reduce the bandwidth of a sparse symmetric matrix.

    The scheme makes use of the following observation. Let y be a
labelled node, and z an unlabelled neighbor of y. To minimize the
bandwidth of the row associated with z, it is apparent that the node z
should be ordered as soon as possible after y. Figure 4.3.1 illustrates
this point.

Figure 4.3.1   Effect on bandwidth of numbering node z after node y, when
they are connected.

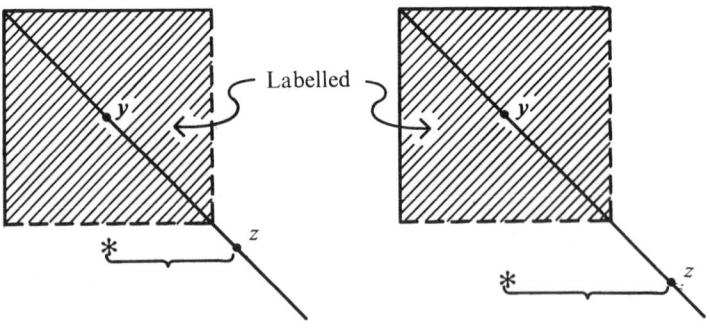

    The Cuthill-McKee scheme may be regarded as a method that
reduces the bandwidth of a matrix via a local minimization of the β_i's.
This suggests that the scheme can be used as a method to reduce the
profile $\sum \beta_i$ of a matrix. George (1971), in his study of the profile

methods, discovered that the ordering obtained by reversing the Cuthill-McKee ordering often turns out to be much superior to the original ordering in terms of profile reduction, although the bandwidth remains unchanged. He called this the *reverse Cuthill-McKee* ordering (RCM). It has since been proved that the reverse scheme is never inferior, as far as envelope storage and envelope operation counts are concerned (Liu 1975).

We describe the RCM algorithm for a connected graph as follows. (The task of determining the starting node in Step 1 is considered in the next section.)

Step 1   Determine a starting node r and assign $x_1 \leftarrow r$.
Step 2   (*Main loop*) For $i = 1, \ldots, N$, find all the unnumbered neighbors of the node x_i and number them in increasing order of degree.
Step 3   (*Reverse ordering*) The reverse Cuthill-McKee ordering is given by $y_1, y_2, \ldots, y_N$ where $y_i = x_{N-i+1}$ for $i = 1, \ldots, N$.

In the case when the graph G^A is disconnected, we can apply the above algorithm to each connected component of the graph. For a given starting node, the algorithm is relatively simple and we go through it in the following example.

Figure 4.3.2   Graph to which the RCM algorithm is to be applied.

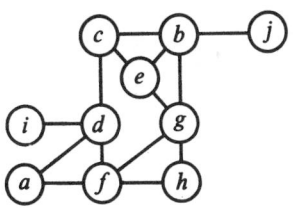

Suppose the node "g" is picked as the starting node, that is $x_1 = g$. Figure 4.3.3 illustrates how nodes are numbered in Step 2 of the algorithm. The resulting reverse Cuthill-McKee ordering is given in Figure 4.3.4, and the envelope size is 22.

The effectiveness of the ordering algorithm depends quite crucially on the choice of the starting node. In the example, if we pick node "a" instead as the starting node, we get a smaller profile of 18. In Section 4.3.3, we present an algorithm which experience has shown to provide a good starting node for the Cuthill-McKee algorithm.

We now establish a rough complexity bound for the execution time of the RCM algorithm, assuming that a starting node is provided. The underlying assumption here is that the execution time of the

Figure 4.3.3   Table showing numbering in Step 2 of the RCM algorithm.

i	Node x_i	Unnumbered neighbors in increasing order of degree
1	g	h, e, b, f
2	h	–
3	e	c
4	b	j
5	f	a, d
6	c	–
7	j	–
8	a	–
9	d	i
10	i	–

Figure 4.3.4   The final ordering and corresponding matrix structure.

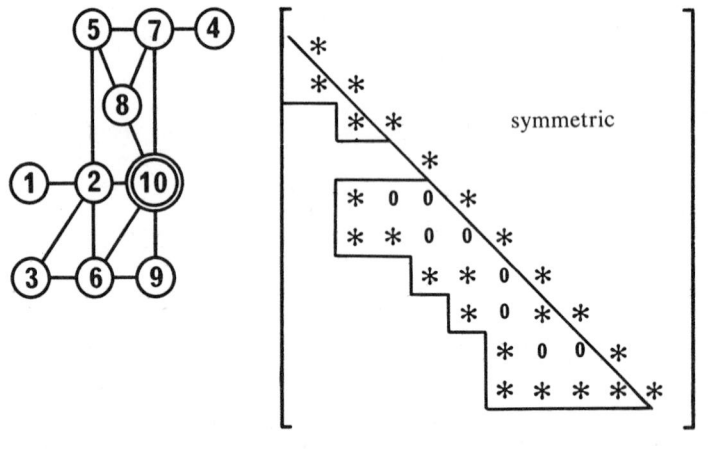

sorting algorithm used is proportional to the number of operations performed, where an operation might be a comparison, or a retrieval of a data item from the adjacency structure used to store the graph.

Theorem 4.3.1

If linear insertion is used for sorting, the time complexity of the RCM algorithm is bounded by $O(m \,|\, E \,|)$, where m is the maximum degree of any node.

Figure 4.3.5 The RCM ordering of the example of Figure 4.3.2, using a different starting node.

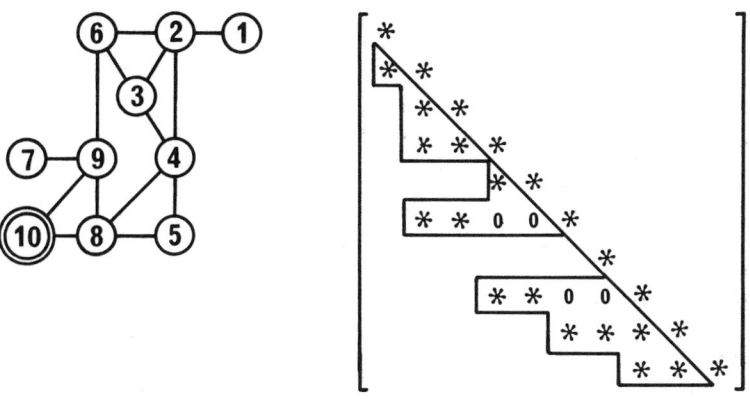

Figure 4.3.6 Diagram showing the effect of reversing the orderings indicated in Figure 4.3.1.

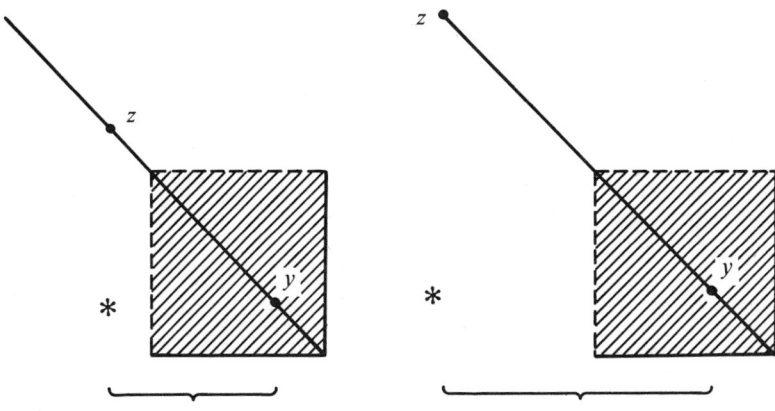

Proof   The major cost is obviously due to Step 2 of the algorithm, since Step 3 can be done in $O(N)$ time. For some constant c, sorting t elements using linear insertion requires ct^2 operations (Aho 1974). Thus, the overall time spent in sorting is less than

$$c \sum_{x \in X} |Deg(x)|^2 \le cm \sum_{x \in X} |Deg(x)| = 2cm |E| .$$

For each index in Step 2, we have to examine the neighbors of node i, in order to retrieve the unnumbered ones for sorting by degree. This sweep through the adjacency structure requires $2|E|$ operations. The computation of the degrees of the nodes requires a further $2|E|$ operations. Thus, the RCM algorithm requires at most

$$4|E| + 2cm|E| + N \text{ operations },$$

where the last term represents the time required to reverse the ordering.   □

4.3.2 Finding a Starting Node

We now turn to the problem of finding a starting node for the RCM algorithm. We consider this problem separately because its solution is useful in connection with several other algorithms we consider in this book. In all cases the objective is to find a pair of nodes which are at maximum or near maximum "distance" apart (defined below). Substantial experience indicates that such nodes are good starting nodes for several ordering algorithms, including the RCM algorithm.

Recall from Section 3.1 that a path of length k from node x_0 to x_k is an ordered set of distinct vertices $(x_0, x_1, \ldots, x_k)$, where $x_i \in Adj(x_{i+1})$ for $0 \le i \le k-1$. The *distance* $d(x,y)$ between two nodes x and y in the connected graph $G = (X,E)$ is simply the length of a shortest path joining nodes x and y. Following Berge (1962), we define the *eccentricity* of a node x to be the quantity

$$\ell(x) = \max\{d(x,y) \mid y \in X\} . \tag{4.3.1}$$

The *diameter* of G is then given by

$$\delta(G) = \max\{\ell(x) \mid x \in X\} ,$$

or equivalently

$$\delta(G) = \max\{d(x,y) \mid x,y \in X\} .$$

A node $x \in X$ is said to be a *peripheral* node if its eccentricity is equal to the diameter of the graph, that is, if $\ell(x) = \delta(G)$. Figure 4.3.7 shows a graph having 8 nodes, with a diameter of 5. The nodes x_2, x_5 and x_7 are peripheral nodes.

Figure 4.3.7   An 8-node graph G with $\delta(G) = 5$.

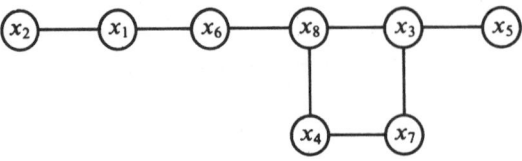

With this terminology established, our objective in this subsection is to describe an efficient heuristic algorithm for finding nodes of high eccentricity. We emphasize that the algorithm is not *guaranteed* to find a peripheral node, or even one that is close to being peripheral. Nevertheless, the nodes found usually do have high eccentricity, and are good starting nodes for the algorithms that employ them. Futhermore, except for some fairly trivial situations, there seems to be no reason to expect that peripheral nodes are any better as starting nodes than those found by this algorithm. Finally, in many situations it is probably too expensive to find peripheral nodes even if it were known to be desirable to use them, since the best known algorithm for finding them has a time complexity bound of $O(|X||E|)$ (Smyth 1974). For most sparse matrix applications this bound would be $O(|X|^2)$. In what follows, we will refer to nodes produced by this algorithm as *pseudo-peripheral* nodes.

We now introduce some notation and terminology which is useful in describing the algorithm. The reader may find it helpful to review the definitions of adjacent set, degree, section graph and connected component, introduced in Section 3.1. A key construct in the algorithm is the *rooted level structure* (Arany 1972).[2] Given a node $x \in X$, the level structure rooted at x is the *partitioning* $\mathcal{L}(x)$ of X satisfying

$$\mathcal{L}(x) = \{L_0(x),\ L_1(x),\ \ldots,\ L_{\ell(x)}(x)\},\qquad(4.3.2)$$

where $\qquad L_0(x) = \{x\}\,,\qquad L_1(x) = Adj(L_0(x))\,,$

and $\quad L_i(x) = Adj(L_{i-1}(x)) - L_{i-2}(x),\ i = 2,\ 3,\ \ldots,\ \ell(x).$ (4.3.3)

The eccentricity $\ell(x)$ of x is called the *length* of $\mathcal{L}(x)$, and the *width* $w(x)$ of $\mathcal{L}(x)$ is defined by

$$w(x) = \max\{|L_i(x)|\ \mid\ 0 \le i \le \ell(x)\}\,.\qquad(4.3.4)$$

In Figure 4.3.8 we show a rooted level structure of the graph of Figure 4.3.7, rooted at the node x_6. Note that $\ell(x_6) = 3$ and $w(x_6) = 3$.

We are now ready to describe the pseudo-peripheral node finding algorithm which is essentially a modification of an algorithm due to Gibbs et al. (1976b). For details on why these modifications were made, see George and Liu (1979b). Using our level structure notation just introduced, the algorithm is as follows.

Step 1 (*Initialization*): Choose an arbitrary node r in X.

Step 2 (*Generate a level structure*): Construct the level structure rooted at r : $\mathcal{L}(r) = \{L_0(r),\ L_1(r),\ \ldots,\ L_{\ell(r)}(r)\}\,.$

---

[2] A *general* level structure is a partitioning $\mathcal{L} = \{L_0, L_1, \ldots, L_\ell\}$ where $Adj(L_0) \subset L_1$, $Adj(L_\ell) \subset L_{\ell-1}$ and $Adj(L_i) \subset L_{i-1} \cup L_{i+1}$, $i = 2,\ 3,\ \ldots,\ \ell-1$.

Figure 4.3.8   A level structure, rooted at x_6, of the graph of Figure 4.3.7.

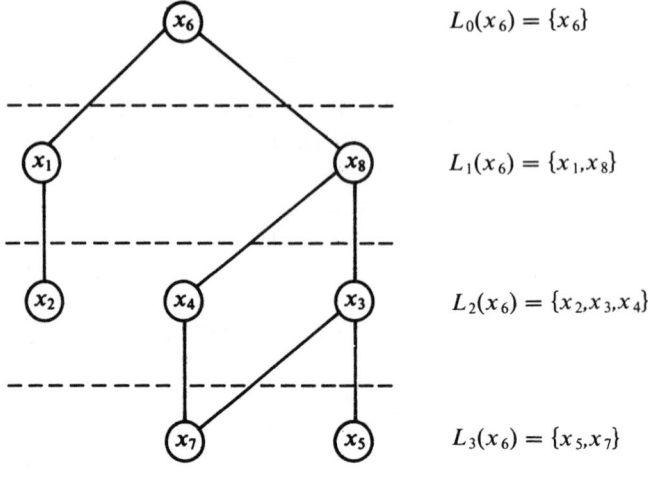

$$L_0(x_6) = \{x_6\}$$

$$L_1(x_6) = \{x_1, x_8\}$$

$$L_2(x_6) = \{x_2, x_3, x_4\}$$

$$L_3(x_6) = \{x_5, x_7\}$$

Step 3   (*Shrink last level*): Choose a node x in $L_{\ell(r)}(r)$ of minimum degree.

Step 4   (*Generate a level structure*):

    a)   Construct the level structure rooted at x:

$$\mathcal{L}(x) = \{L_0(x), \ L_1(x), \ \ldots, \ L_{\ell(x)}(x)\} \ .$$

    b)   If $\ell(x) > \ell(r)$, set $r \leftarrow x$ and go to Step 3.

Step 5   (*Finished*): The node x is a pseudo-peripheral node.

Computer subroutines FNROOT and ROOTLS, which implement this algorithm, are presented and discussed in the next subsection. An example showing the operation of the algorithm is given in Figure 4.3.9. Nodes in level i of the level structures are labelled with the integer i.

4.3.3 Subroutines for Finding a Starting Node

In this subsection we present and describe a pair of subroutines which implement the algorithm of the previous section. In these subroutines, as well as those in Sections 4.3.4 and 4.4.2, several input parameters are the same, and have already been described in Section 3.3. The reader might find it useful to review that section before proceeding.

Figure 4.3.9 An example of the application of the pseudo-peripheral node finding algorithm.

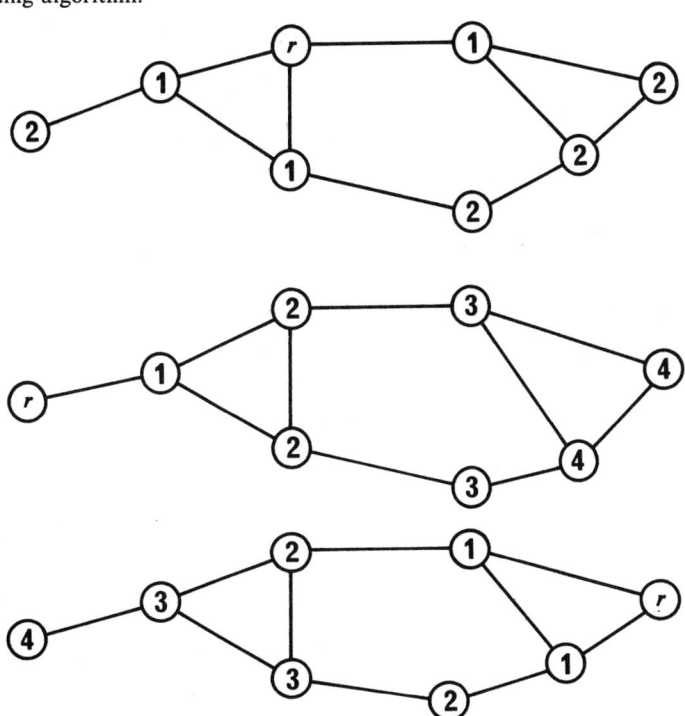

ROOTLS (ROOTed Level Structure)

The purpose of this subroutine is to generate a level structure of the connected component specified by the input parameters ROOT, MASK, XADJ, and ADJNCY, as described in Section 3.3. On exit from the subroutine, the level structure generated is rooted at ROOT, and is contained in the array pair (XLS, LS), with nodes at level k given by LS(j), XLS(k) $\leq j <$ XLS($k + 1$). The number of levels is provided by the variable NLVL. Note that since Fortran does not allow zero subscripts, we cannot have a "zero level," so k here corresponds to level L_{k-1} in the level structure $\mathcal{L}$(ROOT) in section 4.3.2. Thus, NLVL is one greater than the eccentricity of ROOT.

The subroutine finds the nodes level by level; a new level is obtained for each execution of the loop DO 400 .... As each new node is found (in executing the loop DO 300 ...), the node number is placed in the array LS, and its corresponding MASK value is set to zero so it will not be put in LS more than once. After the level structure has been generated, the values of MASK for the nodes in the level structure are reset to 1 (by executing the loop DO 500 ...).

```
C***************************************************************
C***************************************************************
C********      ROOTLS ..... ROOTED LEVEL STRUCTURE      ******
C***************************************************************
C***************************************************************
C
C     PURPOSE - ROOTLS GENERATES THE LEVEL STRUCTURE ROOTED
C        AT THE INPUT NODE CALLED ROOT. ONLY THOSE NODES FOR
C        WHICH MASK IS NONZERO WILL BE CONSIDERED.
C
C     INPUT PARAMETERS -
C        ROOT - THE NODE AT WHICH THE LEVEL STRUCTURE IS TO
C               BE ROOTED.
C        (XADJ, ADJNCY) - ADJACENCY STRUCTURE PAIR FOR THE
C               GIVEN GRAPH.
C        MASK - IS USED TO SPECIFY A SECTION SUBGRAPH. NODES
C               WITH MASK(I)=0 ARE IGNORED.
C
C     OUTPUT PARAMETERS -
C        NLVL - IS THE NUMBER OF LEVELS IN THE LEVEL STRUCTURE.
C        (XLS, LS) - ARRAY PAIR FOR THE ROOTED LEVEL STRUCTURE.
C
C***************************************************************
C
      SUBROUTINE ROOTLS ( ROOT, XADJ, ADJNCY, MASK, NLVL, XLS, LS )
C
C***************************************************************
C
         INTEGER ADJNCY(1), LS(1), MASK(1), XLS(1)
         INTEGER XADJ(1), I, J, JSTOP, JSTRT, LBEGIN,
     1           CCSIZE, LVLEND, LVSIZE, NBR, NLVL,
     1           NODE, ROOT
C
C***************************************************************
C
C        ------------------
C        INITIALIZATION ...
C        ------------------
         MASK(ROOT) = 0
         LS(1) = ROOT
         NLVL = 0
         LVLEND = 0
         CCSIZE = 1
C        ----------------------------------------------------
C        LBEGIN IS THE POINTER TO THE BEGINNING OF THE CURRENT
C        LEVEL, AND LVLEND POINTS TO THE END OF THIS LEVEL.
C        ----------------------------------------------------
  200    LBEGIN = LVLEND + 1
         LVLEND = CCSIZE
         NLVL = NLVL + 1
         XLS(NLVL) = LBEGIN
C        ----------------------------------------------------
C        GENERATE THE NEXT LEVEL BY FINDING ALL THE MASKED
C        NEIGHBORS OF NODES IN THE CURRENT LEVEL.
C        ----------------------------------------------------
         DO 400 I = LBEGIN, LVLEND
            NODE = LS(I)
            JSTRT = XADJ(NODE)
            JSTOP = XADJ(NODE + 1) - 1
            IF ( JSTOP .LT. JSTRT )  GO TO 400
            DO 300 J = JSTRT, JSTOP
               NBR = ADJNCY(J)
               IF (MASK(NBR) .EQ. 0) GO TO 300
                  CCSIZE = CCSIZE + 1
                  LS(CCSIZE) = NBR
                  MASK(NBR) = 0
  300       CONTINUE
```

```
  400     CONTINUE
C         ----------------------------------------
C         COMPUTE THE CURRENT LEVEL WIDTH.
C         IF IT IS NONZERO, GENERATE THE NEXT LEVEL.
C         ----------------------------------------
          LVSIZE = CCSIZE - LVLEND
          IF (LVSIZE .GT. 0 ) GO TO 200
C         -----------------------------------------------------
C         RESET MASK TO ONE FOR THE NODES IN THE LEVEL STRUCTURE.
C         -----------------------------------------------------
          XLS(NLVL+1) = LVLEND + 1
          DO 500 I = 1, CCSIZE
             NODE = LS(I)
             MASK(NODE) = 1
  500     CONTINUE
          RETURN
       END
```

FNROOT (FiNd ROOT)

This subroutine finds a pseudo-peripheral node of a connected component of a given graph, using the algorithm described in Section 4.3.2. The subroutine operates on the connected component specified by the input arguments ROOT, MASK, XADJ, and ADJNCY, as we described in Section 3.3.

The first call to ROOTLS corresponds to Step 2 of the algorithm. If the component consists of a single node or a chain with ROOT as its endpoint, then ROOT is a peripheral node and LS contains its corresponding rooted level structure, so execution terminates. Otherwise, a node of minimum degree in the last level is found (Step 3 of the algorithm; DO 300 ... loop of the subroutine). The new level structure rooted at this node is generated (the call to ROOTLS with label 400) and the termination test (Step 4.b of the algorithm) is performed. If the test fails, control transfers to statement 100 and the procedure is repeated. On exit, ROOT is the node number of the pseudo-peripheral node, and the array pair (XLS, LS) contains the corresponding rooted level structure.

```
C*******************************************************************
C*******************************************************************
C*******    FNROOT ..... FIND PSEUDO-PERIPHERAL NODE    *******
C*******************************************************************
C*******************************************************************
C
C    PURPOSE - FNROOT IMPLEMENTS A MODIFIED VERSION OF THE
C       SCHEME BY GIBBS, POOLE, AND STOCKMEYER TO FIND PSEUDO-
C       PERIPHERAL NODES.  IT DETERMINES SUCH A NODE FOR THE
C       SECTION SUBGRAPH SPECIFIED BY MASK AND ROOT.
C
C    INPUT PARAMETERS -
C       (XADJ, ADJNCY) - ADJACENCY STRUCTURE PAIR FOR THE GRAPH.
C       MASK - SPECIFIES A SECTION SUBGRAPH. NODES FOR WHICH
C              MASK IS ZERO ARE IGNORED BY FNROOT.
C
C    UPDATED PARAMETER -
C       ROOT - ON INPUT, IT (ALONG WITH MASK) DEFINES THE
```

```
C                    COMPONENT FOR WHICH A PSEUDO-PERIPHERAL NODE IS
C                    TO BE FOUND. ON OUTPUT, IT IS THE NODE OBTAINED.
C
C     OUTPUT PARAMETERS -
C        NLVL - IS THE NUMBER OF LEVELS IN THE LEVEL STRUCTURE
C               ROOTED AT THE NODE ROOT.
C        (XLS,LS) - THE LEVEL STRUCTURE ARRAY PAIR CONTAINING
C                   THE LEVEL STRUCTURE FOUND.
C
C     PROGRAM SUBROUTINES -
C        ROOTLS.
C
C*************************************************************
C
      SUBROUTINE FNROOT ( ROOT, XADJ, ADJNCY, MASK, NLVL, XLS, LS )
C
C*************************************************************
C
      INTEGER ADJNCY(1), LS(1), MASK(1), XLS(1)
      INTEGER XADJ(1), CCSIZE, J, JSTRT, K, KSTOP, KSTRT,
     1        MINDEG, NABOR, NDEG, NLVL, NODE, NUNLVL,
     1        ROOT
C
C*************************************************************
C
C        ---------------------------------------------
C        DETERMINE THE LEVEL STRUCTURE ROOTED AT ROOT.
C        ---------------------------------------------
         CALL   ROOTLS ( ROOT, XADJ, ADJNCY, MASK, NLVL, XLS, LS )
         CCSIZE = XLS(NLVL+1) - 1
         IF ( NLVL .EQ. 1 .OR. NLVL .EQ. CCSIZE ) RETURN
C        ---------------------------------------------------
C        PICK A NODE WITH MINIMUM DEGREE FROM THE LAST LEVEL.
C        ---------------------------------------------------
  100    JSTRT = XLS(NLVL)
         MINDEG = CCSIZE
         ROOT = LS(JSTRT)
         IF ( CCSIZE .EQ. JSTRT )  GO TO 400
            DO 300 J = JSTRT, CCSIZE
               NODE = LS(J)
               NDEG = 0
               KSTRT = XADJ(NODE)
               KSTOP = XADJ(NODE+1) - 1
               DO 200 K = KSTRT, KSTOP
                  NABOR = ADJNCY(K)
                  IF ( MASK(NABOR) .GT. 0 )  NDEG = NDEG + 1
  200          CONTINUE
               IF ( NDEG .GE. MINDEG ) GO TO 300
               ROOT = NODE
               MINDEG = NDEG
  300       CONTINUE
C        ---------------------------------------
C        AND GENERATE ITS ROOTED LEVEL STRUCTURE.
C        ---------------------------------------
  400    CALL   ROOTLS ( ROOT, XADJ, ADJNCY, MASK, NUNLVL, XLS, LS )
         IF (NUNLVL .LE. NLVL)  RETURN
         NLVL = NUNLVL
         IF ( NLVL .LT. CCSIZE )  GO TO 100
         RETURN
      END
```

4.3.4 Subroutines for the Reverse Cuthill-McKee Algorithm

In this subsection we describe the three subroutines DEGREE, RCM, and
GENRCM, which together with the subroutines of the previous section
provide a complete implementation for the RCM algorithm described
in Section 4.3.1. The roles of the input parameters ROOT, MASK, XADJ,
ADJNCY, and PERM are as described in Section 3.3. The control relationship among the subroutines is given below.

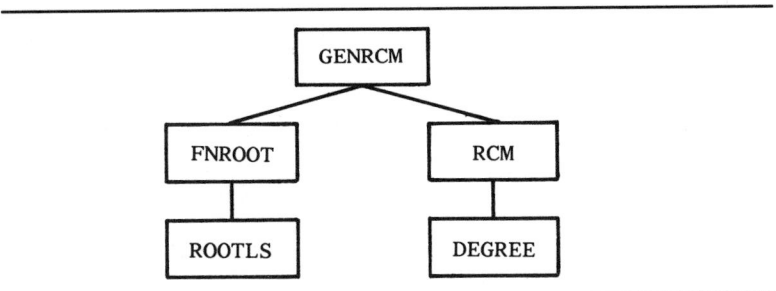

DEGREE
      This subroutine computes the degrees of the nodes in a connected
component of a graph. The subroutine operates on the connected
component specified by the input parameters ROOT, MASK, XADJ, and
ADJNCY.
      Beginning with the first level (containing only ROOT), the degrees
of the nodes are computed one level at a time (loop DO 400 I = ...).
As the neighbors of these nodes are examined (loop
DO 200 J = ...), those which are not already recorded in LS are put
in that array, thus generating the next level of nodes. When a node is
put in LS, its corresponding value of XADJ has its sign changed, so that
the node will only be recorded once. (This function was performed using MASK in the subroutine ROOTLS, but here MASK must be maintained
in its input form so that the degree will be computed correctly). The
variable CCSIZE contains the number of nodes currently in LS. After
all nodes have been found, and their degrees have been computed, the
nodes in LS are used to reset the signs of the corresponding elements of
XADJ to their original values (loop DO 500 I = ...).

```
C***************************************************************
C***************************************************************
C*******       DEGREE ..... DEGREE IN MASKED COMPONENT   ******
C***************************************************************
C***************************************************************
C
C     PURPOSE - THIS ROUTINE COMPUTES THE DEGREES OF THE NODES
C        IN THE CONNECTED COMPONENT SPECIFIED BY MASK AND ROOT.
C        NODES FOR WHICH MASK IS ZERO ARE IGNORED.
C
C     INPUT PARAMETER -
C        ROOT - IS THE INPUT NODE THAT DEFINES THE COMPONENT.
C        (XADJ, ADJNCY) - ADJACENCY STRUCTURE PAIR.
C        MASK - SPECIFIES A SECTION SUBGRAPH.
C
C     OUTPUT PARAMETERS -
C        DEG - ARRAY CONTAINING THE DEGREES OF THE NODES IN
C             THE COMPONENT.
C        CCSIZE-SIZE OF THE COMPONENT SPECIFED BY MASK AND ROOT
C
C     WORKING PARAMETER -
C        LS - A TEMPORARY VECTOR USED TO STORE THE NODES OF THE
C             COMPONENT LEVEL BY LEVEL.
C
C***************************************************************
C
      SUBROUTINE  DEGREE ( ROOT, XADJ, ADJNCY, MASK,
     1                      DEG, CCSIZE, LS )
C
C***************************************************************
C
      INTEGER ADJNCY(1), DEG(1), LS(1), MASK(1)
      INTEGER XADJ(1), CCSIZE, I, IDEG, J, JSTOP, JSTRT,
     1        LBEGIN, LVLEND, LVSIZE, NBR, NODE, ROOT
C
C***************************************************************
C
C
C        ----------------------------------------------------
C        INITIALIZATION ...
C        THE ARRAY XADJ IS USED AS A TEMPORARY MARKER TO
C        INDICATE WHICH NODES HAVE BEEN CONSIDERED SO FAR.
C        ----------------------------------------------------
         LS(1) = ROOT
         XADJ(ROOT) = -XADJ(ROOT)
         LVLEND = 0
         CCSIZE = 1
C        -------------------------------------------------------
C        LBEGIN IS THE POINTER TO THE BEGINNING OF THE CURRENT
C        LEVEL, AND LVLEND POINTS TO THE END OF THIS LEVEL.
C        -------------------------------------------------------
  100    LBEGIN = LVLEND + 1
         LVLEND = CCSIZE
C        ---------------------------------------------------
C        FIND THE DEGREES OF NODES IN THE CURRENT LEVEL,
C        AND AT THE SAME TIME, GENERATE THE NEXT LEVEL.
C        ---------------------------------------------------
         DO 400 I = LBEGIN, LVLEND
            NODE = LS(I)
            JSTRT = -XADJ(NODE)
            JSTOP = IABS(XADJ(NODE + 1)) - 1
            IDEG = 0
            IF ( JSTOP .LT. JSTRT ) GO TO 300
               DO 200 J = JSTRT, JSTOP
                  NBR = ADJNCY(J)
                  IF ( MASK(NBR) .EQ. 0 )  GO TO  200
                     IDEG = IDEG + 1
                     IF ( XADJ(NBR) .LT. 0 ) GO TO 200
```

```
                     XADJ(NBR) = -XADJ(NBR)
                     CCSIZE = CCSIZE + 1
                     LS(CCSIZE) = NBR
      200            CONTINUE
      300            DEG(NODE) = IDEG
      400       CONTINUE
C            -------------------------------------------
C            COMPUTE THE CURRENT LEVEL WIDTH.
C            IF IT IS NONZERO , GENERATE ANOTHER LEVEL.
C            -------------------------------------------
             LVSIZE = CCSIZE - LVLEND
             IF ( LVSIZE .GT. 0 ) GO TO 100
C            -------------------------------------------
C            RESET XADJ TO ITS CORRECT SIGN AND RETURN.
C            -------------------------------------------
             DO 500 I = 1, CCSIZE
                NODE = LS(I)
                XADJ(NODE) = -XADJ(NODE)
      500    CONTINUE
             RETURN
          END
```

RCM (Reverse Cuthill-McKee)

This subroutine applies the RCM algorithm described in Section 4.3.1 to a connected component of a subgraph. It operates on a connected component specified by the input parameters ROOT, MASK, XADJ, and ADJNCY. The starting node is ROOT.

Since the algorithm requires the degrees of the nodes in the component, the first step is to compute those degrees by calling the subroutine DEGREE. The nodes are found and ordered in a level by level fashion; a new level is numbered each time the loop DO 600 I = ... is executed. The loop DO 200 I = ... finds the unnumbered neighbors of a node, and the remainder of the DO 600 loop implements a linear insertion sort to order those neighbors in increasing order of degree. The new ordering is recorded in the array PERM as explained in Section 3.3. The final loop (DO 700 I = ...) reverses the ordering, so that the reverse Cuthill-McKee ordering, rather than the standard Cuthill-McKee ordering is obtained.

Note that just as in the subroutine ROOTLS, MASK(i) is set to zero as node i is recorded. However, unlike ROOTLS, the subroutine RCM does not restore MASK to its original input state. The values of MASK corresponding to the nodes of the connected component that has been numbered remain set to zero on exit from the subroutine.

```
C*************************************************************
C*************************************************************
C********    RCM ..... REVERSE CUTHILL-MCKEE ORDERING   ******
C*************************************************************
C*************************************************************
C
C     PURPOSE - RCM NUMBERS A CONNECTED COMPONENT SPECIFIED BY
C        MASK AND ROOT, USING THE RCM ALGORITHM.
C        THE NUMBERING IS TO BE STARTED AT THE NODE ROOT.
C
C     INPUT PARAMETERS -
C        ROOT - IS THE NODE THAT DEFINES THE CONNECTED
C               COMPONENT AND IT IS USED AS THE STARTING
C               NODE FOR THE RCM ORDERING.
C        (XADJ, ADJNCY) - ADJACENCY STRUCTURE PAIR FOR
C               THE GRAPH.
C
C     UPDATED PARAMETERS -
C        MASK - ONLY THOSE NODES WITH NONZERO INPUT MASK
C               VALUES ARE CONSIDERED BY THE ROUTINE.  THE
C               NODES NUMBERED BY RCM WILL HAVE THEIR
C               MASK VALUES SET TO ZERO.
C
C     OUTPUT PARAMETERS -
C        PERM - WILL CONTAIN THE RCM ORDERING.
C        CCSIZE - IS THE SIZE OF THE CONNECTED COMPONENT
C               THAT HAS BEEN NUMBERED BY RCM.
C
C     WORKING PARAMETER -
C        DEG - IS A TEMPORARY VECTOR USED TO HOLD THE DEGREE
C               OF THE NODES IN THE SECTION GRAPH SPECIFIED
C               BY MASK AND ROOT.
C
C     PROGRAM SUBROUTINES -
C        DEGREE.
C
C*************************************************************
C
      SUBROUTINE  RCM ( ROOT, XADJ, ADJNCY, MASK,
     1                  PERM, CCSIZE, DEG )
C
C*************************************************************
C
      INTEGER ADJNCY(1), DEG(1), MASK(1), PERM(1)
      INTEGER XADJ(1), CCSIZE, FNBR, I, J, JSTOP,
     1        JSTRT, K, L, LBEGIN, LNBR, LPERM,
     1        LVLEND, NBR, NODE, ROOT
C
C*************************************************************
C
C        -----------------------------------------
C        FIND THE DEGREES OF THE NODES IN THE
C        COMPONENT SPECIFIED BY MASK AND ROOT.
C        -----------------------------------------
      CALL  DEGREE ( ROOT, XADJ, ADJNCY, MASK, DEG,
     1               CCSIZE, PERM )
      MASK(ROOT) = 0
      IF ( CCSIZE .LE. 1 ) RETURN
      LVLEND = 0
      LNBR = 1
C        --------------------------------------------------
C        LBEGIN AND LVLEND POINT TO THE BEGINNING AND
C        THE END OF THE CURRENT LEVEL RESPECTIVELY.
C        --------------------------------------------------
  100 LBEGIN = LVLEND + 1
      LVLEND = LNBR
      DO 600 I = LBEGIN, LVLEND
```

```
C          ----------------------------------
C          FOR EACH NODE IN CURRENT LEVEL ...
C          ----------------------------------
           NODE = PERM(I)
           JSTRT = XADJ(NODE)
           JSTOP = XADJ(NODE+1) - 1
C          -----------------------------------------------
C          FIND THE UNNUMBERED NEIGHBORS OF NODE.
C          FNBR AND LNBR POINT TO THE FIRST AND LAST
C          UNNUMBERED NEIGHBORS RESPECTIVELY OF THE CURRENT
C          NODE IN PERM.
C          -----------------------------------------------
           FNBR = LNBR + 1
           DO 200 J = JSTRT, JSTOP
              NBR = ADJNCY(J)
              IF ( MASK(NBR) .EQ. 0 )  GO TO 200
                 LNBR = LNBR + 1
                 MASK(NBR) = 0
                 PERM(LNBR) = NBR
  200      CONTINUE
           IF ( FNBR .GE. LNBR )  GO TO 600
C          --------------------------------------------
C          SORT THE NEIGHBORS OF NODE IN INCREASING
C          ORDER BY DEGREE. LINEAR INSERTION IS USED.
C          --------------------------------------------
           K = FNBR
  300      L = K
              K = K + 1
              NBR = PERM(K)
  400         IF ( L .LT. FNBR )  GO TO 500
                 LPERM = PERM(L)
                 IF ( DEG(LPERM) .LE. DEG(NBR) )  GO TO 500
                    PERM(L+1) = LPERM
                    L = L - 1
                    GO TO 400
  500         PERM(L+1) = NBR
              IF ( K .LT. LNBR )  GO TO 300
  600      CONTINUE
           IF (LNBR .GT. LVLEND) GO TO 100
C          --------------------------------------
C          WE NOW HAVE THE CUTHILL MCKEE ORDERING.
C          REVERSE IT BELOW ...
C          --------------------------------------
           K = CCSIZE/2
           L = CCSIZE
           DO 700 I = 1, K
              LPERM = PERM(L)
              PERM(L) = PERM(I)
              PERM(I) = LPERM
              L = L - 1
  700      CONTINUE
           RETURN
        END
```

GENRCM (GENeral RCM)

This subroutine finds the RCM ordering of a general disconnected graph. It proceeds through the graph, and calls the subroutine RCM to number each connected component. The inputs to the subroutine are the number of nodes (or equations) NEQNS, and the graph in the array pair (XADJ, ADJNCY). The arrays MASK and XLS are working arrays, used by the subroutines FNROOT and RCM, which are called by GENRCM.

The subroutine begins by setting all values of MASK to 1 (loop DO 100 I = ...). It then loops through MASK until it finds an *i* for which MASK(*i*) = 1; node *i* along with MASK, XADJ, and ADJNCY will specify a connected subgraph of the original graph *G*. The subroutines FNROOT and RCM are then called to order the nodes of that subgraph. (Recall that the numbered nodes will have their MASK values set to zero by RCM.) Note that NUM points to the first free position in the array PERM, and is updated after each call to RCM. The actual parameter in GENRCM corresponding to PERM in RCM is PERM(NUM); that is, PERM in RCM corresponds to the last NEQNS - NUM + 1 elements of PERM in GENRCM. Note also that these same elements of PERM are used to store the level structure in FNROOT. They correspond to the array LS in the execution of that subroutine.

After the component is ordered, the search for another *i* for which MASK(*i*) $\neq$ 0 resumes, until either the loop is exhausted, or NEQNS nodes have been numbered.

```
C***************************************************************
C***************************************************************
C*******   GENRCM .....  GENERAL REVERSE CUTHILL MCKEE   ******
C***************************************************************
C***************************************************************
C
C     PURPOSE - GENRCM FINDS THE REVERSE CUTHILL-MCKEE
C        ORDERING FOR A GENERAL GRAPH. FOR EACH CONNECTED
C        COMPONENT IN THE GRAPH, GENRCM OBTAINS THE ORDERING
C        BY CALLING THE SUBROUTINE RCM.
C
C     INPUT PARAMETERS -
C        NEQNS - NUMBER OF EQUATIONS
C        (XADJ, ADJNCY) - ARRAY PAIR CONTAINING THE ADJACENCY
C                STRUCTURE OF THE GRAPH OF THE MATRIX.
C
C     OUTPUT PARAMETER -
C        PERM - VECTOR THAT CONTAINS THE RCM ORDERING.
C
C     WORKING PARAMETERS -
C        MASK - IS USED TO MARK VARIABLES THAT HAVE BEEN
C                NUMBERED DURING THE ORDERING PROCESS. IT IS
C                INITIALIZED TO 1, AND SET TO ZERO AS EACH NODE
C                IS NUMBERED.
C        XLS - THE INDEX VECTOR FOR A LEVEL STRUCTURE.  THE
C                LEVEL STRUCTURE IS STORED IN THE CURRENTLY
C                UNUSED SPACES IN THE PERMUTATION VECTOR PERM.
C
C     PROGRAM SUBROUTINES -
C        FNROOT, RCM.
C
C*************************************************************
C
      SUBROUTINE  GENRCM ( NEQNS, XADJ, ADJNCY, PERM, MASK, XLS )
C
C*************************************************************
C
         INTEGER ADJNCY(1), MASK(1), PERM(1), XLS(1)
         INTEGER XADJ(1), CCSIZE, I, NEQNS, NLVL,
     1           NUM, ROOT
C
```

```
C***************************************************************
C
          DO 100 I = 1, NEQNS
              MASK(I) = 1
    100    CONTINUE
          NUM = 1
          DO 200 I = 1, NEQNS
C         ----------------------------------------
C             FOR EACH MASKED CONNECTED COMPONENT ...
C         ----------------------------------------
              IF (MASK(I) .EQ. 0) GO TO 200
              ROOT = I
C         ----------------------------------------
C             FIRST FIND A PSEUDO-PERIPHERAL NODE ROOT.
C             NOTE THAT THE LEVEL STRUCTURE FOUND BY
C             FNROOT IS STORED STARTING AT PERM(NUM).
C             THEN RCM IS CALLED TO ORDER THE COMPONENT
C             USING ROOT AS THE STARTING NODE.
C         ----------------------------------------
              CALL   FNROOT ( ROOT, XADJ, ADJNCY, MASK,
    1                         NLVL, XLS, PERM(NUM) )
              CALL      RCM ( ROOT, XADJ, ADJNCY, MASK,
    1                         PERM(NUM), CCSIZE, XLS )
              NUM = NUM + CCSIZE
              IF (NUM .GT. NEQNS) RETURN
    200    CONTINUE
          RETURN
       END
```

Exercises

4.3.1)   Let the graph associated with a given matrix be the n by n grid graph. Here is the case when $n = 5$.

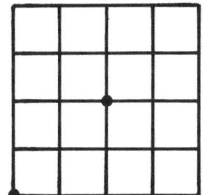

a)   Show that if the reverse Cuthill-McKee algorithm starts at a corner node, the profile is $\frac{2}{3}n^3 + O(n^2)$.

b)   What if the scheme starts at the center node?

4.3.2)   Give an example where the algorithm of Section 4.3.2 will fail to find a peripheral node. Find a large example which is particularly bad, say some significant fraction of $|X|$ from the diameter. The authors do not know of a large example where the execution time will be greater than $O(|E|)$. Can you find one?

4.3.3)   The original pseudo-peripheral node finding algorithm of Gibbs et al. (1976b) did not have a "shrinking step;" Steps 3 and 4 were as follows:

Step 3:   (*Sort the last level*): Sort the nodes in $L_{\ell(r)}(r)$ in order of increasing degree.

Step 4:   (*Test for termination*): For $x \in L_{\ell(r)}(r)$ in order of increasing degree, generate

$$\mathcal{L}(x) = \{L_0(x), L_1(x), \ldots, L_{\ell(x)}(x)\} .$$

If $\ell(x) > \ell(r)$, set $r \leftarrow x$ and go to Step 3.

Give an example to show that the execution time of this algorithm can be greater than $O(|E|)$. Answer the first two questions in Exercise 4.3.2 for this algorithm.

4.3.4)   Suppose we delete Step 3 of the algorithm of Section 4.3.1. The ordering given by $x_1, x_2, \ldots, x_N$ is called the *Cuthill-McKee ordering*. Let A_c be the matrix ordered by this algorithm. Show that

a)   the matrix A_c has the monotone profile property (see Exercise 4.2.2),

b)   in the graph G^{A_c}, for $1 < i \leq N$

$$Adj(\{x_i, \ldots, x_N\}) \subset \{x_{f_i(A_c)}, \ldots, x_{i-1}\} .$$

4.3.5)   Show that $Env(A_r) = Env(A_c)$ if and only if the matrix A_r has the monotone profile property. Here A_r is the matrix ordered by the algorithm of Section 4.3.1, and A_c is as described in Exercise 4.3.4.

4.3.6)   What ensures that the pseudo-peripheral node finding algorithm described in Section 4.3.2 terminates?

4.3.7)   Consider the N by N symmetric positive definite system of equations $Ax = b$ derived from an n by n finite element mesh as follows. The mesh consists of $(n-1)^2$ small squares, as shown in Figure 4.3.10 for $n = 5$, each mesh square has a node at its vertices and midsides, and there is one variable x_i associated with each node. For some labelling of the $N = 3n^2 - 2n$ nodes, the matrix A has the property that $a_{ij} \neq 0$ if and only if x_i and x_j are associated with the same mesh square.

We have a choice of two orderings, α_1 and α_2, as shown in Figure 4.3.10. The orderings are similar in that they both number the nodes mesh line by mesh line. Their difference is essentially that α_1 numbers nodes on each horizontal mesh line and on the vertical lines immediately *above* it at the same

Figure 4.3.10   Two orderings α_1 and α_2 of a 5 by 5 finite element mesh.

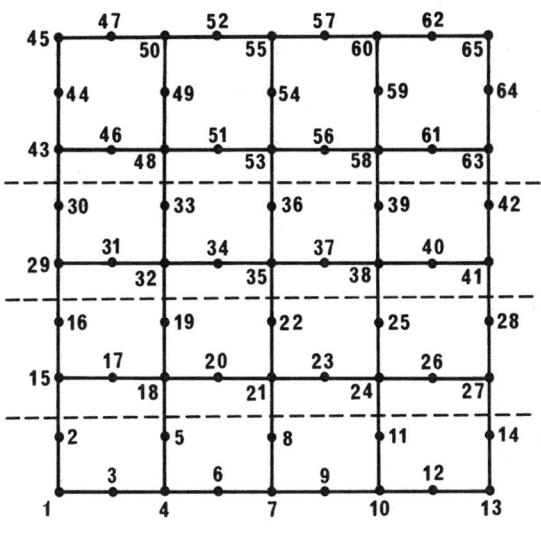

α_1

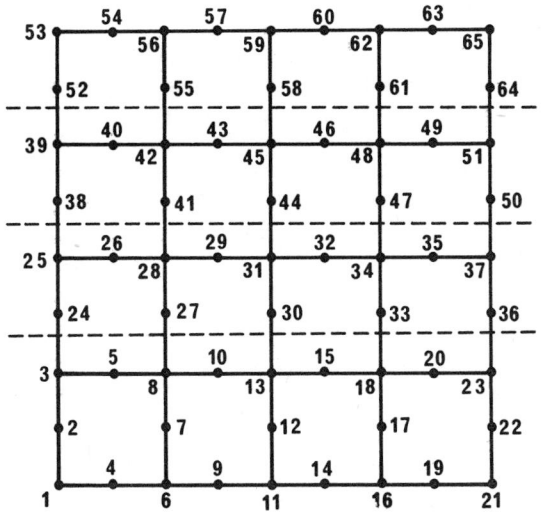

α_2

time, while α_2 numbers nodes on a horizontal line along with nodes on the vertical lines immediately *below* it at the same time, as depicted by the dashed lines in the diagrams.

a) What is the bandwidth of A, for orderings α_1 and α_2?

b) Suppose the envelope method is used to solve $Ax = b$, using orderings α_1 and α_2. Let θ_1 and θ_2 be the corresponding arithmetic operation counts, and let η_1 and η_2 be the corresponding storage requirements. Show that for large n,

$$\begin{aligned}
\theta_1 &= 6n^4 + O(n^3) \\
\theta_2 &= 13.5n^4 + O(n^3) \\
\eta_1 &= 6n^3 + O(n^2) \\
\eta_2 &= 9n^3 + O(n^2) \ .
\end{aligned}$$

Orderings α_1 and α_2 resemble the type of ordering produced by the RCM and standard Cuthill-McKee ordering algorithms respectively; the results above illustrate the substantial differences in storage and operation counts the two orderings can produce. For more details see Liu (1975).

4.3.8) (*King Ordering*) King (1970) has proposed an algorithm for reducing the profile of a symmetric matrix. His algorithm for a connected graph can be described as follows.

Step 1 (*Initialization*) Determine a pseudo-peripheral node r and assign $x_1 \leftarrow r$.

Step 2 (*Main loop*) For $i = 1, \ldots, N - 1$, find a node $y \in Adj(\{x_1, \ldots, x_i\})$ with minimum

$$| Adj(\{x_1, \ldots, x_i, y\})|.$$

Number the node y as x_{i+1}.

Step 3 (*Exit*) The King ordering is given by x_1, $x_2, \ldots, x_N$.

This algorithm reduces the profile by a local minimization of the frontwidth. Implement this algorithm for general disconnected graphs. Run your program on the matrices in test set #1 of Chapter 9. Compare the performance of this algorithm with that of RCM.

4.4 Implementation of the Envelope Method

4.4.1 An Envelope Storage Scheme

The most commonly used storage scheme for the envelope method is the one proposed by Jennings (1966). For each row in the matrix, all the entries from the first nonzero to the diagonal are stored. These row portions are stored in contiguous locations in a one dimensional array. However, we use a modification of this scheme, in which the diagonal entries are stored in a separate vector. An advantage of this variant scheme is that it lends itself readily to the case when A is unsymmetric; this point is pursued in an exercise at the end of this chapter.

The scheme has a main storage array ENV which contains the envelope entries of each row in the matrix. An auxiliary index vector XENV of length N is used to point to the start of each row portion. For uniformity in indexing, we set XENV($N + 1$) to $|Env(A)| + 1$. In this way, the index vector XENV allows us to access any nonzero component conveniently. The mapping from $Env(A)$ to $\{1, 2, \ldots, |Env(A)|\}$ is given by:

$$\{i,j\} \rightarrow \text{XENV}(i + 1) - (i - j).$$

In other words, a component a_{ij} within the envelope region of A is found in ENV(XENV($i + 1$) − ($i - j$)). Figure 4.4.1 illustrates the storage scheme. For example, to retrieve a_{64}, we have

$$\text{XENV}(7) - (6 - 4) = 8$$

so that a_{64} is stored in the 8-th element of the vector ENV.

A more frequently used operation is to retrieve the envelope portion of a row. This can be done conveniently as follows.

```
          :
     JSTRT = XENV(IROW)
     JSTOP = XENV(IROW+1) - 1
     IF (JSTOP.LT.JSTRT) GO TO 200
     DO 100 J = JSTRT, JSTOP
         ELEMNT = ENV(J)
          :
100     CONTINUE
200     :
```

The primary storage of the scheme is $|Env(A)| + N$ and the overhead storage is $N + 1$. The data structure for the storage scheme can be set up in $O(|E|)$ time and the subroutine FNENV, discussed in the next subsection, performs this function.

Figure 4.4.1   Example of the envelope storage scheme.

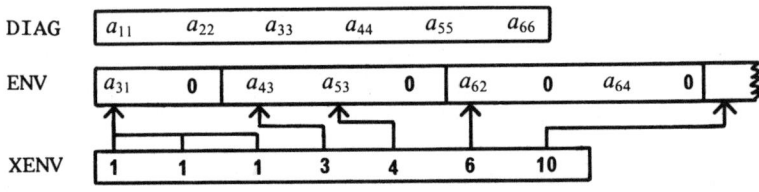

4.4.2   The Storage Allocation Subroutine FNENV *(FiNd-ENVelope)*

In this section we describe the subroutine FNENV.   This subroutine
accepts as input the graph of the matrix A, stored in the array pair
(XADJ, ADJNCY), along with the permutation vector PERM and its
inverse INVP (discussed in Section 3.3). The objective of the subroutine is to compute the components of the array XENV discussed in Section 4.4.1, which is used in connection with storing the factor L of
PAP^T. Also returned is the value ENVSZE, which is the envelope size of
L and equals XENV(NEQNS+1) - 1. Here as before, NEQNS is the
number of equations or nodes.

The subroutine is straightforward and needs little explanation.
The loop DO 200 I = ... processes each row; the index of the first
nonzero in the i-th row (IFIRST) of PAP^T is determined by the loop
DO 100 J = ... At the end of each execution of the loop
DO 100 J = ..., ENVSZE is suitably updated. Note that PERM and
INVP are used since the array pair (XADJ, ADJNCY) stores the
structure of A, but the structure of L we are finding corresponds to
PAP^T.

```
C*************************************************************
C*************************************************************
C***************       FNENV ..... FIND ENVELOPE    ***********
C*************************************************************
C*************************************************************
C
C       PURPOSE - FINDS THE ENVELOPE STRUCTURE OF A PERMUTED
C               MATRIX.
C
C       INPUT PARAMETERS -
C          NEQNS - NUMBER OF EQUATIONS
C          (XADJ, ADJNCY) - ARRAY PAIR CONTAINING THE ADJACENCY
C               STRUCTURE OF THE GRAPH OF THE MATRIX.
C          PERM,INVP - ARRAYS CONTAINING PERMUTATION DATA ABOUT
C               THE REORDERED MATRIX.
C
C       OUTPUT PARAMETERS -
C          XENV - INDEX VECTOR FOR THE LEVEL STRUCTURE
C               TO BE USED TO STORE THE LOWER (OR UPPER)
C               ENVELOPE OF THE REORDERED MATRIX.
C          ENVSZE - IS EQUAL TO XENV(NEQNS+1) - 1.
C          BANDW - BANDWIDTH OF THE REORDERED MATRIX.
C
C*************************************************************
C
        SUBROUTINE  FNENV ( NEQNS, XADJ, ADJNCY, PERM, INVP,
     1                      XENV, ENVSZE, BANDW )
C
C*************************************************************
C
        INTEGER ADJNCY(1), INVP(1), PERM(1)
        INTEGER XADJ(1), XENV(1), BANDW, I, IBAND,
     1          IFIRST, IPERM, J, JSTOP, JSTRT, ENVSZE,
     1          NABOR, NEQNS
C
C*************************************************************
C
        BANDW = 0
        ENVSZE = 1
        DO 200 I = 1, NEQNS
           XENV(I) = ENVSZE
           IPERM = PERM(I)
           JSTRT = XADJ(IPERM)
           JSTOP = XADJ(IPERM + 1) - 1
           IF ( JSTOP .LT. JSTRT ) GO TO 200
C          --------------------------------
C          FIND THE FIRST NONZERO IN ROW I.
C          --------------------------------
           IFIRST = I
           DO 100 J = JSTRT, JSTOP
              NABOR = ADJNCY(J)
              NABOR = INVP(NABOR)
              IF ( NABOR .LT. IFIRST ) IFIRST = NABOR
100        CONTINUE
           IBAND = I - IFIRST
           ENVSZE = ENVSZE + IBAND
           IF ( BANDW .LT. IBAND ) BANDW = IBAND
200     CONTINUE
        XENV(NEQNS+1) = ENVSZE
        ENVSZE = ENVSZE - 1
        RETURN
        END
```

4.5   The Numerical Subroutines ESFCT, ELSLV **and** EUSLV

In this section we describe the subroutines which perform the numerical factorization and solution, using the envelope storage scheme described in Section 4.4.1. We describe the triangular solution subroutines ELSLV (Envelope-Lower-SoLVe) and EUSLV (Envelope-Upper-SoLVe) before the factorization subroutine ESFCT (Envelope-Symmetric-FaCTorization) because ELSLV is used by ESFCT.

4.5.1   The Triangular Solution Subroutines ELSLV *and* EUSLV.

These subroutines carry out the numerical solutions of the lower and upper triangular systems

$$Ly = b$$

and

$$L^T x = y ,$$

respectively, where L is a lower triangular matrix stored as described in Section 4.4.1.

There are several important features of ELSLV which deserve explanation. To begin, the position (IFIRST) of the first nonzero in the right hand side (RHS) is determined. With this initialization, the program then loops (DO 500 I = . . .) over the rows IFIRST, IFIRST+1, . . . , NEQNS of L, using the inner product scheme described in Section 2.2.1. However, the program attempts to exploit strings of zeros in the solution; the variable LAST is used to store the index of the most recently computed *nonzero* component of the solution. (The solution overwrites the input right hand side array RHS.)

The reader is urged to simulate the subroutine's action on the problem described by Figure 4.5.1 to verify that only the nonzeros denoted by ⊞ are actually used by the subroutine ELSLV.

Note that LAST simply allows us to skip certain rows; we still perform some multiplications with zero operands in the DO 300 K . . . loop, but on most machines a test to avoid such a multiplication is more costly than going ahead and doing it.

The test and adjustment of IBAND just preceding the DO 300 . . . loop also requires some explanation. In some circumstances ELSLV is used to solve a lower triangular system where the coefficient matrix to be used is only a *submatrix* of the matrix passed to ELSLV in the array pair (XENV, ENV), as depicted below. Some of the rows of the envelope protrude outside the coefficient matrix to be used, and IBAND is appropriately adjusted to account for them. In the example below, L is actually 16 by 16, and if the system we wish to solve is the submatrix

Figure 4.5.1   Elements of *L* actually used by ELSLV are denoted by ⊠.

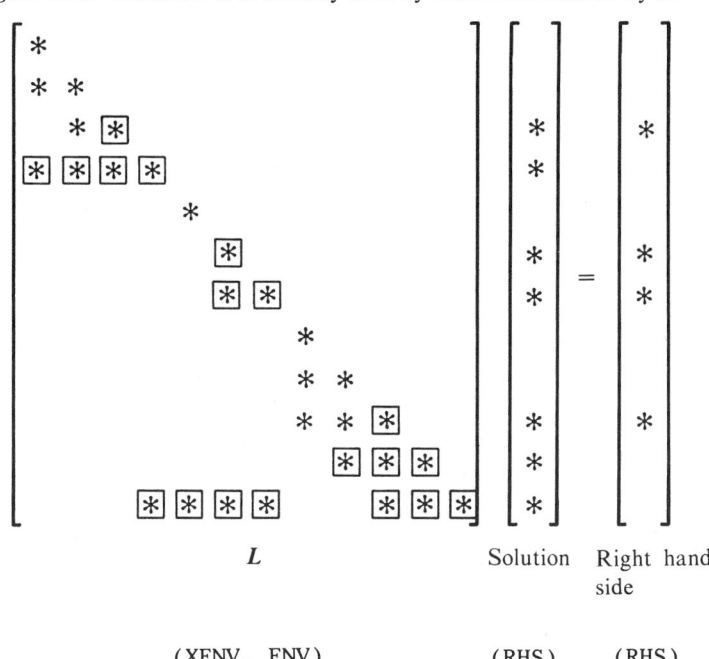

L	Solution	Right hand side
(XENV, ENV)	(RHS)	(RHS)

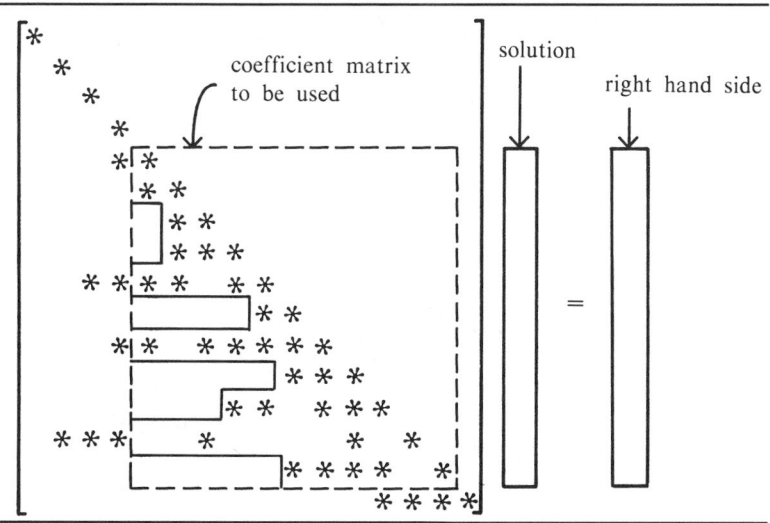

indicated by the 11 by 11 system with right hand side RHS, we would solve it by executing the statement

CALL ELSLV(11, XENV(5), ENV, DIAG(5), RHS)

In the subroutine, XENV(5) is interpreted as XENV(1), XENV(6) becomes XENV(2), etc. This "trick" is used heavily by the subroutine ESFCT, which calls ELSLV.

```
C****************************************************************
C****************************************************************
C*********    ELSLV .....  ENVELOPE LOWER SOLVE    *********
C****************************************************************
C****************************************************************
C
C     PURPOSE - THIS SUBROUTINE SOLVES A LOWER TRIANGULAR
C         SYSTEM L X = RHS.  THE FACTOR L IS STORED IN THE
C         ENVELOPE FORMAT.
C
C     INPUT PARAMETERS -
C         NEQNS - NUMBER OF EQUATIONS.
C         (XENV, ENV) - ARRAY PAIR FOR THE ENVELOPE OF L.
C         DIAG - ARRAY FOR THE DIAGONAL OF L.
C
C     UPDATED PARAMETER -
C         RHS - ON INPUT, IT CONTAINS THE RIGHT HAND VECTOR.
C             ON RETURN, IT CONTAINS THE SOLUTION VECTOR.
C         OPS - DOUBLE PRECISION VARIABLE CONTAINED IN THE
C             LABELLED COMMON BLOCK OPNS.  ITS VALUE IS
C             INCREASED BY THE NUMBER OF OPERATIONS
C             PERFORMED BY THIS SUBROUTINE.
C
C****************************************************************
C
C         SUBROUTINE  ELSLV ( NEQNS, XENV, ENV, DIAG, RHS )
C
C****************************************************************
C
C         DOUBLE PRECISION COUNT, OPS
C         COMMON  /SPKOPS/ OPS
C         REAL DIAG(1), ENV(1), RHS(1), S
C         INTEGER XENV(1), I, IBAND, IFIRST, K, KSTOP,
C     1             KSTRT, L, LAST, NEQNS
C
C****************************************************************
C
C         --------------------------------------------------
C         FIND THE POSITION OF THE FIRST NONZERO IN RHS AND
C         PUT IT IN IFIRST.
C         --------------------------------------------------
          IFIRST = 0
  100     IFIRST = IFIRST + 1
              IF ( RHS(IFIRST) .NE. 0.0E0 )  GO TO 200
              IF ( IFIRST .LT. NEQNS ) GO TO 100
              RETURN
  200     LAST = 0
C         --------------------------------------------------
C         LAST CONTAINS THE POSITION OF THE MOST RECENTLY
C         COMPUTED NONZERO COMPONENT OF THE SOLUTION.
C         --------------------------------------------------
          DO 500 I = IFIRST, NEQNS
              IBAND = XENV(I+1) - XENV(I)
              IF ( IBAND .GE. I )  IBAND = I - 1
              S = RHS(I)
              L = I - IBAND
              RHS(I) = 0.0E0
C         --------------------------------------------------
C         ROW OF THE ENVELOPE IS EMPTY, OR CORRESPONDING
```

```
C               COMPONENTS OF THE SOLUTION ARE ALL ZEROS.
C               -------------------------------------------------
                IF ( IBAND .EQ. 0 .OR. LAST .LT. L )   GO TO 400
                KSTRT = XENV(I+1) - IBAND
                KSTOP = XENV(I+1) - 1
                DO 300 K = KSTRT, KSTOP
                    S = S - ENV(K)*RHS(L)
                    L = L + 1
  300           CONTINUE
                COUNT = IBAND
                OPS = OPS + COUNT
  400           IF ( S .EQ. 0.0E0 )   GO TO 500
                RHS(I) = S/DIAG(I)
                OPS = OPS + 1.0D0
                LAST = I
  500       CONTINUE
            RETURN
        END
```

We now turn to a description of the subroutine EUSLV, which solves the problem $L^Tx = y$, with L stored using the same storage scheme as that used by ELSLV. This means that we have convenient access to the *columns* of L^T, and sparsity can be exploited completely, as discussed in Section 2.2.1, using an outer product form of the computation. The i-th column of L^T is used in the computation only if the i-th element of the solution is nonzero.

Just as in ELSLV, the subroutine EUSLV can be used to solve upper triangular systems involving only a submatrix of L contained in the array pair (XENV, ENV), using techniques analogous to those we described above. The value of IBAND is appropriately adjusted for those columns of L^T that protrude outside the part of L actually being used.

All the subroutines which perform numerical computation contain a labelled COMMON block SPKOPS, which has a single variable OPS. Each subroutine counts the number of operations (multiplications and divisions) it performs, and increments the value of OPS accordingly. Thus, if the user of the subroutines wishes to monitor the number of operations performed, he can make the same common block declaration in his calling program and examine the value of OPS.

The variable OPS has been declared to be double precision to avoid the possibility of serious rounding error in the computation of operation counts. Our subroutines may be used to solve very large systems, so OPS may easily assume values as large as 10^8 or 10^9, even though OPS may be incremented in each subroutine by relatively small numbers. On many computers, if single precision is used, the floating point addition of a small number (say less than 10) to 10^8 will again yield 10^8. (Try it, simulating 6 digit floating point arithmetic!) Using double precision for OPS makes serious rounding error in the operation count very unlikely.

```
C**************************************************************
C**************************************************************
C**********      EUSLV .....  ENVELOPE UPPER SOLVE    *********
C**************************************************************
C**************************************************************
C
C     PURPOSE - THIS SUBROUTINE SOLVES AN UPPER TRIANGULAR
C        SYSTEM U X = RHS.  THE FACTOR U IS STORED IN THE
C        ENVELOPE FORMAT.
C
C     INPUT PARAMETERS -
C        NEQNS - NUMBER OF EQUATIONS.
C        (XENV, ENV) - ARRAY PAIR FOR THE ENVELOPE OF U.
C        DIAG - ARRAY FOR THE DIAGONAL OF U.
C
C     UPDATED PARAMETER -
C        RHS - ON INPUT, IT CONTAINS THE RIGHT HAND SIDE.
C            ON OUTPUT, IT CONTAINS THE SOLUTION VECTOR.
C        OPS - DOUBLE PRECISION VARIABLE CONTAINED IN THE
C            LABELLED COMMON BLOCK OPNS.  ITS VALUE IS
C            INCREASED BY THE NUMBER OF OPERATIONS
C            PERFORMED BY THIS SUBROUTINE.
C
C**************************************************************
C
      SUBROUTINE  EUSLV ( NEQNS, XENV, ENV, DIAG, RHS )
C
C**************************************************************
C
         DOUBLE PRECISION COUNT, OPS
         COMMON  /SPKOPS/ OPS
         REAL DIAG(1), ENV(1), RHS(1), S
         INTEGER XENV(1), I, IBAND, K, KSTOP, KSTRT, L,
     1            NEQNS
C
C**************************************************************
C
         I = NEQNS + 1
  100    I = I - 1
         IF ( I .EQ. 0 )  RETURN
            IF ( RHS(I) .EQ. 0.0E0 )  GO TO 100
            S = RHS(I)/DIAG(I)
            RHS(I) = S
            OPS = OPS + 1.0D0
            IBAND = XENV(I+1) - XENV(I)
            IF ( IBAND .GE. I )  IBAND = I - 1
            IF ( IBAND .EQ. 0 )  GO TO 100
               KSTRT = I - IBAND
               KSTOP = I - 1
               L = XENV(I+1) - IBAND
               DO 200 K = KSTRT, KSTOP
                  RHS(K) = RHS(K) - S*ENV(L)
                  L = L + 1
  200          CONTINUE
               COUNT = IBAND
               OPS = OPS + COUNT
         GO TO 100
      END
```

4.5.2   The Factorization Subroutine ESFCT

In this section we describe some details about the numerical factorization subroutine ESFCT, which computes the Cholesky factorization LL^T of a given matrix A, stored using the envelope storage scheme described in Section 4.4.1. The variant of Cholesky's method used is the bordering method (see Section 2.1.2).

Recall that if A is partitioned as

$$A = \begin{pmatrix} M & u \\ u^T & s \end{pmatrix}$$

where M is the leading principal submatrix of A and $L_M L_M^T$ is its Cholesky factorization, then the factor of A is given by

$$L = \begin{pmatrix} L_M & 0 \\ w^T & t \end{pmatrix},$$

where $L_M w = u$ and $t = (s - w^T w)^{1/2}$. Thus, the Cholesky factor of A can be computed row by row, working with successively larger matrices M, beginning with the one by one matrix a_{11}. The main point of interest in ESFCT concerns the exploitation of the fact that the vectors u are "short" because we are dealing with an envelope matrix.

Referring to Figure 4.5.2, suppose the first $i - 1$ steps of the factorization have been completed, so that the leading $(i - 1) \times (i - 1)$ principal submatrix of A has been factored. (Thus, the statements preceeding the loop DO 300 I = 2, . . . have been executed, and the loop DO 300 I = 2, . . ., has been executed $i - 2$ times.) In order to compute row i of L, we must solve the system of equations $L_M w = u$.

Figure 4.5.2   Sketch showing the way sparsity is exploited in ESFCT; only $\bar{L}$ enters into the computation of $\bar{w}$ from $\bar{u}$.

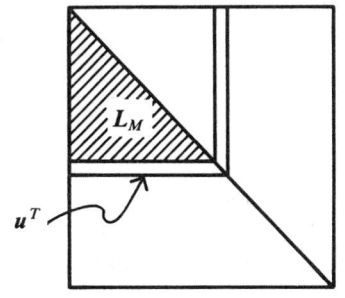
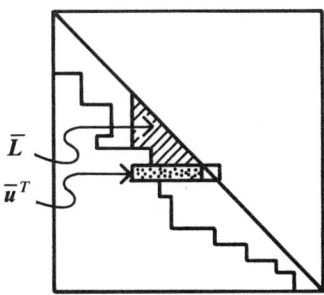

However, it is clear from the picture (and from Lemmas 2.2.1 and

2.2.4) that only part of L_M is involved in the computation, namely that part of L_M labelled $\bar{L}$. Thus ELSLV is called with the size of the triangular system specified as IBAND, the size of $\bar{u}$ (and $\bar{L}$), and IFIRST is the index in L of the first row of $\bar{L}$.

```
C*************************************************************
C*************************************************************
C******    ESFCT ..... ENVELOPE SYMMETRIC FACTORIZATION  ****
C*************************************************************
C*************************************************************
C
C     PURPOSE - THIS SUBROUTINE FACTORS A POSITIVE DEFINITE
C        MATRIX A INTO L*L(TRANSPOSE).  THE MATRIX A IS STORED
C        IN THE ENVELOPE FORMAT.  THE ALGORITHM USED IS THE
C        STANDARD BORDERING METHOD.
C
C     INPUT PARAMETERS -
C        NEQNS - NUMBER OF EQUATIONS.
C        XENV - THE ENVELOPE INDEX VECTOR.
C
C     UPDATED PARAMETERS -
C        ENV - THE ENVELOPE OF L OVERWRITES THAT OF A.
C        DIAG - THE DIAGONAL OF L OVERWRITES THAT OF A.
C        IFLAG - THE ERROR FLAG.  IT IS SET TO 1 IF A ZERO OR
C                NEGATIVE SQUARE ROOT IS DETECTED DURING THE
C                FACTORIZATION.
C
C     PROGRAM SUBROUTINES -
C        ELSLV.
C
C*************************************************************
C
          SUBROUTINE  ESFCT ( NEQNS, XENV, ENV, DIAG, IFLAG )
C
C*************************************************************
C
          DOUBLE PRECISION COUNT, OPS
          COMMON /SPKOPS/ OPS
          REAL DIAG(1), ENV(1), S, TEMP
          INTEGER XENV(1), I, IBAND, IFIRST, IFLAG, IXENV,
     1            J, JSTOP, NEQNS
C
C*************************************************************
C
          IF ( DIAG(1) .LE. 0.0E0 )  GO TO 400
          DIAG(1) = SQRT(DIAG(1))
          IF ( NEQNS .EQ. 1 )  RETURN
C        ---------------------------------------------------
C        LOOP OVER ROWS 2,3,..., NEQNS OF THE MATRIX ....
C        ---------------------------------------------------
          DO 300 I = 2, NEQNS
             IXENV = XENV(I)
             IBAND = XENV(I+1) - IXENV
             TEMP = DIAG(I)
             IF ( IBAND .EQ. 0 )  GO TO 200
             IFIRST = I - IBAND
C           ----------------------------------------
C           COMPUTE ROW I OF THE TRIANGULAR FACTOR.
C           ----------------------------------------
             CALL  ELSLV ( IBAND, XENV(IFIRST), ENV,
     1                     DIAG(IFIRST), ENV(IXENV) )
             JSTOP = XENV(I+1) - 1
             DO 100 J = IXENV, JSTOP
                S = ENV(J)
```

```
                  TEMP = TEMP - S*S
100           CONTINUE
200           IF ( TEMP .LE. 0.0E0 )  GO TO 400
              DIAG(I) = SQRT(TEMP)
              COUNT = IBAND
              OPS = OPS + COUNT
300       CONTINUE
          RETURN
C
C         ------------------------------------------------
C         SET ERROR FLAG  -  NON POSITIVE DEFINITE MATRIX.
C         ------------------------------------------------
400       IFLAG = 1
          RETURN
      END
```

Exercises

4.5.1) Suppose A has symmetric structure but $A \neq A^T$, and assume that Gaussian elimination applied to A is numerically stable without pivoting. The bordering equations for factoring A, analogous to those used by ESFCT in Section 4.5.2, are as follows.

$$A = \begin{pmatrix} M & v \\ u^T & s \end{pmatrix}$$

$$L = \begin{pmatrix} L_M & 0 \\ w^T & 1 \end{pmatrix} \qquad U = \begin{pmatrix} U_M & g \\ 0 & t \end{pmatrix}$$

$$L_M g = v, \qquad U_M^T w = u, \qquad t = s - w^T g .$$

Here L is now *unit* lower triangular (ones on the diagonal), and of course $L \neq U^T$.

a) Using ELSLV as a base, implement a Fortran subroutine EL1SLV that solves unit lower triangular systems stored using the envelope storage scheme.

b) Using ESFCT as a base, implement a Fortran subroutine EFCT that factors A into LU, where L and U^T are stored using the envelope scheme.

c) What subroutines do you need to solve $Ax = b$, where A is as described in this question?   Hints:

    i)   Very few changes in ELSLV and ESFCT are required.

    ii) Your implementation of EFCT should use EL1SLV and ELSLV.

4.5.2) Suppose L and b have the structure shown below, where L is stored in the arrays XENV, ENV, and DIAG, as described in Section 4.4.1. How many arithmetic operations will ELSLV perform in solving $Lx = b$? How many will EUSLV perform in solving $L^T x = b$?

$$
\begin{bmatrix}
* \\
* & * \\
* & * & * \\
& & & * \\
& & & * & * \\
& & & & * & * \\
& & & & & * \\
& & & & & & * \\
& & & * & & * & & * \\
* & * & & & * & & * & * & *
\end{bmatrix}
\begin{bmatrix}
\\
\\
* \\
\\
* \\
\\
* \\
\\
\end{bmatrix}
$$

$$L \qquad\qquad\qquad b$$

4.6 Additional Notes

Our lack of enthusiasm for band orderings is due in part to the fact that we only consider "in core" methods in our book. Band orderings are attractive if auxiliary storage is to be used, since it is quite easy to implement factorization and solution subroutines which utilize auxiliary storage, provided about $\beta(\beta + 1)/2$ main storage locations are available (Felippa 1970). Wilson et al. (1974) describe an out-of-core band-oriented scheme which requires even less storage; their program can execute even if there is only enough storage to hold two columns of the band of L. Another context in which band orderings are important is in the use of so-called minimal storage band methods (Sherman 1975). The basic computational scheme is similar to those which use auxiliary storage, except that the columns of L are computed, used, and then "thrown away," instead of being written on auxiliary storage. The parts of L needed later are recomputed.

Several other algorithms for producing low profile orderings have been proposed. Levy (1971) describes an algorithm which picks nodes to number on the basis of minimum increase in the envelope size. King (1970) has proposed a similar scheme, except that the candidates for numbering are restricted to those having at least one numbered neighbor, and therefore requires a starting node. More recently, several algorithms more closely related to the one described in this chapter have been proposed (Gibbs et al. 1976b, Gibbs 1976c).

Several researchers have described "frontal" or "wavefront" techniques to exploit the variation in the bandwidth when using auxiliary storage (Melosh 1969, Irons 1970). These schemes require only about $\omega(\omega + 1)/2$ main storage locations rather than $\beta(\beta + 1)/2$ for the band schemes, although the programs tend to be substantially more complicated as a result.   These ideas have been proposed in the context of solving finite element equations, and a second novel feature the methods have is that the equations are generated and solved *in tandem*.

It has been shown that given a starting node, the RCM algorithm can be implemented to run in $O(|E|)$ time (Chan 1979). Since each edge of the graph must be examined at least once, this new method is apparently optimal.

A set of subroutines which are similar to ELSLV, EUSLV, and ESFCT is provided in Eisenstat (1974).

5/ General Sparse Methods

5.0 Introduction

In this chapter we consider methods which, unlike those of Chapter 4, attempt to exploit all the zero elements in the triangular factor L of A. The ordering algorithm we study in this chapter is called the *minimum degree algorithm* (Rose 1972a). It is a heuristic algorithm for finding an ordering for A which suffers low fill when it is factored. This algorithm has been used widely in industrial applications, and enjoys a good reputation. The computer implementations of the allocation and numerical subroutines are adapted from those of the Yale Sparse Matrix Package (Eisenstat 1981).

5.1 Symmetric Factorization

Let A be a symmetric sparse matrix. The *nonzero structure* of A is defined by

$$Nonz(A) = \{\{i,j\} \mid a_{ij} \neq 0 \text{ and } i \neq j\}.$$

Suppose the matrix is factored into LL^T using the Cholesky factorization algorithm. The *filled matrix* $F(A)$ of A is the matrix sum $L + L^T$. When the matrix under study is clear from context, we use F rather than $F(A)$. Its corresponding structure is then

$$Nonz(F) = \{\{i,j\} \mid \ell_{ij} \neq 0 \text{ and } i \neq j\}.$$

Recall that throughout our book, we assume that exact numerical cancellation does not occur, so for a given nonzero structure $Nonz(A)$, the corresponding $Nonz(F)$ is completely determined. That is, $Nonz(F)$ is independent of the numerical quantities in A.

This no-cancellation assumption immediately implies that

$$Nonz(A) \subset Nonz(F),$$

and the *fill* of the matrix A can then be defined as

$$Fill(A) = Nonz(F) - Nonz(A).$$

For example, consider the matrix in Figure 5.1.1, where fill-in entries are indicated by ⊛. The corresponding sets are given by

$$Nonz\,(A\,) = \big\{\{1, 5\}, \{1, 8\}, \{2, 4\}, \{2, 5\}, \{3, 8\}, \{4, 7\}, \{5, 6\}, \{6, 8\}, \{8, 9\}\big\}$$

$$Fill\,(A\,) = \big\{\{4, 5\}, \{5, 7\}, \{5, 8\}, \{6, 7\}, \{7, 8\}\big\}\,.$$

In the next section, we shall consider how $Fill\,(A\,)$ can be obtained from $Nonz\,(A\,)$.

Figure 5.1.1   A matrix example of *Nonz* and *Fill*.

$$\begin{bmatrix}
① & & & & * & & * & \\
& ② & & * & * & & & \\
& & ③ & & & & & * \\
& * & & ④ & ⊛ & & * & \\
* & * & & ⊛ & ⑤ & * & ⊛ & ⊛ \\
& & & & * & ⑥ & ⊛ & * \\
& & & * & ⊛ & ⊛ & ⑦ & ⊛ \\
* & & * & & ⊛ & * & ⊛ & ⑧ & * \\
& & & & & & * & ⑨
\end{bmatrix}$$

5.1.1   Elimination Graph Model

We now relate the application of symmetric Gaussian elimination to A, to corresponding changes in its graph G^A. Recall from Chapter 2 that the first step of the outer product version of the algorithm applied to an N by N symmetric positive definite matrix $A = A_0$ can be described by the equation:

$$A = A_0 = H_0 = \begin{pmatrix} d_1 & v_1^T \\ v_1 & \overline{H}_1 \end{pmatrix} \tag{5.1.1}$$

$$= \begin{pmatrix} \sqrt{d_1} & 0 \\ \dfrac{v_1}{\sqrt{d_1}} & I_{N-1} \end{pmatrix} \begin{pmatrix} 1 & 0 \\ 0 & H_1 \end{pmatrix} \begin{pmatrix} \sqrt{d_1} & v_1^T/\sqrt{d_1} \\ 0 & I_{N-1} \end{pmatrix}$$

$$= L_1 A_1 L_1^T,$$

where

$$H_1 = \overline{H}_1 - v_1 v_1^T / d_1 \; . \tag{5.1.2}$$

The basic step is then recursively applied to H_1, H_2, and so on. Making the usual assumption that exact cancellation does not occur, equation (5.1.2) implies that the jk-th entry of H_1 is nonzero if the corresponding entry in $\overline{H}_1$ is already nonzero, or if *both* $(v_1)_j \neq 0$ and $(v_1)_k \neq 0$. Of course both situations may prevail, but when only the latter one does, some *fill-in* occurs. This phenomenon is illustrated pictorially in Figure 5.1.2. After the first step of the factorization is completed, we are left with the matrix H_1 to factor.

Figure 5.1.2   Pictorial illustration of fill-in in the outer-product formulation.

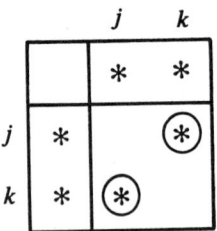

Following Parter (1961) and Rose (1972a), we now establish a correspondence between the transformation of H_0 to H_1 and the corresponding changes to their respective graphs. As usual, we denote the graphs of $H_0(=A)$ and H_1 by G^{H_0} and G^{H_1} respectively, and for convenience we denote the node $\alpha(i)$ by x_i, where α is the labelling of G^A implied by A. Now as shown in the example of Figure 5.1.3, the graph of H_1 is obtained from that of H_0 by :

1)   deleting node x_1 and its incident edges
2)   adding edges to the graph so that nodes in $Adj(x_1)$ are pairwise adjacent in G^{H_1}.

The recipe is due to Parter (1961).

Thus, as observed by Rose, symmetric Gaussian elimination can be interpreted as generating a sequence of *elimination graphs*

$$G_i^\alpha = G^{H_i} = (X_i^\alpha, E_i^\alpha), \; i = 1, 2, \ldots, N-1 \; ,$$

where G_i^α is obtained from G_{i-1}^α according to the procedure described above. When α is clear from context, we use G_i instead of G_i^α. The example in Figure 5.1.3 illustrates this vertex elimination operation. The darker lines depict edges added during the factorization. For example, the elimination of the node x_2 in the graph G_1 generates three fill-in edges $\{x_3,x_4\}$, $\{x_4,x_6\}$, $\{x_3,x_6\}$ in G_2 since $\{x_3, x_4, x_6\}$ is the adjacent set of x_2 in G_1.

Let L be the triangular factor of the matrix A. Define the filled graph of G^A to be the symmetric graph $G^F = (X^F,E^F)$, where $F = L + L^T$. Here the edge set E^F consists of all the edges in E^A together with all the edges added during the factorization. Obviously, $X^F = X^A$. The edge sets E^F and E^A are related by the following lemma due to Parter (1961). Its proof is left as an exercise.

Lemma 5.1.1

The unordered pair $\{x_i,x_j\} \in E^F$ if and only if $\{x_i,x_j\} \in E^A$ or $\{x_i,x_k\} \in E^F$ and $\{x_k,x_j\} \in E^F$ for some $k < \min\{i,j\}$.

The notion of elimination graphs allows us to interpret the step by step elimination process as a sequence of graph transformations. Moreover, the set of edges added in the elimination graphs corresponds to the set of fill-ins. Thus, for the example in Figure 5.1.3, the structures of the corresponding matrix $F = L + L^T$ and the filled graph G^F are given in Figure 5.1.4.

Thus, we see that the filled graph G^F can be easily constructed from the sequence of elimination graphs. Finding G^F is important because it contains the structure of L. We need to know it if we intend to use a storage scheme which exploits all the zeros in L.

5.1.2   Modelling Elimination By Reachable Sets

Section 5.1.1 defines the sequence of elimination graphs

$$G_0 \rightarrow G_1 \rightarrow \cdots \rightarrow G_{N-1}$$

and provides a recursive characterization of the edge set E^F. It is often helpful, both in theoretical and computational terms, to have characterizations of G_i and E^F directly in terms of the *original* graph G^A. Our objective in this section is to provide such characterizations using the notion of *reachable sets*.

Let us first study the way the fill edge $\{x_4,x_6\}$ is formed in the example of Figure 5.1.3. In G_1, there is the path

$$(x_4,x_2,x_6) \ ,$$

so that when x_2 is eliminated, the edge $\{x_4,x_6\}$ is created. However, the edge $\{x_2,x_6\}$ is not present in the original graph; it is formed from the path

$$(x_2,x_1,x_6)$$

Figure 5.1.3   The sequence of elimination graphs.

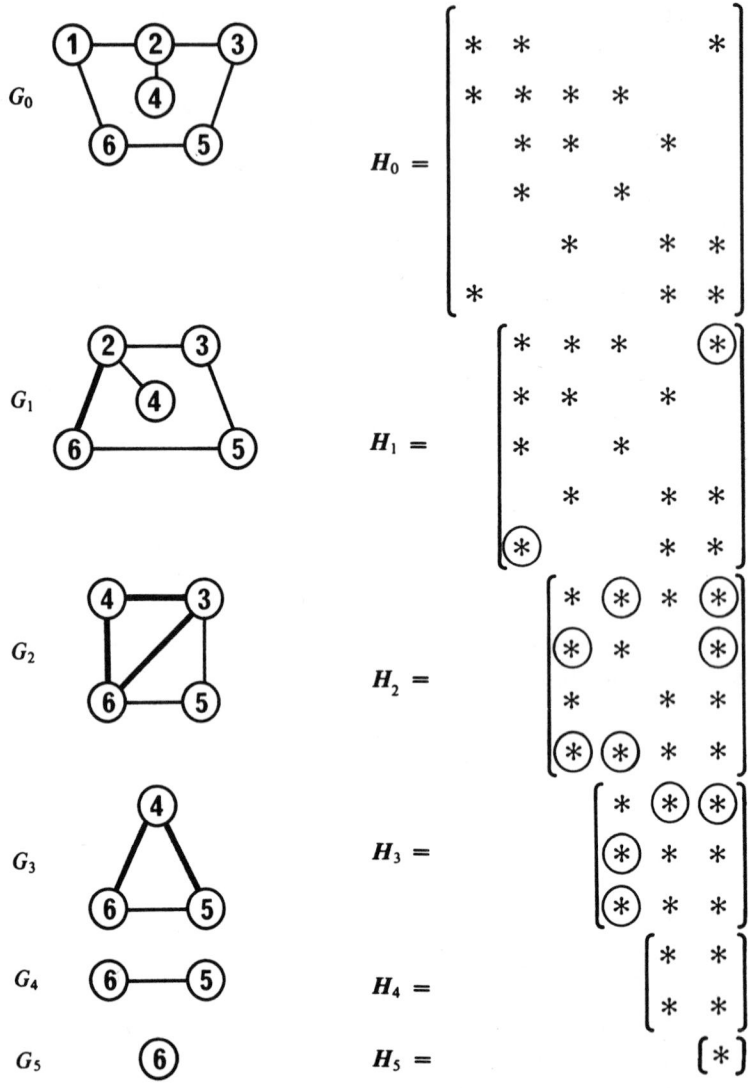

when x_1 is eliminated from G_0. On combining the two, we see that the path (x_4, x_2, x_1, x_6) in the original graph is really responsible for the filled edge $\{x_4, x_6\}$. This motivates the use of reachable sets, which we now introduce (George 1980).

Let S be a subset of the node set and $x \notin S$. The node x is said to be *reachable from* a node y *through* S if there exists a path $(y, v_1, \ldots, v_k, x)$ from y to x such that $v_i \in S$ for $1 \le i \le k$. Note

Figure 5.1.4   The filled graph and matrix of the example in Figure 5.1.3.

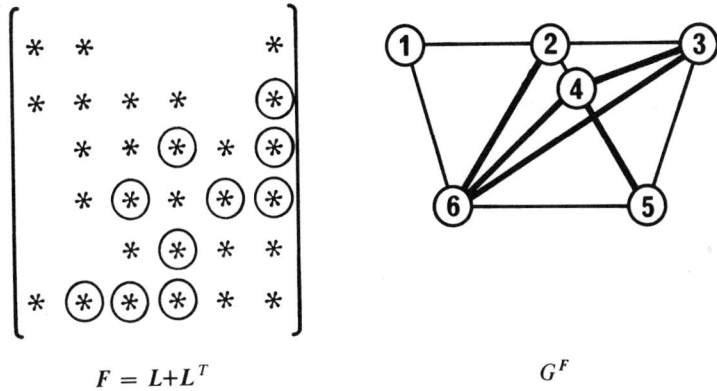

$$F = L + L^T$$ $$G^F$$

that k can be zero, so that any adjacent node of y not in S is reachable from y through S.

The reachable set of y through S, denoted by $Reach\,(y,S)$, is then defined to be

$$Reach\,(y,S) = \{\, x \notin S \mid x \text{ is reachable from } y \text{ through } S \,\}. \quad (5.1.3)$$

To illustrate the notion of reachable sets, we consider the example in Figure 5.1.5. If $S = \{s_1,s_2,s_3,s_4\}$, we have

$$Reach\,(y,S) = \{a,b,c\} \;,$$

since we can find the following paths through S:

$$(y,s_2,s_4,a)\;,$$
$$(y,b)\;,$$
$$(y,s_1,c)\;.$$

Figure 5.1.5   Example to illustrate the reachable set concept.

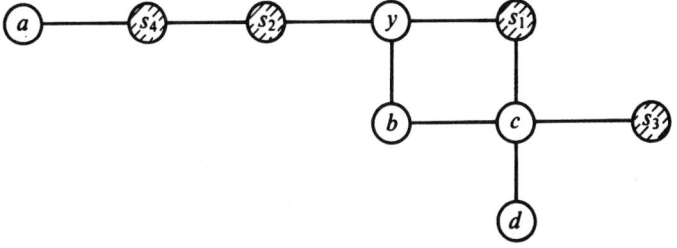

The following theorem characterizes the filled graph by reachable sets.

Theorem 5.1.2

$$E^F = \left\{ \{x_i, x_j\} \mid x_j \in Reach\,(x_i, \{x_1, x_2, \ldots, x_{i-1}\}) \right\} .$$

Proof   Assume $x_j \in Reach\,(x_i, \{x_1, \ldots, x_{i-1}\})$. By definition, there exists a path $(x_i, y_1, \ldots, y_t, x_j)$ in G^A with $y_k \in \{x_1, \ldots, x_{i-1}\}$ for $1 \leq k \leq t$. If $t = 0$ or $t = 1$, the result follows immediately from Lemma 5.1.1. If $t > 1$, a simple induction on t, together with Lemma 5.1.1 shows that $\{x_i, x_j\} \in E^F$.

Conversely, assume $\{x_i, x_j\} \in E^F$ and $i \leq j$. The proof is by induction on the subscript i. The result is true for $i = 1$, since $\{x_i, x_j\} \in E^F$ implies $\{x_i, x_j\} \in E^A$. Suppose the assertion is true for subscripts less than i. If $\{x_i, x_j\} \in E^A$, there is nothing to prove. Otherwise, by Lemma 5.1.1, there exists a $k < \min\{i, j\}$ such that $\{x_i, x_k\} \in E^F$ and $\{x_j, x_k\} \in E^F$. By the inductive assumption, a path can be found from x_i to x_j passing through x_k in the section graph $G^A(\{x_1, \ldots, x_k\} \cup \{x_i, x_j\})$ which implies that $x_j \in Reach\,(x_i, \{x_1, \ldots, x_{i-1}\})$.   □

In terms of the matrix, the set $Reach\,(x_i, \{x_1, \ldots, x_{i-1}\})$ is simply the set of row subscripts that correspond to nonzero entries in the column vector L_{*i}. For example, let the graph of Figure 5.1.5 be ordered as shown in Figure 5.1.6.

Figure 5.1.6   A labelling of the graph of Figure 5.1.5.

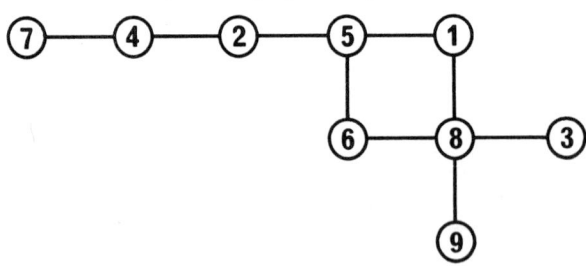

If $S_i = \{x_1, \ldots, x_i\}$, it is not difficult to see from the definition of reachable set that

$$Reach\,(x_1,S_0) = \{x_5,x_8\}$$
$$Reach\,(x_2,S_1) = \{x_4,x_5\}$$
$$Reach\,(x_3,S_2) = \{x_8\}$$
$$Reach\,(x_4,S_3) = \{x_5,x_7\}$$
$$Reach\,(x_5,S_4) = \{x_6,x_7,x_8\}$$
$$Reach\,(x_6,S_5) = \{x_7,x_8\}$$
$$Reach\,(x_7,S_6) = \{x_8\}$$
$$Reach\,(x_8,S_7) = \{x_9\}$$
$$Reach\,(x_9,S_8) = \varnothing\;.$$

It then follows from Theorem 5.1.2 that the structure of the corresponding L is given by

We have thus characterized the structure of L directly in terms of the structure of A. More importantly, there is a convenient way of characterizing the elimination graphs introduced in Section 5.1.1 in terms of reachable sets. Let $G_0,\ G_1,\ \ldots,\ G_{N-1}$ be the sequence of elimination graphs as defined by the nodes $x_1,\ x_2,\ \ldots,\ x_N$, and consider the graph $G_i = (X_i,E_i)$. We then have

Theorem 5.1.3

Let y be a node in the elimination graph $G_i = (X_i,E_i)$. The set of nodes adjacent to y in G_i is given by

$$Reach\ (y, \{x_1, \ldots, x_i\})$$

where the *Reach* operator is applied to the *original* graph G_0.

Proof   The proof can be done by induction on i.   □

Let us re-examine the example in Figure 5.1.3. Consider the graphs G_0 and G_2.

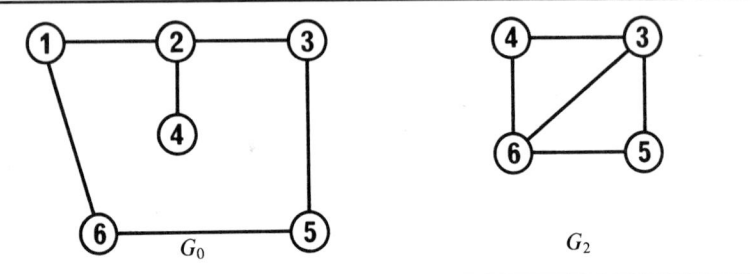

Let $S_2 = \{x_1, x_2\}$. It is clear that

$$Reach\ (x_3, S_2) = \{x_4, x_5, x_6\}\ ,$$
$$Reach\ (x_4, S_2) = \{x_3, x_6\}\ ,$$
$$Reach\ (x_5, S_2) = \{x_3, x_6\}\ ,$$

and

$$Reach\ (x_6, S_2) = \{x_3, x_4, x_5\}\ ,$$

since we have paths

$$(x_3, x_2, x_4)\ ,$$
$$(x_3, x_2, x_1, x_6),$$

and

$$(x_4, x_2, x_1, x_6)$$

in the graph G_0. These reach sets are precisely the adjacent sets in the graph G_2.

The importance of reachable sets in sparse elimination lies in Theorem 5.1.3. Given a graph $G = (X, E)$ and an elimination sequence $x_1, x_2, \ldots, x_N$, the whole elimination process can be described implicitly by this sequence and the *Reach* operator. This can be regarded as an *implicit model* for elimination, as opposed to the *explicit model* using elimination graphs (Section 5.1.1).

Exercises

5.1.1)   For any nonzero structure $Nonz(A)$, can you always find a matrix A^* so that its filled matrix F^* has identical *logical* and *numerical* nonzero structures? Why?

5.1.2)   Consider the star graph with 7 nodes (Figure 4.2.3). Assuming that the centre node is numbered first, determine the sequence of elimination graphs.

5.1.3)   For a given labelled graph $G^A = (X^A, E^A)$, show that

$$Reach(x_i, \{x_1, x_2, \ldots, x_{i-1}\}) \subset Adj(\{x_1, x_2, \ldots, x_i\}),$$

and hence conclude that $Fill(A) \subset Env(A)$.

5.1.4)   Show that the section graph

$$G^A(Reach(x_i, \{x_1, \ldots, x_{i-1}\}) \cup \{x_i\})$$

is a clique in the filled graph G^F.

5.1.5)   (Rose 1972a) A graph is *triangulated* if for every cycle $(x_1, x_2, \ldots, x_\ell, x_1)$ of length $\ell > 3$, there is an edge joining two non-consecutive vertices in the cycle. (Such an edge is called a *chord* of the cycle.) Show that the following conditions are equivalent.

a)   the graph G^A is triangulated
b)   there exists a permutation matrix P such that $Fill(PAP^T) = \emptyset$.

5.1.6)   Show that the graph $G^{F(A)}$ is triangulated. Give a permutation P such that $Fill(PF(A)P^T) = \emptyset$. Hence, or otherwise, show that $Nonz(F(A)) = Nonz(F(F(A)))$.

5.1.7)   Let $S \subset T$ and $y \notin T$. Show that

$$Reach(y, S) \subset Reach(y, T) \cup T.$$

5.1.8)   Let $y \notin S$. Define the *neighborhood set* of y in S to be

$$Nbrhd(y, S) =$$

$$\{s \in S \mid s \text{ is reachable from } y \text{ through a subset of } S\}.$$

Let $x \notin S$. Show that, if

$$Adj(x) \subset Reach(y, S) \cup Nbrhd(y, S) \cup \{y\},$$

then
a) $Nbrhd(x, S) \subset Nbrhd(y, S)$
b) $Reach(x, S) \subset Reach(y, S) \cup \{y\}$.

5.1.9)   Prove Theorem 5.1.3.

5.2 Computer Representation of Elimination Graphs

As discussed in Section 5.1, Gaussian elimination on a sparse symmetric linear system can be modelled by the sequence of elimination graphs. In this section, we study the representation and transformation of elimination graphs on a computer. These issues are important in the implementation of general sparse methods.

5.2.1 Explicit and Implicit Representations

Elimination graphs are, after all, symmetric graphs so that they can be represented explicitly using one of the storage schemes described in Section 3.2. However, what concerns us is that the implementation should be tailored for elimination, so that the transformation from one elimination graph to the next in the sequence can be performed easily.

Let us review the transformation steps. Let G_i be the elimination graph obtained from eliminating the node x_i from G_{i-1}. The adjacency structure of G_i can be obtained as follows.

Step 1 Determine the adjacent set $Adj_{G_{i-1}}(x_i)$ in G_{i-1}.

Step 2 Remove the node x_i and its adjacent list from the adjacency structure.

Step 3 For each node $y \in Adj_{G_{i-1}}(x_i)$, the new adjacent set of y in G_i is given by merging the subsets

$$Adj_{G_{i-1}}(y) - \{x_i\} \quad \text{and} \quad Adj_{G_{i-1}}(x_i) - \{y\} .$$

The above is an algorithmic formulation of the recipe by Parter (Section 5.1.1) to effect the transformation. There are two points that should be mentioned about the implementation. First, the space used to store $Adj_{G_{i-1}}(x_i)$ in the adjacency structure can be re-used after Step 2. Secondly, the *explicit* adjacency structure of G_i may require much more space than that of G_{i-1}. For example, in the star graph of N nodes (Figure 4.2.3), if the centre node is to be numbered first and $G_0 = (X_0, E_0)$ and $G_1 = (X_1, E_1)$ are the corresponding elimination graphs, it is easy to show that (see Exercise 5.1.2)

$$|E_0| = O(N)$$

and

$$|E_1| = O(N^2) .$$

In view of these observations a very flexible data structure has to be used in the explicit implementation to allow for the dynamic change in the structure of the elimination graphs. The *adjacency linked list* structure described in Section 3.2 is a good candidate.

Any explicit computer representation has two disadvantages. First, the flexibility in the data structure often requires significant overhead in storage and execution time. Secondly, the maximum amount of storage required is *unpredictable*. Enough storage is needed for the largest elimination graph G_i that occurs.[1] This may exceed greatly the storage requirement for the original G_0. Futhermore, this maximum storage requirement is not known until the end of the entire elimination process.

Theorem 5.1.3 provides another way to represent elimination graphs. They can be stored *implicitly* using the original graph G and the eliminated subset S_i. The set of nodes adjacent to y in G_i can then be retrieved by generating the reachable set:

$$Reach\,(y, S_i)$$

on the original graph. This implicit representation does not have any of the disadvantages of the explicit method. It has a small and predictable storage requirement and it preserves the adjacency structure of the given graph.

However, the amount of work required to determine reachable sets can be intolerably large, especially at the later stages of elimination when $|S_i|$ is large. In the next section, we shall consider another model which is more suitable for computer implementation, but still retains many of the advantages of using reachable sets.

5.2.2   Quotient Graph Model

Let us first consider elimination on the graph given in Figure 5.1.6. After the elimination of the nodes x_1, x_2, x_3, x_4, x_5 the corresponding elimination graph is given in Figure 5.2.1. Shaded nodes are those that have been eliminated.

Let $S = \{x_1,\ x_2,\ x_3,\ x_4,\ x_5\}$. In the implicit model, to discover that $x_6 \in Reach\,(x_7, S)$, the path

$$(x_7,\ x_4,\ x_2,\ x_5,\ x_6)$$

has to be traversed. Similarly, $x_8 \in Reach\,(x_7, S)$ because of the path

$$(x_7,\ x_4,\ x_2,\ x_5,\ x_1,\ x_8)\ .$$

Note that the lengths of the two paths are 4 and 5 respectively.

---

[1] Here "largest" refers to the number of edges, rather than the number of nodes.

Figure 5.2.1   A graph example and its elimination graph.

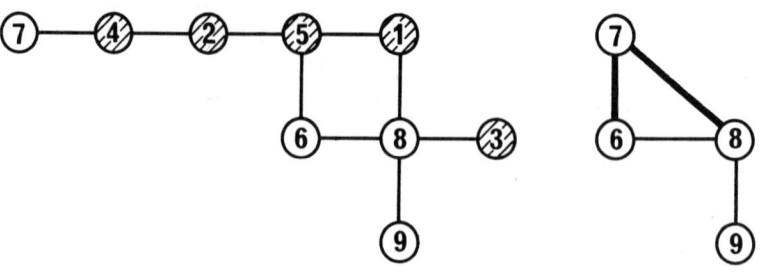

We make two observations:

a)   the amount of work to generate reachable sets can be reduced if the lengths of paths to uneliminated nodes are shortened.

b)   if these paths are shortened to the extreme case, we get the explicit elimination graphs which have undesirable properties as mentioned in the previous section.

We look for a compromise. By coalescing connected eliminated nodes, we obtain a new graph structure that serves our purpose. For example, in Figure 5.2.1, there are two connected components in the graph $G(S)$, whose node sets are

$$\{x_1,\ x_2,\ x_4,\ x_5\} \text{ and } \{x_3\}\ .$$

By forming two "supernodes," we obtain the graph as given in Figure 5.2.2.

Figure 5.2.2   Graph formed by coalescing connected eliminated nodes.

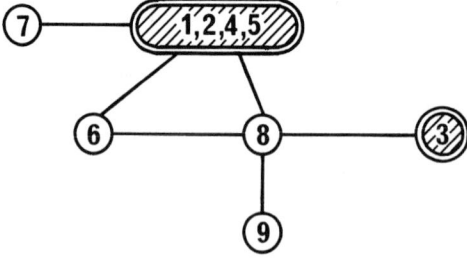

For convenience, we set $\bar{x}_5 = \{x_1,\ x_2,\ x_4,\ x_5\}$ and $\bar{x}_3 = \{x_3\}$ to denote these connected components in S. With this new graph, we note that the paths

$$(x_7,\ \bar{x}_5,\ x_6)$$

and

$$(x_7, \ \overline{x}_5, \ x_8)$$

are of length *two* and they lead us from the node x_7 to x_6 and x_8 respectively.

In general, if we adopt this strategy all such paths are of length less than or equal to two. This has the obvious advantage over the reachable set approach on the original graph, where paths can be of arbitrary lengths (less than N).

What is then its advantage over the explicit elimination graph approach? In the next section we shall show that this approach can be implemented *in-place*; that is, it requires no more space than the original graph structure. In short, this new graph structure can be used to generate reachable sets (or adjacent sets in the elimination graph) quite efficiently and yet it requires a fixed amount of storage.

To formalize this model for elimination, we introduce the notion of *quotient graphs*. Let $G = (X,E)$ be a given graph and let P be a partition on its node set X:

$$P = \{Y_1, \ Y_2, \ldots, \ Y_p\} \ .$$

That is, $\bigcup\limits_{k=1}^{p} Y_k = X$ and $Y_i \cap Y_j = \varnothing$ for $i \neq j$. We define the *quotient graph* of G with respect to P to be the graph $(P, \mathcal{E})$, where $\{Y_i, Y_j\} \in \mathcal{E}$ if and only if $Adj(Y_i) \cap Y_j \neq \varnothing$. Often, we denote this graph by G/P.

For example, the graph in Figure 5.2.2 is the quotient graph of the one in Figure 5.2.1 with respect to the partitioning

$$\{x_1, \ x_2, \ x_4, \ x_5\}, \ \{x_3\}, \ \{x_6\}, \ \{x_7\}, \ \{x_8\}, \ \{x_9\} \ .$$

The notion of quotient graphs will be treated in more detail in Chapter 6 where partitioned matrices are considered. Here, we study its role in modelling elimination. The new model represents the elimination process as a sequence of quotient graphs.

Let $G = (X,E)$ be a given graph and consider a stage in the elimination where S is the set of eliminated nodes. We now associate a quotient graph with respect to this set S as motivated by the example in Figure 5.2.2. Define the set

$$\mathcal{C}(S) = \tag{5.2.1}$$

$$\{C \subset S \mid G(C) \text{ is a connected component in the subgraph } G(S)\} \ ,$$

and the partitioning on X,

$$\overline{\mathcal{C}}(S) = \{\{y\} \mid y \in X - S\} \cup \mathcal{C}(S) \ . \tag{5.2.2}$$

This uniquely defines the quotient graph

$$G/\overline{\mathcal{C}}(S) ,$$

which can be viewed as the graph obtained by coalescing connected sets in S. Figure 5.2.2 is the resulting quotient graph for $S = \{x_1, x_2, x_3, x_4, x_5\}$.

We now study the relevance of quotient graphs in elimination. Let $x_1, x_2, \ldots, x_N$ be the sequence of node elimination in the given graph G. As before, let

$$S_i = \{x_1, x_2, \ldots, x_i\} , \quad 1 \le i \le N .$$

For each i, the subset S_i induces the partitioning $\overline{\mathcal{C}}(S_i)$ and the corresponding quotient graph

$$\mathcal{G}_i = G/\overline{\mathcal{C}}(S_i) = (\overline{\mathcal{C}}(S_i),\mathcal{E}_i) . \tag{5.2.3}$$

In this way, we obtain a sequence of quotient graphs

$$\mathcal{G}_1 \rightarrow \mathcal{G}_2 \rightarrow \cdots \rightarrow \mathcal{G}_N$$

from the node elimination sequence. Figure 5.2.3 shows the sequence for the graph example of Figure 5.2.1. For notational convenience, we use y instead of $\{y\}$ for such "supernodes" in $\overline{\mathcal{C}}(S_i)$.

The following theorem shows that quotient graphs of the form (5.2.3) are indeed representations of elimination graphs.

Theorem 5.2.1
For $y \in X - S_i$,

$$Reach_G(y,S_i) = Reach_{\mathcal{G}_i}(y,\mathcal{C}(S_i)) .$$

Proof   Consider $u \in Reach_G(y,S_i)$. If the nodes y and u are adjacent in G, so are y and u in $\mathcal{G}_i$. Otherwise, there exists a path

$$y, s_1, \ldots, s_t, u$$

in G where $\{s_1, \ldots, s_t\} \subset S_i$. Let $G(C)$ be the connected component in $G(S_i)$ containing $\{s_1\}$. Then we have the path

$$y, C, u$$

in $\mathcal{G}_i$ so that $u \in Reach_{\mathcal{G}_i}(y,\mathcal{C}(S_i))$.

Conversely, consider any $u \in Reach_{\mathcal{G}_i}(y,\mathcal{C}(S_i))$. There exists a path

$$y, C_1, \ldots, C_t, u$$

in $\mathcal{G}_i$ where $\{C_1, \ldots, C_t\} \subset \mathcal{C}(S_i)$. If $t = 0$, y and u are adjacent in the original graph G. If $t > 0$, by definition of connected components,

Figure 5.2.3   A sequence of quotient graphs.

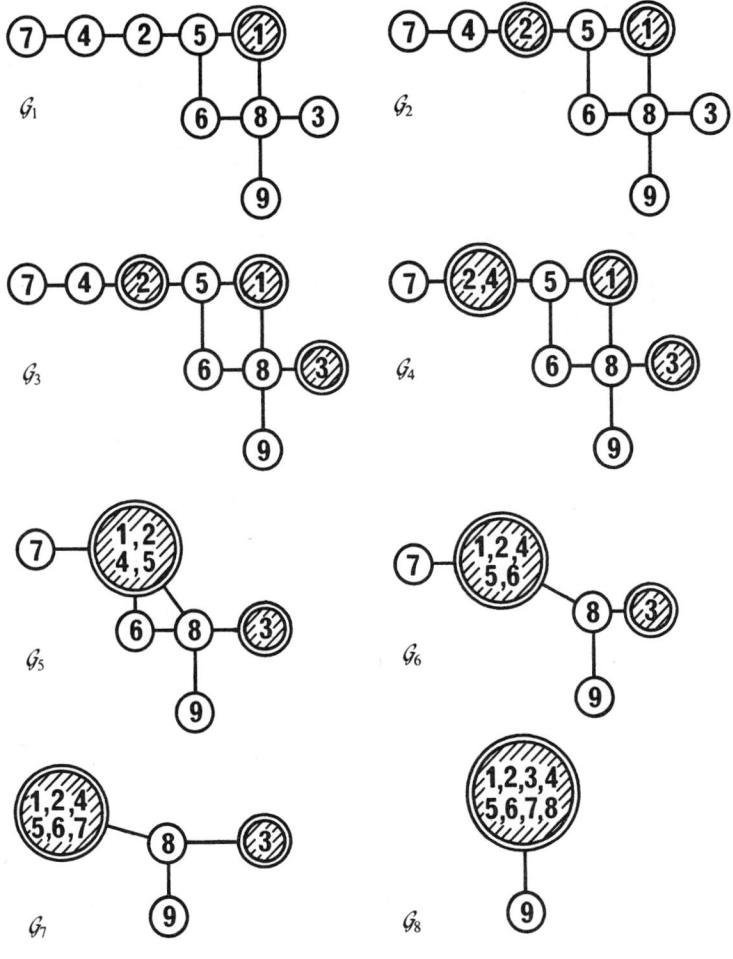

t cannot be greater than one; that is, the path must be

$$y, C, u ,$$

so that we can obtain a path from y to u through C in the graph G. Hence

$$u \in Reach_G(y, S_i) .$$

$\square$

The determination of reachable sets in the quotient graph $\mathcal{G}_i$ is straightforward. For a given node $y \notin \mathcal{C}(S_i)$, the following algorithm returns the set $Reach_{\mathcal{G}_i}(y, \mathcal{C}(S_i))$.

Step 1   (*Initialization*) $R \leftarrow \emptyset$.
Step 2   (*Find reachable nodes*)
        For $x \in Adj_{\mathcal{G}_i}(y)$
           If $x \in \mathcal{C}(S_i)$
                then $R \leftarrow R \cup Adj_{\mathcal{G}_i}(x)$
                else $R \leftarrow R \cup \{x\}$.
Step 3   (*Exit*) The reachable set is given in R.

The connection between elimination graphs and quotient graphs (5.2.3) is quite obvious. Indeed, we can obtain the structure of the elimination graph G_i from that of $\mathcal{G}_i$ by the simple algorithm below.

Step 1   Remove supernodes in $\mathcal{C}(S_i)$ and their incident edges from the quotient graph.
Step 2   For each $C \in \mathcal{C}(S_i)$, add edges to the quotient graph so that all adjacent nodes of C are pairwise adjacent in the elimination graph.

To illustrate the idea, consider the transformation of $\mathcal{G}_4$ to G_4 for the example in Figure 5.2.3. The elimination graph G_4 is given in Figure 5.2.4.

Figure 5.2.4   From quotient graph to elimination graph.

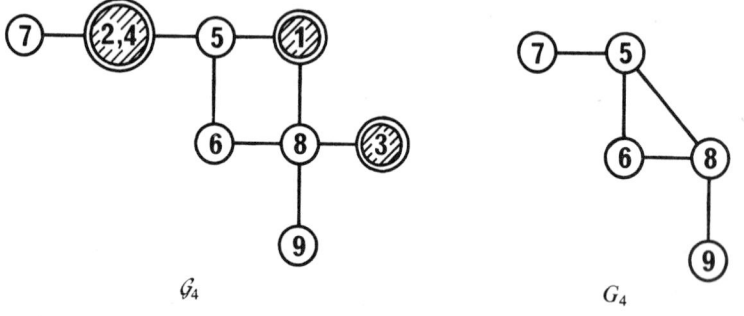

In terms of implicitness, the quotient graph model lies in between the reachable set approach and the elimination graph model as a vehicle for representing the elimination process.

| Reachable set on original graph | $\rightarrow$ | Quotient graph | $\rightarrow$ | Elimination graph |

The correspondence between the three models is summarized in the following table.

Table 5.2.1

	Implicit Model	Quotient Model	Explicit Model
Represen-	S_1	$\mathcal{G}_1$	G_1
tation	S_2	$\mathcal{G}_2$	G_2
	.	.	.
	.	.	.
	.	.	.
	S_{N-1}	$\mathcal{G}_{N-1}$	G_{N-1}
Adjacency	$Reach\,(y,S_i)$	$Reach_{\mathcal{G}_i}(y,\mathcal{C}(S_i))$	$Adj_{G_i}(y)$

5.2.3   Implementation of the Quotient Graph Model

Consider the quotient graph $\mathcal{G} = G/\overline{\mathcal{C}}(S)$ induced by the eliminated set S. For notational convenience, if $s \in S$, we use the notation $\overline{s}$ to denote the connected component in the subgraph $G(S)$, containing the node s. For example, in the quotient graph of Figure 5.2.2,

$$\overline{x}_5 = \overline{x}_1 = \overline{x}_2 = \overline{x}_4 = \{x_1,\ x_2,\ x_4,\ x_5\}\ .$$

On the other hand, for a given $C \in \mathcal{C}(S)$, we can select any node x from C and use x as a *representative* for C, that is, $\overline{x} = C$. Before we discuss the choice of representative in the implementation, we establish some results that can be used to show that the model can be implemented *in-place*; that is, in the space provided by the adjacency structure of the original graph.

Lemma 5.2.2
Let $G = (X,E)$ and $C \subset X$ where $G(C)$ is a connected subgraph. Then

$$\sum_{x \in C} |\,Adj(x)\,| \ge |\,Adj(C)\,| + 2(|\,C\,| - 1)\ .$$

Proof   Since $G(C)$ is connected, there are at least $|\,C\,| - 1$ edges in the subgraph. These edges are counted twice in $\sum_{x \in C} |\,Adj(x)\,|$ and hence the result.   □

Let $x_1, x_2, \ldots, x_N$ be the node sequence and $S_i = \{x_1, \ldots, x_i\}$, $1 \le i \le N$. For $1 \le i \le N$, let

$$\mathcal{G}_i = G/\overline{\mathcal{C}}(S_i) = (\overline{\mathcal{C}}(S_i), \mathcal{E}_i) .$$

Lemma 5.2.3
Let $y \in X - S_i$. Then

$$| Adj_G(y) | \ge | Adj_{\mathcal{G}_i}(y) | .$$

Proof   This follows from the inequality

$$| Adj_{\mathcal{G}_i}(y) | \ge | Adj_{\mathcal{G}_{i+1}}(y) |$$

for $y \in X - S_{i+1}$. The problem of verifying this inequality is left as an exercise.   □

Theorem 5.2.4

$$\max_{1 \le i \le N} | \mathcal{E}_i | \le | E | .$$

Proof   Consider the quotient graphs $\mathcal{G}_i$ and $\mathcal{G}_{i+1}$. If x_{i+1} is isolated in the subgraph $G(S_{i+1})$, clearly $| \mathcal{E}_{i+1} | = | \mathcal{E}_i |$. Otherwise the node x_{i+1} is merged with some components in S_i to form a new component in S_{i+1}. The results of Lemmas 5.2.2 and 5.2.3 apply, so that

$$| \mathcal{E}_{i+1} | < | \mathcal{E}_i | .$$

Hence, in all cases,

$$| \mathcal{E}_{i+1} | \le | \mathcal{E}_i |$$

and the result follows.   □

Theorem 5.2.4 shows that the sequence of quotient graphs produced by elimination requires no more space than the original graph structure. On coalescing a connected set C into a supernode, we know from Lemma 5.2.2 that there are enough storage locations for $Adj(C)$ from those of $Adj(x)$, $x \in C$. Moreover, for $|C| > 1$, there is a surplus of $2(|C| - 1)$ locations, which can be used for links or pointers.

Figure 5.2.5 is an illustration of the data structure used to represent $Adj_{\mathcal{G}}(C)$, in the quotient graph $\mathcal{G}$, where $C = \{a, b, c\}$. Here, zero signifies the end of the neighbor list in $\mathcal{G}$.

Note that in the example, the node "a" is chosen to be the representative for $C = \{a, b, c\}$. In the computer implementation, it is important to choose a *unique* representative for each $C \in \mathcal{C}(S)$, so that any reference to C can be made through its representative.

Figure 5.2.5   Data structure for quotient graph.

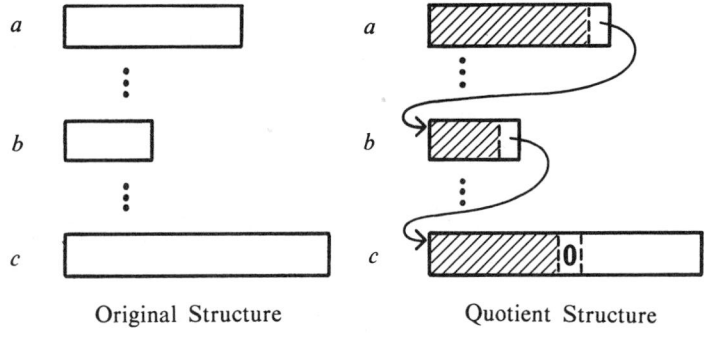

Original Structure   Quotient Structure

Let $x_1, x_2, \ldots, x_N$ be the node sequence, and $C \in \mathcal{C}(S)$. We choose the node $x_r \in C$ to be the representative of C, where

$$r = \max\{j \mid x_j \in C\} . \qquad (5.2.4)$$

That is, x_r is the node in C last eliminated.

So far, we have described the data structure of the quotient graphs and how to represent supernodes. Another important aspect in the implementation of the quotient graph model for elimination is the transformation of quotient graphs due to node elimination. Let us now consider how the adjacency structure of $\mathcal{G}_i$ can be obtained from that of $\mathcal{G}_{i-1}$ when the node x_i is eliminated. The following algorithm performs the transformation.

Step 1   (*Preparation*)   Determine the sets

$$T = Adj_{\mathcal{G}_{i-1}}(x_i) \cap \mathcal{C}(S_{i-1})$$

$$R = Reach_{\mathcal{G}_{i-1}}(x_i, \mathcal{C}(S_{i-1})) .$$

Step 2   (*Form new supernode and partitioning*)   Form

$$\bar{x}_i = \{x_i\} \cup T$$

$$\mathcal{C}(S_i) = (\mathcal{C}(S_{i-1}) - T) \cup \{\bar{x}_i\} .$$

Step 3   (*Update adjacency*) $Adj_{\mathcal{G}_i}(\bar{x}_i) = R$
For $y \in R$, $Adj_{\mathcal{G}_i}(y) = \{\bar{x}_i\} \cup Adj_{\mathcal{G}_{i-1}}(y) - (T \cup \{x_i\})$.

Let us apply this algorithm to transform $\mathcal{G}_4$ to $\mathcal{G}_5$ in the example of Figure 5.2.3. In $\mathcal{G}_4$

$$\mathcal{C}(S_4) = \{\bar{x}_1, \bar{x}_3, \bar{x}_4\} .$$

On applying Step 1 to the node x_5, we obtain

$$T = \{\bar{x}_1, \ \bar{x}_4\}$$

and

$$R = \{x_6, \ x_7, \ x_8\} \ .$$

Therefore, the new "supernode" is given by

$$\bar{x}_5 = \{x_5\} \cup \bar{x}_1 \cup \bar{x}_4 = \{x_1, \ x_2, \ x_4, \ x_5\} \ .$$

and the new partitioning is

$$\mathcal{C}(S_5) = \{\bar{x}_3, \ \bar{x}_5\} \ .$$

Finally, in Step 3 the adjacency sets are updated and we get

$$Adj_{\mathcal{G}_5}(x_6) = \{\bar{x}_5, \ x_8\}$$

$$Adj_{\mathcal{G}_5}(x_7) = \{\bar{x}_5\}$$

$$Adj_{\mathcal{G}_5}(x_8) = \{\bar{x}_3, \ \bar{x}_5, \ x_6, \ x_9\} \ ,$$

and

$$Adj_{\mathcal{G}_5}(\bar{x}_5) = R = \{x_6, \ x_7, \ x_8\} \ .$$

The effect of the quotient graph transformation on the data structure can be illustrated by an example. Consider the example of Figure 5.2.3, where we assume that the adjacency structure is represented as shown in Figure 5.2.6.

Figure 5.2.6   Adjacency representation.

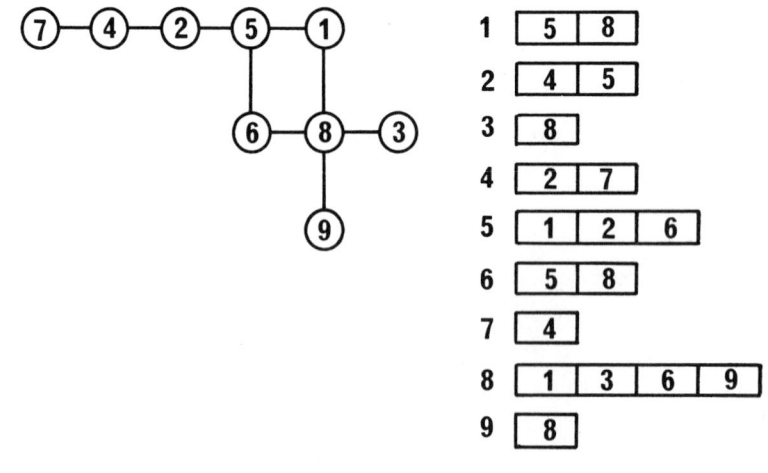

Figure 5.2.7 shows some important steps in producing quotient graphs for this example. The adjacency structure remains unchanged when the quotient graphs $\mathcal{G}_1$, $\mathcal{G}_2$ and $\mathcal{G}_3$ are formed. To transform $\mathcal{G}_3$ to $\mathcal{G}_4$, the nodes x_2 and x_4 are to be coalesced, so that in $\mathcal{G}_4$, the new adjacent set of node x_4 contains that of the subset $\{x_2, x_4\}$ in the original graph, namely $\{x_5, x_7\}$. Here, the last location for the adjacent set of x_4 is used as a link. Note also that in the adjacent list of node x_5, the neighbor x_2 has been changed to x_4 in $\mathcal{G}_4$ since node x_4 becomes the representative of the component subset $\{x_2, x_4\}$.

The representations for $\mathcal{G}_5$ and $\mathcal{G}_6$ in this storage mode are also included in Figure 5.2.7.

This way of representing quotient graphs for elimination will be used in the implementation of the minimum degree ordering algorithm, to be discussed in the Section 5.4.

Exercises

5.2.1) a) Design and implement a subroutine called REACH which can be used to determine the reachable set of a given node ROOT through a subset S. The subset is given by an array SFLAG, where a node i belongs to S if SFLAG(i) is nonzero. Describe the parameters of the subroutine and any auxiliary storage you require.

   b) Suppose a graph is stored in the array pair (XADJ, ADJNCY). For any given elimination sequence, use the subroutine REACH to print out the adjacency structures of the sequence of elimination graphs.

5.2.2) Let $\bar{\mathcal{C}}(S_i)$ be as defined in (5.2.2) and show that $| \bar{\mathcal{C}}(S_{i+1}) | \le | \bar{\mathcal{C}}(S_i) |$.

5.2.3) Prove the inequality that appears in the proof of Lemma 5.2.3.

5.2.4) Let $\mathcal{X} = \{ C \mid C \in \mathcal{C}(S_i)$ for some $i \}$. Show that $| \mathcal{X} | = N$.

5.2.5) Let $C \in \mathcal{C}(S_i)$, and $\bar{x}_r = C$ where

$$r = \max\{ j \mid x_j \in C \} .$$

Show that
   a) $Adj_G(C) = Reach_G(x_r, S_i)$.
   b) $Reach_G(x_r, S_i) = Reach_G(x_r, S_{r-1})$.

5.2.6) Display the sequence $\{\mathcal{G}_i\}$ of quotient graphs for the star graph of 7 nodes, where the centre node is numbered first.

Figure 5.2.7 An in-place quotient graph transformation.

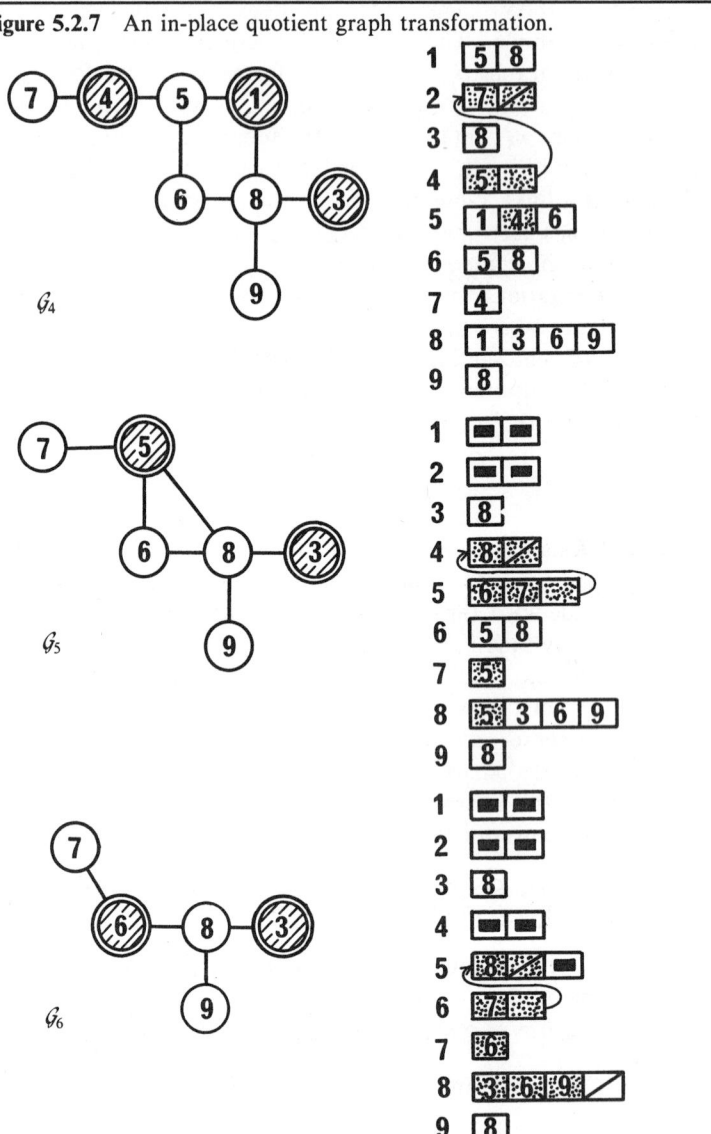

5.3   The Minimum Degree Ordering Algorithm

Let A be a given symmetric matrix and let P be a permutation matrix. Although the nonzero structures of A and PAP^T are different, their sizes are the same: $|Nonz(A)| = |Nonz(PAP^T)|$. However, the crucial point is that there may be a dramatic difference between $|Nonz(F(A))|$ and $|Nonz(F(PAP^T))|$ for some permutation P. The example in Figure 4.2.3 illustrates this fact.

Ideally, we want to find a permutation P^* that minimizes the size of the nonzero structure of the filled matrix:

$$|Nonz(F(P^*AP^{*T}))| = \min_P |Nonz(F(PAP^T))| \ .$$

So far, there is no efficient algorithm for getting such an optimal P^* for a general symmetric matrix. Indeed, the problem when A is unsymmetric has been shown to be very difficult—a so-called NP complete problem (Rose 1975). Thus, we have to rely on heuristics which will produce an ordering P with an acceptably small but not necessarily minimum $|Nonz(F(PAP^T))|$.

By far the most popular fill-reducing scheme used is the *minimum degree* algorithm (Tinney 1969), which corresponds to the Markowitz scheme (Markowitz 1957) for unsymmetric matrices. The scheme is based on the following observation, which is depicted in Figure 5.3.1.

Figure 5.3.1   Motivation of the minimum degree algorithm.

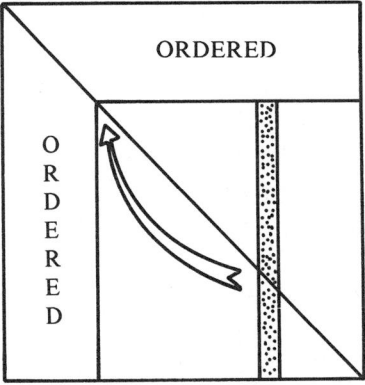

Suppose $\{x_1, \ldots, x_{i-1}\}$ have been labelled. The number of nonzeros in the filled graph for these columns is fixed. In order to reduce the number of nonzeros in the i-th column, it is apparent that in the submatrix remaining to be factored, the column with the fewest nonzeros should be moved to become column i. In other words, the scheme may be regarded as a method that reduces the fill of a matrix by a local minimization of $\eta(L_{*i})$ in the factored matrix.

5.3.1   The Basic Algorithm

The minimum degree algorithm can be most easily described in terms of ordering a symmetric graph. Let $G_0 = (X,E)$ be an unlabelled graph. Using the elimination graph model, the basic algorithm is as follows.

Step 1   (*Initialization*) $i \leftarrow 1$.

Step 2   (*Minimum degree selection*) In the elimination graph $G_{i-1} = (X_{i-1}, E_{i-1})$, choose a node x_i of minimum degree in G_{i-1}.

Step 3   (*Graph transformation*) Form the new elimination graph $G_i = (X_i, E_i)$ by eliminating the node x_i from G_{i-1}.

Step 4   (*Loop or stop*) $i \leftarrow i+1$. If $i > |X|$, stop. Otherwise, go to Step 2.

As an illustration of the algorithm, we consider the graph in Figure 5.3.2.

Figure 5.3.2   A minimum degree ordering for a graph.

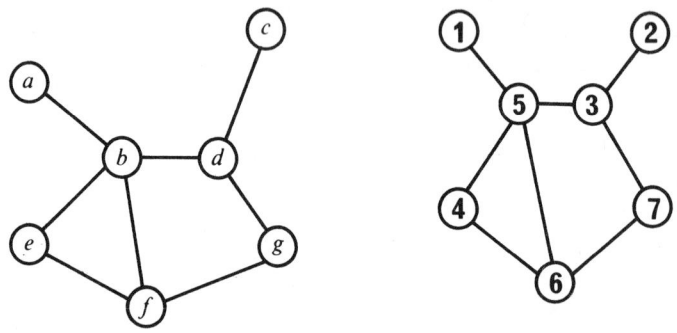

The way the minimum degree algorithm is carried out for this example is shown step by step in Figure 5.3.3. Notice that there can be more than one node with the minimum degree at a particular step. Here we break the ties arbitrarily. However, different tie-breaking strategies give different versions of the minimum degree algorithm.

Figure 5.3.3   Numbering in the minimum degree algorithm.

i	Elimination Graph G_{i-1}	Node Selected	Minimum Degree
1	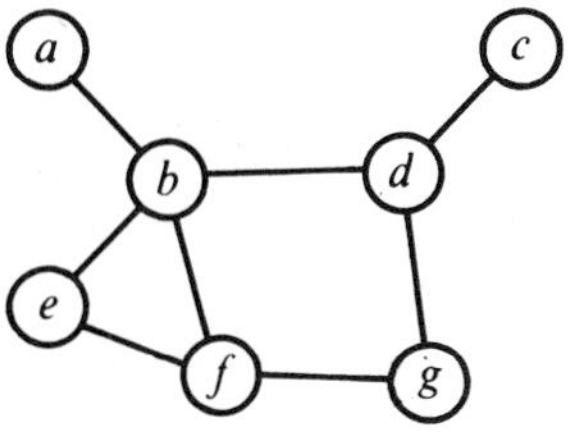	a	1
2	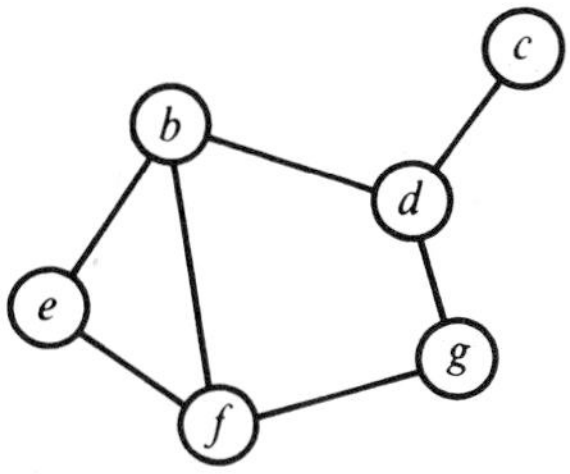	c	1
3	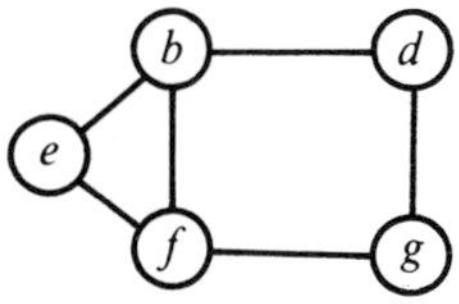	d	2
4	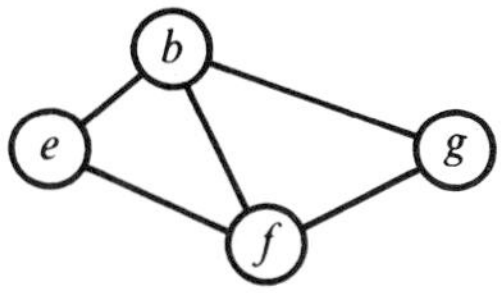	e	2
5	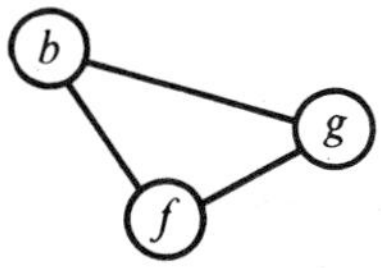	b	2
6		f	1
7		g	0

5.3.2 Description of the Minimum Degree Algorithm Using Reachable Sets

The use of elimination graphs in the minimum degree algorithm provides the mechanism by which we select the next node to be numbered. Each step of the algorithm involves a graph transformation, which is the most expensive part of the algorithm in terms of implementation. These transformations can be eliminated if we can provide an alternative way to compute the degrees of the nodes in the elimination graph.

Theorem 5.1.3 provides a mechanism for achieving this through the use of reachable sets. With this connection, we can restate the minimum degree algorithm as follows.

Step 1   (*Initialization*) $S \leftarrow \emptyset$. $Deg(x) \leftarrow |Adj(x)|$, for $x \in X$.

Step 2   (*Minimum degree selection*) Pick a node $y \in X - S$ where $Deg(y) = \min\limits_{x \in X - S} Deg(x)$. Number the node y next and set $T \leftarrow S \cup \{y\}$.

Step 3   (*Degree update*) $Deg(u) \leftarrow |Reach(u,T)|$ for $u \in X - T$.

Step 4   (*Loop or stop*) If $T = X$, stop. Otherwise, set $S \leftarrow T$ and go to Step 2.

This approach uses the original graph structure throughout the entire process. Indeed, the algorithm can be carried out with only the adjacency structure

$$G_0 = (X,E) .$$

It is appropriate here to point out that in the degree update step of the algorithm, it is not necessary to recompute the sizes of the reachable sets for every node in $X - T$, since most of them remain unchanged. This observation is formalized in the following lemma. Its proof follows from the definition of reachable sets and is left as an exercise.

Lemma 5.3.1

Let $y \notin S$ and $T = S \cup \{y\}$. Then $Reach(x,T)$

$$= \begin{cases} Reach(x,S) & \text{for } x \notin Reach(y,S) \\ Reach(x,S) \cup Reach(y,S) - \{x,y\} & \text{otherwise.} \end{cases}$$

In the example of Figure 5.3.3, consider the stage when node d is being eliminated.

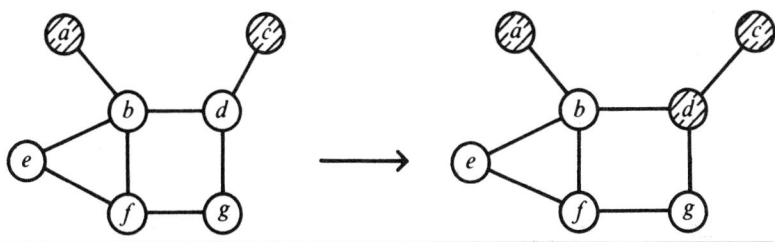

We have $S = \{a,c\}$, so that $Reach(d,S) = \{b,g\}$. Therefore, the elimination of d only affects the degrees of the nodes b and g. By this observation, Step 3 in the algorithm can be restated as

Step 3   (*Degree update*)

$$Deg(u) \leftarrow |\, Reach(u,T)|\,, \quad \text{for} \quad u \in Reach(y,S)\,.$$

Corollary 5.3.2
Let y, S, T be as in Lemma 5.3.1. For $x \in X - T$,

$$|\, Reach(x,T)| \geq |\, Reach(x,S)| - 1\,.$$

Proof   The result follows directly from Lemma 5.3.1.   □

5.3.3   An Enhancement

As the algorithm stands, one node is numbered each time the loop is executed. However, when a node y of minimum degree is found at Step 2, it is often possible to detect that a subset of nodes may automatically be numbered next, without carrying out any minimum degree search.

Let us begin the study by introducing an equivalence relation. Consider a stage in the elimination process, where S is the set of eliminated nodes. Two nodes $x,y \in X - S$ are said to be *indistinguishable with respect to elimination*[2] if

$$Reach(x,S) \cup \{x\} = Reach(y,S) \cup \{y\}\,. \qquad (5.3.1)$$

Consider the graph example in Figure 5.3.4. The subset S contains 36 shaded nodes. (This is an actual stage that occurs when the minimum degree algorithm is applied to this graph.) We note that the nodes a, b and c are indistinguishable with respect to elimination, since $Reach(a,S) \cup \{a\}$, $Reach(b,S) \cup \{b\}$ and $Reach(c,S) \cup \{c\}$ are all equal to

---

[2] Henceforth, it should be understood that nodes referred to as "indistinguishable" are *indistinguishable with respect to elimination*.

$$\{a,\ b,\ c,\ d,\ e,\ f,\ g,\ h,\ j,\ k\}\ .$$

There are two more groups that can be identified as indistinguishable. They are

$$\{j,k\}\ ,$$

and

$$\{f,g\}\ .$$

Figure 5.3.4   A graph example on indistinguishable nodes.

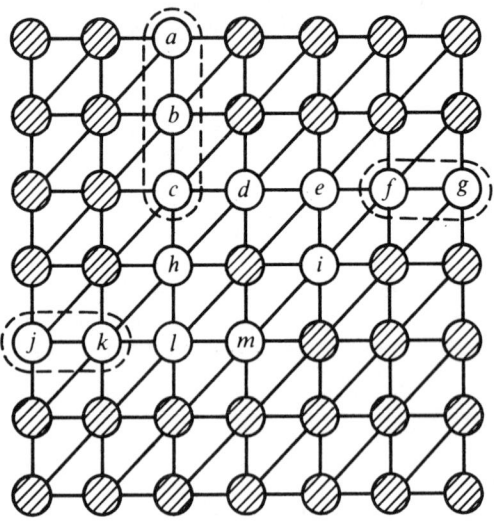

We now study the implication of this equivalence relation and its role in the minimum degree algorithm. As we shall see later, this notion can be used to speed up the execution of the minimum degree algorithm.

Theorem 5.3.3
Let $x,y \in X - S$. If

$$Reach\,(x,S) \cup \{x\} = Reach\,(y,S) \cup \{y\}\ ,$$

then for all $X - \{x,y\} \supset T \supset S$,

$$Reach\,(x,T) \cup \{x\} = Reach\,(y,T) \cup \{y\}\ .$$

Proof   Obviously, $x \in Reach\,(y,S) \subset Reach\,(y,T) \cup T$,   (see Exercise 5.1.7) so that $x \in Reach\,(y,T)$. We now want to show that $Reach\,(x,T) \subset Reach\,(y,T) \cup \{y\}$. Consider $z \in Reach\,(x,T)$. There exists a path

$$(x, s_1, \ldots, s_t, z)$$

where $\{s_1, \ldots, s_t\} \subset T$. If all $s_i \in S$, there is nothing to prove. Otherwise, let s_i be the first node in $\{s_1, \ldots, s_t\}$ not in S, that is

$$s_i \in Reach(x,S) \cap T .$$

This implies $s_i \in Reach(y,S)$ and hence $z \in Reach(y,T)$. Together, we have

$$Reach(x,T) \cup \{x\} \subset Reach(y,T) \cup \{y\} .$$

The inclusion in the other direction follows from symmetry, yielding the result.   □

Corollary 5.3.4
Let x,y be indistinguishable with respect to the subset S. Then for $T \supset S$,

$$| Reach(x,T)| = | Reach(y,T)| .$$

In other words, if two nodes become indistinguishable at some stage of the elimination, they remain indistinguishable until one of them is eliminated. Moreover, the following theorem shows that they can be eliminated *together* in the minimum degree algorithm.

Theorem 5.3.5
If two nodes become indistinguishable at some stage in the minimum degree algorithm. then they can be eliminated together in the algorithm.

Proof   Let x,y be indistinguishable after the elimination of the subset S. Assume that x becomes a node of minimum degree after the set $T \supset S$ has been eliminated, that is,

$$| Reach(x,T)| \le | Reach(z,T)| \text{ for all } z \in X - T .$$

Then, by Corollary 5.3.4,

$$| Reach(y,T \cup \{x\})| = | Reach(y,T) - \{x\}|$$
$$= | Reach(y,T)| - 1$$
$$= | Reach(x,T)| - 1 .$$

Therefore, for all $z \in X - T \cup \{x\}$, by Corollary 5.3.2,

$$| Reach(y,T \cup \{x\})| \le | Reach(z,T)| - 1$$
$$\le | Reach(z,T \cup \{x\})| .$$

In other words, after the elimination of the node x, the node y becomes a node of minimum degree.   □

These observations can be exploited in the implementation of the minimum degree algorithm. After carrying out a minimum degree search to determine the next node $y \in X - S$ to eliminate, we can number immediately after y the set of nodes indistinguishable from y.

In addition, in the degree update step, by virtue of Corollary 5.3.4, work can be reduced since indistinguishable nodes have the same degree in the elimination graphs. Once nodes are identified as being indistinguishable, they can be "glued" together and treated as a single supernode thereafter.

For example, Figure 5.3.5 shows two stages in the eliminations where supernodes are formed from indistinguishable nodes. For simplicity, the eliminated nodes are not shown. After the elimination of the indistinguishable set $\{a, b, c\}$, all the nodes have identical reachable sets so that they can be merged into one.

Figure 5.3.5 Indistinguishable nodes in two stages of elimination for the example in Figure 5.3.4.

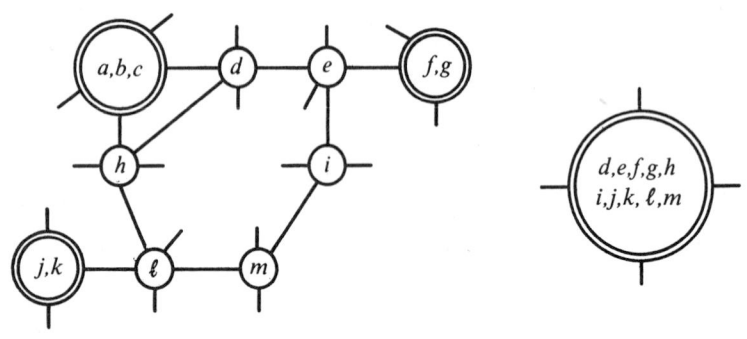

In general, to identify indistinguishable nodes via the definition (5.3.1) is time consuming. Since the enhancement does not require the merging of *all possible* indistinguishable nodes, we look for some simple, easily-implemented condition. In what follows, a condition is presented which experience has shown to be very effective. In most cases, it identifies all indistinguishable nodes.

Let $G = (X, E)$ and S be the set of eliminated nodes. Let $G(C_1)$ and $G(C_2)$ be two connected components in the subgraph $G(S)$; that is,

$$C_1, C_2 \in \mathcal{C}(S) .$$

Lemma 5.3.6
Let $R_1 = Adj(C_1)$, and $R_2 = Adj(C_2)$. If $y \in R_1 \cap R_2$, and

$$Adj(y) \subset R_1 \cup R_2 \cup C_1 \cup C_2$$

then $Reach(y,S) \cup \{y\} = R_1 \cup R_2$.

Proof   Let $x \in R_1 \cup R_2$. Assume $x \in R_1 = Adj(C_1)$. Since $G(C_1)$ is a connected component in $G(S)$, we can find a path from y to x through $C_1 \subset S$. Therefore, $x \in Reach(y,S) \cup \{y\}$.

On the other hand, $y \in R_1 \cup R_2$ by definition. Moreover, if $x \in Reach(y,S)$, there exists a path from y to x through S:

$$y, \, s_1, \, s_2, \, \ldots, \, s_t, \, x \, .$$

If $t = 0$, then $x \in Adj(y) - S \subset R_1 \cup R_2$. Otherwise, if $t > 0$, $s_1 \in Adj(y) \cap S \subset C_1 \cup C_2$. This means $\{s_1, \ldots, s_t\}$ is a subset of either C_1 or C_2 so that $x \in R_1 \cup R_2$. Hence $Reach(y,S) \cup \{y\} \subset R_1 \cup R_2$.   □

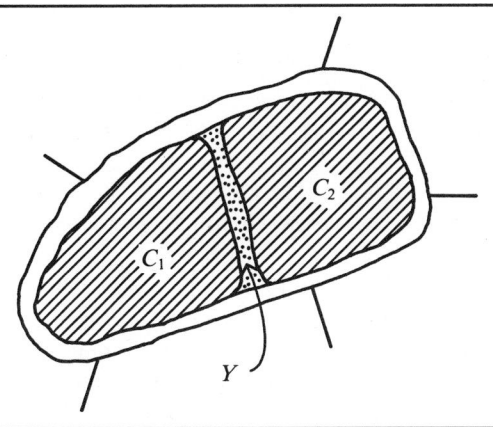

Theorem 5.3.7
Let C_1, C_2 and R_1, R_2 be as in Lemma 5.3.6. Then the nodes in

$$Y = \{y \in R_1 \cap R_2 | \, Adj(y) \subset R_1 \cup R_2 \cup C_1 \cup C_2\} \qquad (5.3.2)$$

are indistinguishable with respect to the eliminated subset S.

Proof   It follows from Lemma 5.3.6.   □

Corollary 5.3.8
For $y \in Y$,

$$|Reach(y,S)| = |R_1 \cup R_2| - 1 \, .$$

Theorem 5.3.7 can be used to merge indistinguishable nodes in the intersection of the two reachable sets R_1 and R_2. The test can be simply done by inspecting the adjacent set of nodes in the intersection $R_1 \cap R_2$.

This notion of indistinguishable nodes can be applied to the minimum degree algorithm. The new enhanced algorithm can be stated as follows.

Step 1   (*Initialization*) $S \leftarrow \emptyset$,

$$Deg(x) = |Adj(x)|, \text{ for } x \in X .$$

Step 2   (*Selection*) Pick a node $y \in X - S$ such that

$$Deg(y) = \min_{x \in X - S} Deg(x) .$$

Step 3   (*Elimination*) Number the nodes in

$$Y = \{x \in X - S \mid x \text{ is indistinguishable from } y\}$$

next in the ordering.

Step 4   (*Degree update*) For $u \in Reach(y,S) - Y$

$$Deg(u) = |Reach(u, S \cup Y)|$$

and identify indistinguishable nodes in the set $Reach(y,S) - Y$.

Step 5   (*Loop or stop*) Set $S \leftarrow S \cup Y$. If $S = X$, stop. Otherwise, go to Step 2.

5.3.4   Implementation of the Minimum Degree Algorithm

The implementation of the minimum degree algorithm presented here incorporates the notion of indistinguishable nodes as described in the previous sections. Nodes identified as indistinguishable are merged together to form a supernode. They will be treated essentially *as one node* in the remainder of the algorithm. They share the same adjacent set, have the same degree, and can be eliminated together in the algorithm. In the implementation, this supernode will be referenced by a representative of the set.

The algorithm requires the determination of reachable sets for degree update. The quotient graph model (Section 5.2.2) is used for this purpose to improve the overall efficiency of the algorithm. In effect, eliminated connected nodes are merged together and the computer representation of the sequence of quotient graphs (Section 5.2.3) is utilized.

It should be emphasized that the idea of *quotient* (or merging nodes into supernodes) is applied here in two different contexts.

a) *eliminated* connected nodes to facilitate the determination of reachable sets.

b) *uneliminated indistinguishable* nodes to speed up elimination.

This is illustrated in Figure 5.3.6. It shows how the graph of Figure 5.3.4 is stored conceptually in this implementation by the two forms of quotient. The shaded double-circled nodes denote supernodes that have been eliminated, while blank double-circled supernodes represent those formed from indistinguishable nodes.

Figure 5.3.6   A quotient graph formed from two types of supernodes.

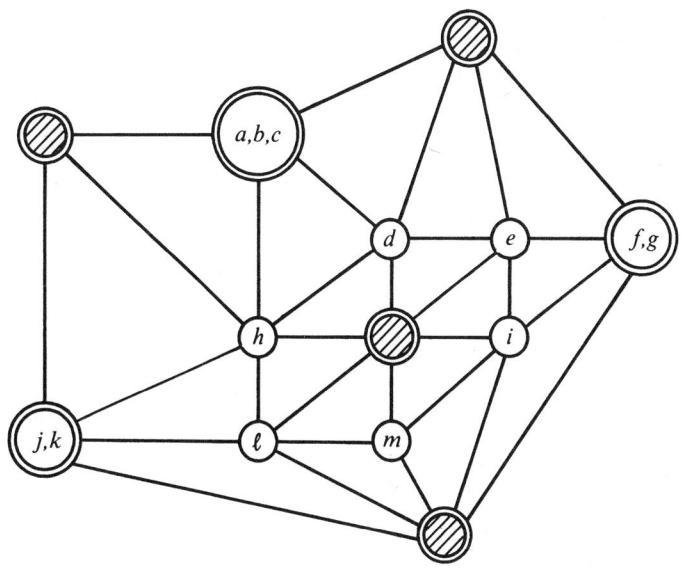

In this subsection, we describe a set of subroutines, which implement the minimum degree algorithm as presented earlier. Some of the parameters used are the same as those discussed in Chapter 3. We shall briefly review them here and readers are referred to Section 3.3 for details.

The graph $G = (X, E)$ is stored using the integer array pair (XADJ, ADJNCY), and the number of variables in X is given by NEQNS. The resulting minimum degree ordering is stored in the vector PERM, while INVP returns the inverse of this ordering.

This collection of subroutines requires some working vectors to implement the quotient graph model and the notion of indistinguishable nodes. The current degrees of the nodes in the (implicit) elimination graph are kept in the array DEG. The DEG value for nodes that have been eliminated is set to -1.

In the representation of the sequence of quotient graphs, connected eliminated nodes are merged to form a supernode. As mentioned in Section 5.3.2, for the purpose of reference, it is sufficient to pick a *representative* from the supernode. If $G(C)$ is such a connected component, we always choose the node $x \in C$ last eliminated to represent C. This implies that the remaining nodes in C can be ignored in subsequent quotient graphs.

The same remark applies to indistinguishable groups of uneliminated nodes. For each group, only the representative will be considered in the present quotient structure.

The working vector MARKER is used to mark those nodes that can be ignored in the adjacency structure. The MARKER values for such nodes are set to -1. This vector is also used temporarily to facilitate the generation of reachable sets.

Two more arrays QSIZE and QLINK are used to completely specify indistinguishable supernodes. If node i is the representative, the number of nodes in this supernode is given by QSIZE(i) and the nodes are given by

$$i, \text{QLINK}(i), \text{QLINK}(\text{QLINK}(i)), \ldots .$$

Figure 5.3.7 illustrates the use of the vectors QSIZE, QLINK and MARKER. The nodes $\{2, 5, 8\}$ form an indistinguishable supernode represented by node 2. Thus, the MARKER values of 5 and 8 are -1. On the other hand, $\{3, 6, 9\}$ forms an eliminated supernode. Its representative is node 9 so that MARKER(3) and MARKER(6) are -1.

Figure 5.3.7   Illustration of the role of QLINK, QSIZE and MARKER working vectors.

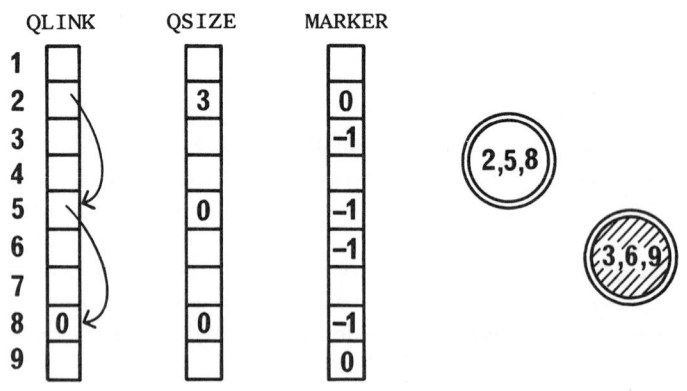

There are five subroutines in this set, namely GENQMD, QMDRCH, QMDQT, QMDUPD, and QMDMRG. Their control relationship is as shown in Figure 5.3.8. They are described in detail below.

Figure 5.3.8   Control relation of subroutines for the minimum degree algorithm.

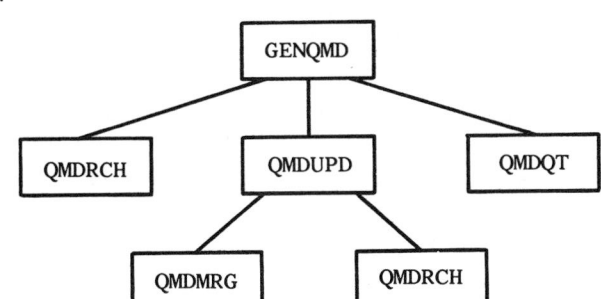

GENQMD (GENeral Quotient Minimum Degree algorithm)

The purpose of this subroutine is to find the minimum degree ordering for a general disconnected graph. It operates on the input graph as given by NEQNS and (XADJ, ADJNCY), and returns the ordering in the vectors PERM and INVP. On return, the adjacency structure will be destroyed because it is used by the subroutine to store the sequence of quotient graph structures.

The subroutine begins by initializing the working arrays QSIZE, QLINK, MARKER and the DEG vector. It then prepares itself for the main loop of the algorithm. In the main loop the subroutine first determines a node of minimum degree by the technique of threshold searching. It keeps two variables THRESH and MINDEG. Any node with its current degree equal to the value of THRESH is one with minimum degree in the elimination graph. The variable MINDEG keeps the lowest degree greater than the threshold value THRESH, and it is used to update the value of THRESH.

Having found a node NODE of minimum degree, GENQMD then determines the reachable set of NODE through eliminated supernodes by calling the subroutine QMDRCH. The set is contained in the vector RCHSET and its size in RCHSZE. The nodes indistinguishable from NODE are then retrieved via the vector QLINK, and numbered (eliminated).

Next, the nodes in the reachable set have their degree updated and at the same time more indistinguishable nodes are identified. In the program, this is done by calling the subroutine QMDUPD. Afterwards, the threshold value is also updated.

Before the program loops back for the next node of minimum degree, the quotient graph transformation is performed by the subroutine QMDQT. The program exits when all the nodes in the graph have been numbered.

```
C****************************************************************
C****************************************************************
C********      GENQMD .....  QUOT MIN DEGREE ORDERING    *****
C****************************************************************
C****************************************************************
C
C      PURPOSE - THIS ROUTINE IMPLEMENTS THE MINIMUM DEGREE
C         ALGORITHM.  IT MAKES USE OF THE IMPLICIT REPRESENT-
C         ATION OF THE ELIMINATION GRAPHS BY QUOTIENT GRAPHS,
C         AND THE NOTION OF INDISTINGUISHABLE NODES.
C         CAUTION - THE ADJACENCY VECTOR ADJNCY WILL BE
C         DESTROYED.
C
C      INPUT PARAMETERS -
C         NEQNS - NUMBER OF EQUATIONS.
C         (XADJ, ADJNCY) - THE ADJACENCY STRUCTURE.
C
C      OUTPUT PARAMETERS -
C         PERM - THE MINIMUM DEGREE ORDERING.
C         INVP - THE INVERSE OF PERM.
C
C      WORKING PARAMETERS -
C         DEG - THE DEGREE VECTOR. DEG(I) IS NEGATIVE MEANS
C               NODE I HAS BEEN NUMBERED.
C         MARKER - A MARKER VECTOR, WHERE MARKER(I) IS
C               NEGATIVE MEANS NODE I HAS BEEN MERGED WITH
C               ANOTHER NODE AND THUS CAN BE IGNORED.
C         RCHSET - VECTOR USED FOR THE REACHABLE SET.
C         NBRHD - VECTOR USED FOR THE NEIGHBORHOOD SET.
C         QSIZE - VECTOR USED TO STORE THE SIZE OF
C               INDISTINGUISHABLE SUPERNODES.
C         QLINK - VECTOR TO STORE INDISTINGUISHABLE NODES,
C               I, QLINK(I), QLINK(QLINK(I)) ... ARE THE
C               MEMBERS OF THE SUPERNODE REPRESENTED BY I.
C
C      PROGRAM SUBROUTINES -
C         QMDRCH, QMDQT, QMDUPD.
C
C****************************************************************
C
C
       SUBROUTINE  GENQMD ( NEQNS, XADJ, ADJNCY, PERM, INVP, DEG,
      1                     MARKER, RCHSET, NBRHD, QSIZE, QLINK,
      1                     NOFSUB )
C
C****************************************************************
C
       INTEGER ADJNCY(1), PERM(1), INVP(1), DEG(1), MARKER(1),
      1        RCHSET(1), NBRHD(1), QSIZE(1), QLINK(1)
       INTEGER XADJ(1), INODE, IP, IRCH, J, MINDEG, NDEG,
      1        NEQNS, NHDSZE, NODE, NOFSUB, NP, NUM, NUMP1,
      1        NXNODE, RCHSZE, SEARCH, THRESH
C
C****************************************************************
C
C      ----------------------------------------------------------
C      INITIALIZE DEGREE VECTOR AND OTHER WORKING VARIABLES.
C      ----------------------------------------------------------
       MINDEG = NEQNS
       NOFSUB = 0
       DO 100 NODE = 1, NEQNS
          PERM(NODE) = NODE
          INVP(NODE) = NODE
          MARKER(NODE) = 0
          QSIZE(NODE)  = 1
          QLINK(NODE)  = 0
          NDEG = XADJ(NODE+1) - XADJ(NODE)
```

```
                 DEG(NODE) = NDEG
                 IF ( NDEG .LT. MINDEG ) MINDEG = NDEG
      100     CONTINUE
                 NUM = 0
C           ------------------------------------------------------
C           PERFORM THRESHOLD SEARCH TO GET A NODE OF MIN DEGREE.
C           VARIABLE SEARCH POINTS TO WHERE SEARCH SHOULD START.
C           ------------------------------------------------------
      200     SEARCH = 1
                 THRESH = MINDEG
                 MINDEG = NEQNS
      300     NUMP1 = NUM + 1
                 IF ( NUMP1 .GT. SEARCH )   SEARCH = NUMP1
                 DO 400 J = SEARCH, NEQNS
                     NODE = PERM(J)
                     IF ( MARKER(NODE) .LT. 0 )  GOTO 400
                         NDEG = DEG(NODE)
                         IF ( NDEG .LE. THRESH )  GO TO 500
                         IF ( NDEG .LT. MINDEG )  MINDEG = NDEG
      400     CONTINUE
                 GO TO 200
C           ------------------------------------------------------
C           NODE HAS MINIMUM DEGREE. FIND ITS REACHABLE SETS BY
C           CALLING QMDRCH.
C           ------------------------------------------------------
      500     SEARCH = J
                 NOFSUB = NOFSUB + DEG(NODE)
                 MARKER(NODE) = 1
                 CALL QMDRCH (NODE, XADJ, ADJNCY, DEG, MARKER,
     1                        RCHSZE, RCHSET, NHDSZE, NBRHD )
C           ------------------------------------------------------
C           ELIMINATE ALL NODES INDISTINGUISHABLE FROM NODE.
C           THEY ARE GIVEN BY NODE, QLINK(NODE), ....
C           ------------------------------------------------------
                 NXNODE = NODE
      600     NUM = NUM + 1
                 NP  = INVP(NXNODE)
                 IP  = PERM(NUM)
                 PERM(NP) = IP
                 INVP(IP) = NP
                 PERM(NUM) = NXNODE
                 INVP(NXNODE) = NUM
                 DEG(NXNODE) = - 1
                 NXNODE = QLINK(NXNODE)
                 IF (NXNODE .GT. 0) GOTO 600
C
                 IF ( RCHSZE .LE. 0 )  GO TO 800
C           ------------------------------------------------------
C           UPDATE THE DEGREES OF THE NODES IN THE REACHABLE
C           SET AND IDENTIFY INDISTINGUISHABLE NODES.
C           ------------------------------------------------------
                 CALL  QMDUPD ( XADJ, ADJNCY, RCHSZE, RCHSET, DEG,
     1                          QSIZE, QLINK, MARKER,
     1                          RCHSET(RCHSZE+1), NBRHD(NHDSZE+1) )
C           ------------------------------------------------------
C           RESET MARKER VALUE OF NODES IN REACH SET.
C           UPDATE THRESHOLD VALUE FOR CYCLIC SEARCH.
C           ALSO CALL QMDQT TO FORM NEW QUOTIENT GRAPH.
C           ------------------------------------------------------
                 MARKER(NODE) = 0
                 DO 700 IRCH = 1, RCHSZE
                     INODE = RCHSET(IRCH)
                     IF ( MARKER(INODE) .LT. 0 )  GOTO 700
                         MARKER(INODE) = 0
                         NDEG = DEG(INODE)
                         IF ( NDEG .LT. MINDEG )  MINDEG = NDEG
                         IF ( NDEG .GT. THRESH )  GOTO 700
                             MINDEG = THRESH
```

```
                          THRESH = NDEG
                          SEARCH = INVP(INODE)
      700             CONTINUE
                      IF ( NHDSZE .GT. 0 )  CALL  QMDQT ( NODE, XADJ,
          1               ADJNCY, MARKER, RCHSZE, RCHSET, NBRHD )
      800       IF ( NUM .LT. NEQNS )  GO TO 300
              RETURN
            END
```

QMDRCH (Quotient MD ReaCHable set)

This subroutine determines the reachable set of a given node ROOT through the set of eliminated nodes. The adjacency structure is assumed to be stored in the quotient graph format as described in Section 5.2.3. On exit, the reachable set determined is placed in the vector RCHSET and its size is given by RCHSZE. As a byproduct, the set of eliminated supernodes adjacent to ROOT is returned in the set NBRHD with its size NHDSZE. Nodes in these two sets will have their MARKER values set to nonzero.

This is an exact implementation of the algorithm in Section 5.2.2. After initialization, the loop DO 600 ... considers each neighbor of the node ROOT. If the neighbor is a representative of an eliminated supernode, its own adjacent set in the quotient graph is included into the reachable set in the DO 500 ... loop. Otherwise, the neighbor itself is included.

```
C***************************************************************
C***************************************************************
C*********        QMDRCH ..... QUOT MIN DEG REACH SET    ********
C***************************************************************
C***************************************************************
C
C     PURPOSE - THIS SUBROUTINE DETERMINES THE REACHABLE SET OF
C        A NODE THROUGH A GIVEN SUBSET.   THE ADJACENCY STRUCTURE
C        IS ASSUMED TO BE STORED IN A QUOTIENT GRAPH FORMAT.
C
C     INPUT PARAMETERS -
C        ROOT - THE GIVEN NODE NOT IN THE SUBSET.
C        (XADJ, ADJNCY) - THE ADJACENCY STRUCTURE PAIR.
C        DEG - THE DEGREE VECTOR.  DEG(I) LT 0 MEANS THE NODE
C              BELONGS TO THE GIVEN SUBSET.
C
C     OUTPUT PARAMETERS -
C        (RCHSZE, RCHSET) - THE REACHABLE SET.
C        (NHDSZE, NBRHD) - THE NEIGHBORHOOD SET.
C
C     UPDATED PARAMETERS -
C        MARKER - THE MARKER VECTOR FOR REACH AND NBRHD SETS.
C                 GT 0 MEANS THE NODE IS IN REACH SET.
C                 LT 0 MEANS THE NODE HAS BEEN MERGED WITH
C                 OTHERS IN THE QUOTIENT OR IT IS IN NBRHD SET.
C
C***************************************************************
C
      SUBROUTINE  QMDRCH ( ROOT, XADJ, ADJNCY, DEG, MARKER,
          1                RCHSZE, RCHSET, NHDSZE, NBRHD )
C
C***************************************************************
```

```
C
              INTEGER  ADJNCY(1),  DEG(1),  MARKER(1),
       1              RCHSET(1),  NBRHD(1)
              INTEGER  XADJ(1),  I,  ISTRT,  ISTOP,  J,  JSTRT,  JSTOP,
       1              NABOR,  NHDSZE,  NODE,  RCHSZE,  ROOT
C
C****************************************************************
C
C
C             ---------------------------------------------
C             LOOP THROUGH THE NEIGHBORS OF ROOT IN THE
C             QUOTIENT GRAPH.
C             ---------------------------------------------
              NHDSZE = 0
              RCHSZE = 0
              ISTRT = XADJ(ROOT)
              ISTOP = XADJ(ROOT+1) - 1
              IF ( ISTOP .LT. ISTRT )  RETURN
                 DO 600 I = ISTRT, ISTOP
                    NABOR =  ADJNCY(I)
                    IF ( NABOR .EQ. 0 ) RETURN
                    IF ( MARKER(NABOR) .NE. 0 )  GO TO 600
                       IF ( DEG(NABOR) .LT. 0 )    GO TO 200
C                      ------------------------------------
C                      INCLUDE NABOR INTO THE REACHABLE SET.
C                      ------------------------------------
                       RCHSZE = RCHSZE + 1
                       RCHSET(RCHSZE) = NABOR
                       MARKER(NABOR) = 1
                       GO TO 600
C                ------------------------------------
C                NABOR HAS BEEN ELIMINATED. FIND NODES
C                REACHABLE FROM IT.
C                ------------------------------------
  200                MARKER(NABOR) = -1
                     NHDSZE = NHDSZE + 1
                     NBRHD(NHDSZE) = NABOR
  300                JSTRT = XADJ(NABOR)
                     JSTOP = XADJ(NABOR+1) - 1
                     DO 500 J = JSTRT, JSTOP
                        NODE = ADJNCY(J)
                        NABOR = - NODE
                        IF (NODE) 300, 600, 400
  400                   IF ( MARKER(NODE) .NE. 0 )  GO TO 500
                        RCHSZE = RCHSZE + 1
                        RCHSET(RCHSZE) = NODE
                        MARKER(NODE) = 1
  500                CONTINUE
  600             CONTINUE
                  RETURN
              END
```

QMDQT (Quotient MD Quotient graph Transformation)

This subroutine performs the quotient graph transformation on the adjacency structure (XADJ, ADJNCY). The new eliminated supernode contains the node ROOT and the nodes in the array NBRHD, and it will be represented by ROOT in the new structure. Its adjacent set in the new quotient graph is given in (RCHSZE, RCHSET).

After initialization, the new adjacent set in (RCHSZE, RCHSET) will be placed in the adjacency list of ROOT in the structure (DO 200 ...). If there is not enough space, the program will use the space provided by the nodes in the set NBRHD. We know from

Section 5.2.3 that there are always enough storage locations.

Before exit, the representative node ROOT is added to the neighbor list of each node in RCHSET. This is done in the DO 600 ... loop.

```
C***************************************************************
C***************************************************************
C*******       QMDQT   .....  QUOT MIN DEG QUOT TRANSFORM   ******
C***************************************************************
C***************************************************************
C
C     PURPOSE - THIS SUBROUTINE PERFORMS THE QUOTIENT GRAPH
C         TRANSFORMATION AFTER A NODE HAS BEEN ELIMINATED.
C
C     INPUT PARAMETERS -
C         ROOT - THE NODE JUST ELIMINATED. IT BECOMES THE
C             REPRESENTATIVE OF THE NEW SUPERNODE.
C         (XADJ, ADJNCY) - THE ADJACENCY STRUCTURE.
C         (RCHSZE, RCHSET) - THE REACHABLE SET OF ROOT IN THE
C             OLD QUOTIENT GRAPH.
C         NBRHD - THE NEIGHBORHOOD SET WHICH WILL BE MERGED
C             WITH ROOT TO FORM THE NEW SUPERNODE.
C         MARKER - THE MARKER VECTOR.
C
C     UPDATED PARAMETER -
C         ADJNCY - BECOMES THE ADJNCY OF THE QUOTIENT GRAPH.
C
C***************************************************************
C
        SUBROUTINE  QMDQT ( ROOT, XADJ, ADJNCY, MARKER,
     1                      RCHSZE, RCHSET, NBRHD )
C
C***************************************************************
C
        INTEGER ADJNCY(1), MARKER(1), RCHSET(1), NBRHD(1)
        INTEGER XADJ(1), INHD, IRCH, J, JSTRT, JSTOP, LINK,
     1          NABOR, NODE, RCHSZE, ROOT
C
C***************************************************************
C
        IRCH = 0
        INHD = 0
        NODE = ROOT
  100   JSTRT = XADJ(NODE)
        JSTOP = XADJ(NODE+1) - 2
        IF ( JSTOP .LT. JSTRT ) GO TO 300
C         -------------------------------------------------
C         PLACE REACH NODES INTO THE ADJACENT LIST OF NODE
C         -------------------------------------------------
        DO 200 J = JSTRT, JSTOP
           IRCH = IRCH + 1
           ADJNCY(J) = RCHSET(IRCH)
           IF ( IRCH .GE. RCHSZE ) GOTO 400
  200   CONTINUE
C         -------------------------------------------------
C         LINK TO OTHER SPACE PROVIDED BY THE NBRHD SET.
C         -------------------------------------------------
  300   LINK = ADJNCY(JSTOP+1)
        NODE = - LINK
        IF ( LINK .LT. 0 ) GOTO 100
           INHD = INHD + 1
           NODE = NBRHD(INHD)
           ADJNCY(JSTOP+1) = - NODE
           GO TO 100
C         -------------------------------------------------
C         ALL REACHABLE NODES HAVE BEEN SAVED.  END THE ADJ LIST.
```

```
C              ADD ROOT TO THE NBR LIST OF EACH NODE IN THE REACH SET.
C              ------------------------------------------------------------
   400     ADJNCY(J+1) = 0
           DO 600 IRCH = 1, RCHSZE
              NODE = RCHSET(IRCH)
              IF ( MARKER(NODE) .LT. 0 )   GOTO 600
                 JSTRT = XADJ(NODE)
                 JSTOP = XADJ(NODE+1) - 1
                 DO 500 J = JSTRT, JSTOP
                    NABOR = ADJNCY(J)
                    IF ( MARKER(NABOR) .GE. 0 ) GO TO 500
                       ADJNCY(J) = ROOT
                       GOTO 600
   500        CONTINUE
   600     CONTINUE
           RETURN
         END
```

QMDUPD (Quotient MD UPDate)

   This subroutine performs the degree update step in the minimum degree algorithm. The nodes whose new degrees are to be determined are given by the pair (NLIST, LIST). The subroutine also merges indistinguishable nodes in this subset by using Theorem 5.3.7.

   The first loop DO 200 ... and the call to the subroutine QMDMRG determine groups of indistinguishable nodes in the given set. They will be merged together and have their degrees updated.

   For those nodes not being merged, the loop DO 600 ... determines their new degrees by calling the subroutine QMDRCH. The vectors RCHSET and NBRHD are used as temporary arrays.

```
C*****************************************************************
C*****************************************************************
C**********     QMDUPD ..... QUOT MIN DEG UPDATE     *********
C*****************************************************************
C*****************************************************************
C
C     PURPOSE - THIS ROUTINE PERFORMS DEGREE UPDATE FOR A SET
C        OF NODES IN THE MINIMUM DEGREE ALGORITHM.
C
C     INPUT PARAMETERS -
C        (XADJ, ADJNCY) - THE ADJACENCY STRUCTURE.
C        (NLIST, LIST) - THE LIST OF NODES WHOSE DEGREE HAS TO
C           BE UPDATED.
C
C     UPDATED PARAMETERS -
C        DEG - THE DEGREE VECTOR.
C        QSIZE - SIZE OF INDISTINGUISHABLE SUPERNODES.
C        QLINK - LINKED LIST FOR INDISTINGUISHABLE NODES.
C        MARKER - USED TO MARK THOSE NODES IN REACH/NBRHD SETS.
C
C     WORKING PARAMETERS -
C        RCHSET - THE REACHABLE SET.
C        NBRHD -  THE NEIGHBORHOOD SET.
C
C     PROGRAM SUBROUTINES -
C        QMDMRG.
C
C*****************************************************************
C
```

```
          SUBROUTINE    QMDUPD ( XADJ, ADJNCY, NLIST, LIST, DEG,
        1                        QSIZE, QLINK, MARKER, RCHSET, NBRHD )
C
C****************************************************************
C
          INTEGER   ADJNCY(1), LIST(1), DEG(1), MARKER(1),
        1           RCHSET(1), NBRHD(1), QSIZE(1), QLINK(1)
          INTEGER   XADJ(1), DEG0, DEG1, IL, INHD, INODE, IRCH,
        1           J, JSTRT, JSTOP, MARK, NABOR, NHDSZE, NLIST,
        1           NODE, RCHSZE, ROOT
C
C****************************************************************
C
C         ----------------------------------------------------
C         FIND ALL ELIMINATED SUPERNODES THAT ARE ADJACENT
C         TO SOME NODES IN THE GIVEN LIST. PUT THEM INTO
C         (NHDSZE, NBRHD). DEG0 CONTAINS THE NUMBER OF
C         NODES IN THE LIST.
C         ----------------------------------------------------
          IF ( NLIST .LE. 0 ) RETURN
          DEG0 = 0
          NHDSZE = 0
          DO 200 IL = 1, NLIST
             NODE = LIST(IL)
             DEG0 = DEG0 + QSIZE(NODE)
             JSTRT = XADJ(NODE)
             JSTOP = XADJ(NODE+1) - 1
             DO 100 J = JSTRT, JSTOP
                NABOR = ADJNCY(J)
                IF ( MARKER(NABOR) .NE. 0  .OR.
        1            DEG(NABOR) .GE. 0 ) GO TO 100
                MARKER(NABOR) = - 1
                NHDSZE = NHDSZE + 1
                NBRHD(NHDSZE) = NABOR
  100        CONTINUE
  200     CONTINUE
C         -----------------------------------------------
C         MERGE INDISTINGUISHABLE NODES IN THE LIST BY
C         CALLING THE SUBROUTINE QMDMRG.
C         -----------------------------------------------
          IF ( NHDSZE .GT. 0 )
        1     CALL QMDMRG ( XADJ, ADJNCY, DEG, QSIZE, QLINK,
        1                   MARKER, DEG0, NHDSZE, NBRHD, RCHSET,
        1                   NBRHD(NHDSZE+1) )
C         -----------------------------------------------------
C         FIND THE NEW DEGREES OF THE NODES THAT HAVE NOT BEEN
C         MERGED.
C         -----------------------------------------------------
          DO 600 IL = 1, NLIST
             NODE = LIST(IL)
             MARK = MARKER(NODE)
             IF ( MARK .GT. 1  .OR.  MARK .LT. 0 ) GO TO 600
                MARKER(NODE) = 2
                CALL QMDRCH ( NODE, XADJ, ADJNCY, DEG, MARKER,
        1                     RCHSZE, RCHSET, NHDSZE, NBRHD )
                DEG1 = DEG0
                IF ( RCHSZE .LE. 0 ) GO TO 400
                   DO 300 IRCH = 1, RCHSZE
                      INODE = RCHSET(IRCH)
                      DEG1 = DEG1 + QSIZE(INODE)
                      MARKER(INODE) = 0
  300              CONTINUE
  400           DEG(NODE) = DEG1 - 1
                IF ( NHDSZE .LE. 0 ) GO TO 600
                DO 500 INHD = 1, NHDSZE
                   INODE = NBRHD(INHD)
                   MARKER(INODE) = 0
  500           CONTINUE
```

```
600     CONTINUE
        RETURN
   END
```

QMDMRG (Quotient MD MeRGe)

This subroutine implements a check for the condition (5.3.2) to determine indistinguishable nodes. Let C_1, C_2, R_1, R_2 and Y be as in Lemma 5.3.6. The subroutine assumes that C_1 and R_1 have already been determined elsewhere. Nodes in R_1 have their MARKER values set to 1.

There may be more than one C_2 input to QMDMRG. They are contained in (NHDSZE, NBRHD), where each NBRHD(i) specifies one eliminated supernode (that is, connected component).

The loop DO 1400 ... applies the condition on each given connected component. It first determines the set $R_2 - R_1$ in (RCHSZE, RCHSET) and the intersection set $R_2 \cap R_1$ in (NOVRLP, OVRLP) in the loop DO 600 .... For each node in the intersection, the condition (5.3.2) is tested in the loop DO 1100 .... If the condition is satisfied, the node is included in the merged supernode by placing it in the QLINK vector. The size of the new supernode is also computed.

```
C************************************************************
C************************************************************
C*********     QMDMRG .....  QUOT MIN DEG MERGE     *******
C************************************************************
C************************************************************
C
C     PURPOSE - THIS ROUTINE MERGES INDISTINGUISHABLE NODES IN
C               THE MINIMUM DEGREE ORDERING ALGORITHM.
C               IT ALSO COMPUTES THE NEW DEGREES OF THESE
C               NEW SUPERNODES.
C
C     INPUT PARAMETERS -
C        (XADJ, ADJNCY) - THE ADJACENCY STRUCTURE.
C        DEG0 - THE NUMBER OF NODES IN THE GIVEN SET.
C        (NHDSZE, NBRHD) - THE SET OF ELIMINATED SUPERNODES
C               ADJACENT TO SOME NODES IN THE SET.
C
C     UPDATED PARAMETERS -
C        DEG - THE DEGREE VECTOR.
C        QSIZE - SIZE OF INDISTINGUISHABLE NODES.
C        QLINK - LINKED LIST FOR INDISTINGUISHABLE NODES.
C        MARKER - THE GIVEN SET IS GIVEN BY THOSE NODES WITH
C               MARKER VALUE SET TO 1.  THOSE NODES WITH DEGREE
C               UPDATED WILL HAVE MARKER VALUE SET TO 2.
C
C     WORKING PARAMETERS -
C        RCHSET - THE REACHABLE SET.
C        OVRLP -  TEMP VECTOR TO STORE THE INTERSECTION OF TWO
C               REACHABLE SETS.
C
C************************************************************
C
        SUBROUTINE QMDMRG ( XADJ, ADJNCY, DEG, QSIZE, QLINK,
      1                     MARKER, DEG0, NHDSZE, NBRHD, RCHSET,
```

```
     1                         OVRLP )
C
C***************************************************************
C
          INTEGER  ADJNCY(1), DEG(1), QSIZE(1), QLINK(1),
     1             MARKER(1), RCHSET(1), NBRHD(1), OVRLP(1)
          INTEGER  XADJ(1), DEG0, DEG1, HEAD, INHD, IOV, IRCH,
     1             J, JSTRT, JSTOP, LINK, LNODE, MARK, MRGSZE,
     1             NABOR, NHDSZE, NODE, NOVRLP, RCHSZE, ROOT
C
C***************************************************************
C
C
C          ------------------
C          INITIALIZATION ...
C          ------------------
          IF ( NHDSZE .LE. 0 )   RETURN
          DO 100 INHD = 1, NHDSZE
             ROOT = NBRHD(INHD)
             MARKER(ROOT) = 0
 100      CONTINUE
C          ------------------------------------------------------
C          LOOP THROUGH EACH ELIMINATED SUPERNODE IN THE SET
C          (NHDSZE, NBRHD).
C          ------------------------------------------------------
          DO 1400 INHD = 1, NHDSZE
             ROOT = NBRHD(INHD)
             MARKER(ROOT) = - 1
             RCHSZE = 0
             NOVRLP = 0
             DEG1   = 0
 200         JSTRT  = XADJ(ROOT)
             JSTOP  = XADJ(ROOT+1) - 1
C             ------------------------------------------------------
C             DETERMINE THE REACHABLE SET AND ITS INTERSECT-
C             ION WITH THE INPUT REACHABLE SET.
C             ------------------------------------------------------
             DO 600 J = JSTRT, JSTOP
                NABOR = ADJNCY(J)
                ROOT  = - NABOR
                IF (NABOR)   200, 700, 300
 300            MARK = MARKER(NABOR)
                IF ( MARK ) 600, 400, 500
 400               RCHSZE = RCHSZE + 1
                   RCHSET(RCHSZE) = NABOR
                   DEG1 = DEG1 + QSIZE(NABOR)
                   MARKER(NABOR) = 1
                   GOTO 600
 500               IF ( MARK .GT. 1 )   GOTO 600
                   NOVRLP = NOVRLP + 1
                   OVRLP(NOVRLP) = NABOR
                   MARKER(NABOR) = 2
 600         CONTINUE
C             ------------------------------------------------------
C             FROM THE OVERLAPPED SET, DETERMINE THE NODES
C             THAT CAN BE MERGED TOGETHER.
C             ------------------------------------------------------
 700         HEAD = 0
             MRGSZE = 0
             DO 1100 IOV = 1, NOVRLP
                NODE = OVRLP(IOV)
                JSTRT = XADJ(NODE)
                JSTOP = XADJ(NODE+1) - 1
                DO 800 J = JSTRT, JSTOP
                   NABOR = ADJNCY(J)
                   IF ( MARKER(NABOR) .NE. 0 )   GOTO 800
                      MARKER(NODE) = 1
                      GOTO 1100
 800            CONTINUE
```

```
C           -------------------------------------------
C                   NODE BELONGS TO THE NEW MERGED SUPERNODE.
C                   UPDATE THE VECTORS QLINK AND QSIZE.
C           -------------------------------------------
            MRGSZE = MRGSZE + QSIZE(NODE)
            MARKER(NODE) = - 1
            LNODE = NODE
 900        LINK   = QLINK(LNODE)
            IF ( LINK .LE. 0 )   GOTO 1000
                LNODE = LINK
                GOTO 900
1000        QLINK(LNODE) = HEAD
            HEAD = NODE
1100     CONTINUE
         IF ( HEAD .LE. 0 )   GOTO 1200
            QSIZE(HEAD) = MRGSZE
            DEG(HEAD) = DEG0 + DEG1 - 1
            MARKER(HEAD) = 2
C           --------------------
C           RESET MARKER VALUES.
C           --------------------
1200        ROOT = NBRHD(INHD)
            MARKER(ROOT) = 0
            IF ( RCHSZE .LE. 0 )   GOTO 1400
            DO 1300 IRCH = 1, RCHSZE
                NODE = RCHSET(IRCH)
                MARKER(NODE) = 0
1300        CONTINUE
1400     CONTINUE
         RETURN
      END
```

Exercises

5.3.1) Let x_i be the node selected from G_{i-1} in the minimum degree algorithm. Let $y \in Adj_{G_{i-1}}(x_i)$ with

$$Deg_{G_i}(y) = Deg_{G_{i-1}}(x_i) - 1 \; .$$

Show that y is a node of minimum degree in G_i.

5.3.2) Let x_i and G_{i-1} be as in Exercise 5.3.1, and $y \in Adj_{G_{i-1}}(x_i)$. Prove that if

$$Adj_{G_{i-1}}(y) \subset Adj_{G_{i-1}}(x_i) \cup \{x_i\}$$

then y is a node of minimum degree in G_i.

5.4   Sparse Storage Schemes

5.4.1   The Uncompressed Scheme

The data structure for the general sparse methods should only store (logical) nonzeros of the factored matrix. The scheme discussed here is oriented to the inner-product formulation of the factorization algorithm (see Section 2.1.2) and can be found in, for example, Gustavson (1972) and Sherman (1975).

Figure 5.4.1   A 7 by 7 matrix A and its factor L.

$$
A = \begin{pmatrix}
a_{11} & & & \text{symmetric} & & & \\
a_{21} & a_{22} & & & & & \\
& & a_{33} & & & & \\
a_{41} & & & a_{44} & & & \\
& & a_{53} & a_{54} & a_{55} & & \\
& & a_{63} & & & a_{66} & \\
& & & & a_{75} & a_{76} & a_{77}
\end{pmatrix}
, \quad
L = \begin{pmatrix}
\ell_{11} & & & & & & \\
\ell_{21} & \ell_{22} & & & & & \\
& & \ell_{33} & & & & \\
\ell_{41} & \ell_{42} & & \ell_{44} & & & \\
& & \ell_{53} & \ell_{54} & \ell_{55} & & \\
& & \ell_{63} & & \ell_{65} & \ell_{66} & \\
& & & & \ell_{75} & \ell_{76} & \ell_{77}
\end{pmatrix}
$$

The scheme has a main storage array LNZ which contains all the nonzero entries in the lower triangular factor. A storage location is provided for each *logical* nonzero in the factor. The nonzeros in L, excluding the diagonal, are stored column after column in LNZ. An accompanying vector NZSUB is provided, which gives the row subscripts of the nonzeros. In addition, an index vector XLNZ is used to point to the start of nonzeros in each column in LNZ (or equivalently NZSUB). The diagonal entries are stored separately in the vector DIAG.

To access a nonzero component a_{ij} or ℓ_{ij}, there is no direct method of calculating the corresponding index in the vector LNZ. Some testing on the subscripts in NZSUB has to be done. The following portion of a program can be used for that purpose. Note that any entry not represented by the data structure is zero.

```
            KSTRT = XLNZ(J)
            KSTOP = XLNZ(J+1) - 1
            AIJ = 0.0
            IF (KSTOP.LT.KSTRT) GO TO 300
                DO 100 K = KSTRT, KSTOP
                    IF (NZSUB(K).EQ.I) GO TO 200
100             CONTINUE
                GO TO 300
200             AIJ = LNZ(K)
300             :
```

Figure 5.4.2 Uncompressed data storage scheme for the matrix and its factor in Figure 5.4.1.

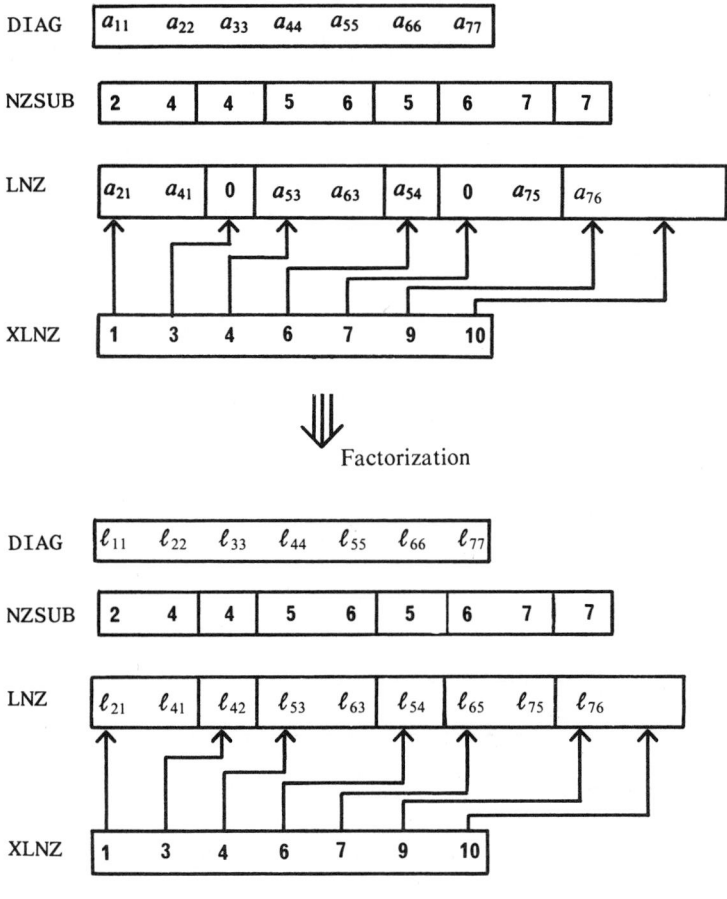

Factorization

Although this scheme is not particularly well suited for random access of nonzero entries, it lends itself quite readily to sparse factorization and solution. The primary storage of the scheme is $|Nonz(F)| + N$ for the vectors LNZ and DIAG, and the overhead storage is $|Nonz(F)| + N$ for NZSUB and XLNZ.

5.4.2 Compressed Scheme

This scheme, which is a modification of the uncompressed scheme, is due to Sherman (1975). The motivation can be provided by considering the minimum degree ordering as discussed in Section 5.3.3. We saw that it was possible to simultaneously number or eliminate a *set Y* of nodes. The nodes in Y satisfy the indistinguishable condition

$$Reach\,(x,S\,) \cup \{x\} = Reach\,(y,S\,) \cup \{y\}\ ,$$

for all $x,y \in Y$. In terms of the matrix factor L, this means all the row subscripts below the block corresponding to Y are identical, as shown in Figure 5.4.3.

Figure 5.4.3   Motivation of the compressed storage scheme.

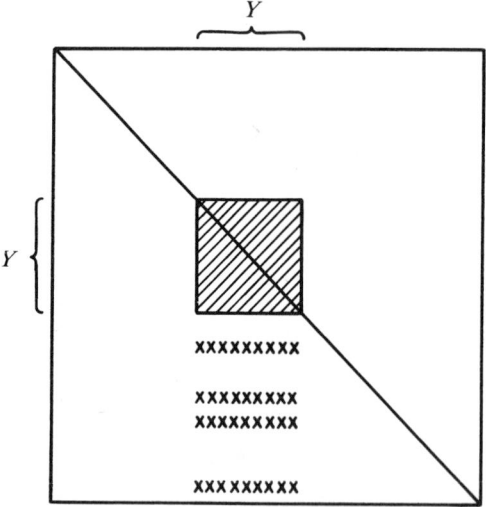

If the structure is stored using the uncompressed scheme, the row subscripts of all but the first column in this block are final subsequences of that of the previous column. Naturally, the subscript vector NZSUB can be compressed so that redundant information is not stored. It is done by removing the row subscripts for a column if they appear as a final subsequence of the previous column.

In exchange for the compression, we need to have an auxiliary index vector XNZSUB which points to the start of row subscripts in NZSUB for each column. The compressed scheme for the example in Figure 5.4.1 is shown in Figure 5.4.4.

In this case, the way to access a nonzero entry in the (i,j)-th position is as follows.

Figure 5.4.4 Compressed storage scheme for the matrix in Figure 5.4.1.

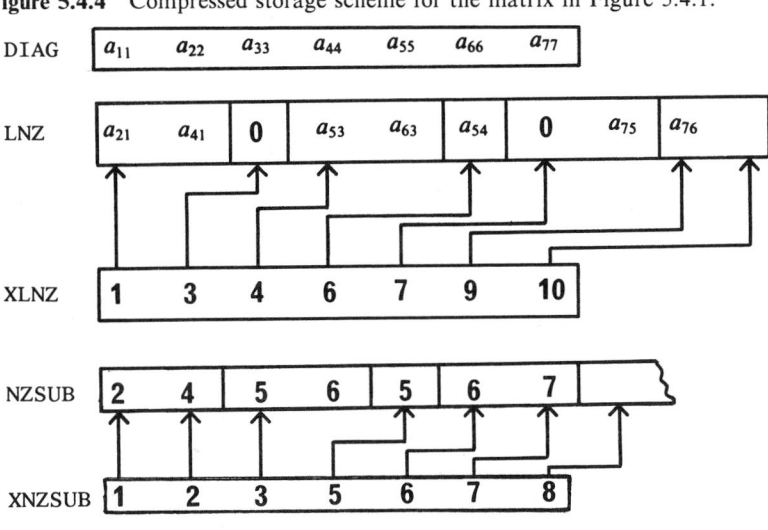

```
        KSTRT = XLNZ(J)
        KSTOP = XLNZ(J+1) - 1
        AIJ = 0.0
        IF (KSTOP.LT.KSTRT) GO TO 300
            KSUB = XNZSUB(J)
            DO 100 K = KSTRT, KSTOP
                IF(NZSUB(KSUB).EQ.I) GO TO 200
                KSUB = KSUB + 1
100         CONTINUE
            GO TO 300
200     AIJ = LNZ(K)
300         :
```

In the compressed scheme, the primary storage remains the same, but the overhead storage is changed and it is less than or equal to $|Nonz(F)| + 2N$. The example given is too small to bring out the significance of the compressed scheme. In Table 5.4.1, we provide some numbers that are obtained from nine larger problems which comprise one of the test sets considered in Chapter 9. The ordering which was used in generating these results was provided by a minimum degree algorithm similar to the one described in the previous section. Typically, for problems of this size and larger, the overhead storage is reduced by at least fifty percent, compared to the uncompressed scheme.

Table 5.4.1 Comparison of uncompressed and compressed storage schemes. The primary storage is equal to the overhead for the uncompressed scheme.

Number of Equations	$|Nonz(A)|$	$|Nonz(F)|$	Overhead for Uncompressed	Overhead for Compressed
936	2664	13870	14806	6903
1009	2928	19081	20090	8085
1089	3136	18626	19715	8574
1440	4032	19047	20487	10536
1180	3285	14685	15865	8436
1377	3808	16793	18170	9790
1138	3156	15592	16730	8326
1141	3162	15696	16837	8435
1349	3876	23726	25075	10666

5.4.3 On Symbolic Factorization

As its name implies, *symbolic factorization* is the process of simulating the numerical factorization of a given matrix A in order to obtain the zero-nonzero structure of its factor L. Since the numerical values of the matrix components are of no significance in this connection, the problem can be conveniently studied using a graph theory approach.

Let $G^\alpha = (X^\alpha, E)$ be an ordered graph, where $|X^\alpha| = N$ and for convenience let $\alpha(i) = x_i$. In view of Theorem 5.1.2, symbolic factorization may be regarded as determination of the sets

$$Reach(x_i, \{x_1, \ldots, x_{i-1}\}) \text{ for } i = 1, \ldots, N.$$

Define $S_i = \{x_1, \ldots, x_i\}$.

We prove the following result about reachable sets.

Lemma 5.4.1

$$Reach(x_i, S_{i-1}) = Adj(x_i) \cup$$

$$\left(\bigcup_k \{Reach(x_k, S_{k-1}) \mid x_i \in Reach(x_k, S_{k-1})\} \right) - S_i .$$

Proof   Let $j > i$. Then by Lemma 5.1.1 and Theorem 5.1.2

$$x_j \in Reach(x_i, S_{i-1}) \text{ iff } \{x_i, x_j\} \in E^F,$$

iff $\{x_i, x_j\} \in E^A$, or $\{x_i, x_k\} \in E^F$ and $\{x_j, x_k\} \in E^F$

for some $k < \min\{i,j\}$,

iff $x_j \in Adj(x_i)$ or $x_i, x_j \in Reach(x_k, S_{k-1})$ for some k .

The lemma then follows. □

Lemma 5.4.1 suggests an algorithm for finding the reachable sets (and hence the structure of the factor L). It may be described as follows.

Step 1   (*Initialization*) For $k = 1, \ldots, N$,

$$Reach(x_k, S_{k-1}) \leftarrow Adj(x_k) - S_{k-1} .$$

Step 2   (*Symbolic factorization*)
      For $k = 1, \ldots, N$,
        if $x_i \in Reach(x_k, S_{k-1})$ then
          $Reach(x_i, S_{i-1}) \leftarrow Reach(x_i, S_{i-1}) \cup$
          $Reach(x_k, S_{k-1}) - S_i.$

A pictorial illustration of the scheme is shown in Figure 5.4.5. This scheme is hardly satisfactory, since it essentially simulates the entire factorization, and its cost will be proportional to the operation count as given in Theorem 2.1.2. Let us look into possible ways of improving the efficiency of the algorithm.

Figure 5.4.5   Merging of reachable sets to obtain $Reach(x_i, S_{i-1})$.

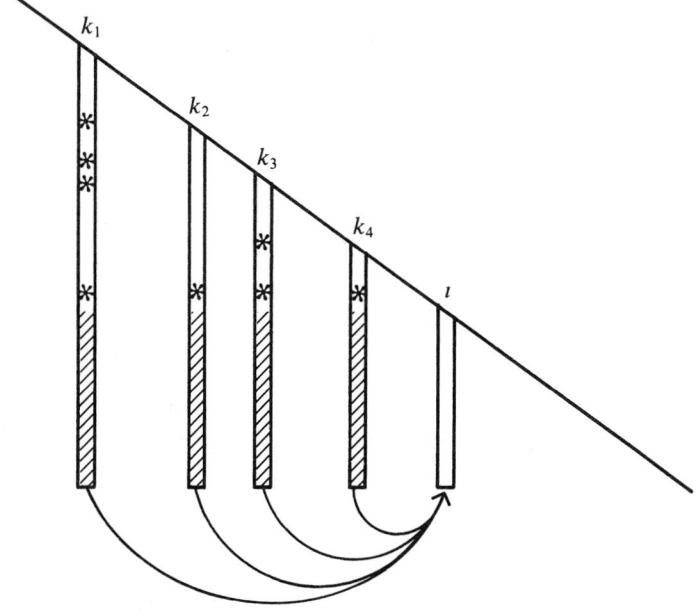

Consider the stage when the set $S_{i-1} = \{x_1, \ldots, x_{i-1}\}$ has been eliminated. For the purpose of this discussion, assume that x_i has two connected components in $G(S_{i-1})$ adjacent to it. Let their node sets be C_1 and C_2.

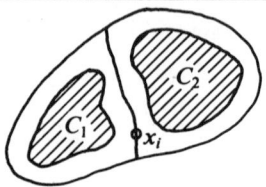

In this case, it can be seen that

$$Reach(x_i, S_{i-1}) = Adj(x_i) \cup Adj(C_1) \cup Adj(C_2) - S_i .$$

However, representatives x_{r_1} and x_{r_2} can be chosen from C_1 and C_2 respectively so that

$$Adj(C_1) = Reach(x_{r_1}, S_{r_1-1})$$

and

$$Adj(C_2) = Reach(x_{r_2}, S_{r_2-1}) .$$

(See Exercise 5.2.5.) Indeed, the representative is given by (5.2.4); specifically, the node in the component last eliminated. In this way, the reachable set can be written as

$$Reach(x_i, S_{i-1}) = Adj(x_i) \cup$$
$$Reach(x_{r_1}, S_{r_1-1}) \cup Reach(x_{r_2}, S_{r_2-1}) - S_i .$$

Thus, instead of having to merge many reachable sets as given in Lemma 5.4.1, we can select representatives. The ideas presented below are motivated from this observation.

For $k = 1, \ldots, N$, define

$$m_k = \min\{j \mid x_j \in Reach(x_k, S_{k-1}) \cup \{x_k\}\} . \qquad (5.4.1)$$

In terms of the matrix, m_k is the subscript of the first nonzero in the column vector L_{*k} excluding the diagonal component.

Lemma 5.4.2

$$Reach(x_k, S_{k-1}) \subset Reach(x_{m_k}, S_{m_k-1}) \cup \{x_{m_k}\} .$$

Proof   For any $x_i \in Reach(x_k, S_{k-1})$, then $k < m_k \leq i$. If $i = m_k$, there is nothing to prove. Otherwise, by Lemma 5.1.1 and

Theorem 5.1.2,

$$x_i \in Reach \, (x_{m_k}, S_{m_k - 1}) \; .$$

$\square$

Lemma 5.4.2 has the following important implication. For $x_i \in Reach \, (x_k, S_{k-1})$ and $i > m_k$, it is redundant to consider $Reach \, (x_k, S_{k-1})$ in determining $Reach \, (x_i, S_{i-1})$ in the algorithm, since all the reachable nodes via x_k can be found in $Reach \, (x_{m_k}, S_{m_k - 1})$. Thus, it is sufficient to merge the reachable sets of some representative nodes. Figure 5.4.6 shows the improvement on the example in Figure 5.4.5. Lemma 5.4.1 can be improved as follows.

Theorem 5.4.3

$$Reach \, (x_i, S_{i-1}) = Adj \, (x_i) \cup$$

$$\left(\bigcup_k \{ Reach \, (x_k, S_{k-1}) \mid m_k = i \} \right) - S_i \; .$$

Proof   Consider any x_k with $x_i \in Reach \, (x_k, S_{k-1})$. Putting $m(k) = m_k$, we have an ascending sequence of subscripts bounded above by i:

$$k < m(k) < m(m(k)) < m^3(k) < \; \cdots \; < i \; .$$

There exists an integer p such that $m^{p+1}(k) = i$. It follows from Lemma 5.4.2 that

$$Reach \, (x_k, S_{k-1}) - S_i \subset Reach \, (x_{m(k)}, S_{m(k)-1}) - S_i$$

$$\subset \; \cdots$$

$$\subset Reach \, (x_{m^p(k)}, S_{m^p(k)-1}) - S_i \; .$$

The result then follows. $\square$

Consider the determination of $Reach \, (x_5, S_4)$ in the graph example of Figure 5.4.7. If Lemma 5.4.1 is used, we see that the sets

$$Reach \, (x_1, S_0) \; ,$$

$$Reach \, (x_2, S_1) \; ,$$

and

$$Reach \, (x_4, S_3)$$

have to be merged with $Adj \, (x_5)$. On the other hand, by Theorem 5.4.3, it is sufficient to consider

Figure 5.4.6   Improvement in merging reachable sets for $Reach(x_i, S_{i-1})$.

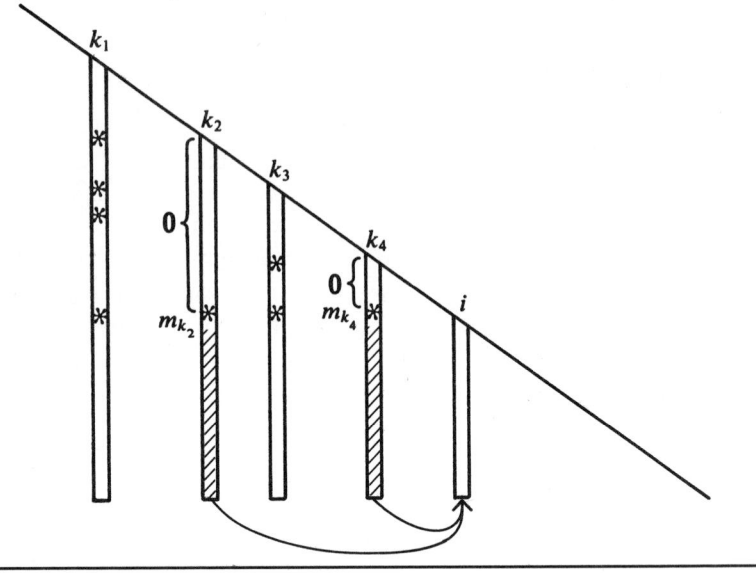

Figure 5.4.7   Illustration of m_k.

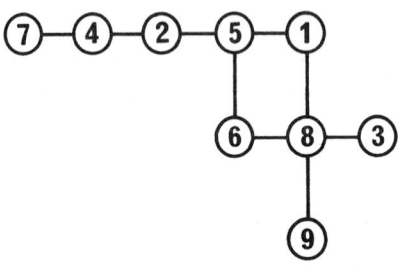

k	m_k	$Reach(x_k, S_{k-1})$
1	5	x_5, x_8
2	4	x_4, x_5
3	8	x_8
4	5	x_5, x_7

$$Reach(x_1, S_0)$$

and

$$Reach(x_4, S_3) \ .$$

Note that $m_2 = 4$ and

$$Reach(x_2, S_1) = \{x_4, x_5\} \subset Reach(x_4, S_3) \cup \{x_4\} \ .$$

The symbolic factorization algorithm can now be refined as:

Step 1   (*Initialization*)
        For $k = 1, \ldots, N$,
$$Reach(x_k, S_{k-1}) = Adj(x_k) - S_k \ .$$
Step 2   (*Symbolic Factorization*)
        For $k = 1, \ldots, N$,
          if $Reach(x_k, S_{k-1}) \neq \emptyset$ then do
$$m = \min\{j \mid x_j \in Reach(x_k, S_{k-1})\}$$
$$Reach(x_m, S_{m-1}) \leftarrow Reach(x_m, S_{m-1}) \cup$$
$$(Reach(x_k, S_{k-1}) - \{x_m\} \ .$$

Theorem 5.4.4
Symbolic factorization can be performed in $O(|E^F|)$ operations.

Proof   For each column k, the value m_k is unique. This means that $Reach(x_k, S_{k-1})$ is accessed only when $Reach(x_{m_k}, S_{m_k-1})$ is being determined. That is, the set $Reach(x_k, S_{k-1})$ is examined exactly once throughout the entire process. Moreover, the union of two reachable sets can be performed in time proportional to the sum of their sizes (see Exercise 5.4.3). Therefore, symbolic factorization can be done in $O(|E^F|)$ operations, where $|E^F| = \sum_{k=1}^{N} |Reach(x_k, S_{k-1})|$.   □

5.4.4   Storage Allocation for the Compressed Scheme and the Subroutine SMBFCT

In this section, we describe the subroutine which performs symbolic factorization as described in the previous section. The result of the process is a data structure for the compressed scheme of Section 5.4.2.

    The implementation is due to Eisenstat, et al. (1981), and can be found in the Yale Sparse Matrix Package. Essentially, it implements the refined algorithm as described in the previous section, with a rearrangement of the order in which reachable sets are merged together.

Step 1   (*Initialization*) For $i = 1, \ldots, N$, $R_i = \emptyset$.
Step 2   (*Symbolic Factorization*)
        For $i = 1, \ldots, N$,
          $Reach(x_i, S_{i-1}) = Adj(x_i) - S_i$
          For $k \in R_i$ do
$$Reach(x_i, S_{i-1}) \leftarrow Reach(x_i, S_{i-1}) \cup$$
$$Reach(x_k, S_{k-1}) - S_i$$
$$m = \min\{j \mid x_j \in Reach(x_i, S_{i-1})\}$$
$$R_m \leftarrow R_m \cup \{x_i\} \ .$$

In this algorithm, the set R_i is used to accumulate the representatives whose reachable sets affect that of x_i. There is an immediate result from Theorem 5.4.3 that can be used to speed up the algorithm. Moreover, it is useful in setting up the compressed storage scheme.

Corollary 5.4.5
If there is only one $m_k = i$, and

$$Adj(x_i) - S_i \subset Reach(x_k, S_{k-1})$$

then $Reach(x_i, S_{i-1}) = Reach(x_k, S_{k-1}) - \{x_i\}$.

The subroutine SMBFCT accepts as input the graph of the matrix stored in the array pair (XADJ, ADJNCY), together with the permutation vector PERM and its inverse INVP. The objective of the subroutine is to set up the data structure for the compressed sparse scheme; that is, to compute the compressed subscript vector NZSUB and the index vectors XLNZ and XNZSUB. Also returned are the values MAXLNZ and MAXSUB which contain the number of off-diagonal nonzeros in the triangular factor and the number of subscripts for the compressed scheme respectively.

Three working vectors RCHLNK, MRGLNK and MARKER are used by the subroutine SMBFCT. The vector RCHLNK is used to facilitate the merging of the reachable sets, while the vector MRGLNK is used to keep track of the non-overlapping representative sets $\{R_i\}$ as introduced above. The vector MARKER is used to detect the condition as given in Corollary 5.4.5.

The subroutine begins by initializing the working vectors MRGLNK and MARKER. It then executes the main loop finding the reachable set for each node. The set $Adj(x_k) - S_k$ is first determined and assigned to the vector RCHLNK. At the same time, the condition in Corollary 5.4.5 is tested. If it is satisfied, the merging of reachable sets can be skipped. Otherwise, based on the information in MRGLNK, previous reachable sets are merged into RCHLNK. With the new reachable set completely formed in RCHLNK, the subroutine checks for possible compression of subscripts and sets up the corresponding portion of the data structure accordingly. Finally, it updates the vector MRGLNK to reflect the changes in the sets $\{R_i\}$.

By merging a set of carefully selected reachable sets, the subroutine SMBFCT is able to find a new reachable set in a very efficient manner. Since the number of subscripts required in the compressed scheme is not known beforehand, the size of the vector NZSUB may not be large enough to accommodate all the subscripts. In that case, the subroutine will abort and the error flag FLAG will be set to 1.

```
C***************************************************************
C***************************************************************
C********    SMBFCT .... SYMBOLIC FACTORIZATION    *******
C***************************************************************
C***************************************************************
C
C       PURPOSE - THIS ROUTINE PERFORMS SYMBOLIC FACTORIZATION
C          ON A PERMUTED LINEAR SYSTEM AND IT ALSO SETS UP THE
C          COMPRESSED DATA STRUCTURE FOR THE SYSTEM.
C
C       INPUT PARAMETERS -
C          NEQNS - NUMBER OF EQUATIONS.
C          (XADJ, ADJNCY) - THE ADJACENCY STRUCTURE.
C          (PERM, INVP) - THE PERMUTATION VECTOR AND ITS INVERSE.
C
C       UPDATED PARAMETERS -
C          MAXSUB - SIZE OF THE SUBSCRIPT ARRAY NZSUB.  ON RETURN,
C                   IT CONTAINS THE NUMBER OF SUBSCRIPTS USED
C
C       OUTPUT PARAMETERS -
C          XLNZ - INDEX INTO THE NONZERO STORAGE VECTOR LNZ.
C          (XNZSUB, NZSUB) - THE COMPRESSED SUBSCRIPT VECTORS.
C          MAXLNZ - THE NUMBER OF NONZEROS FOUND.
C          FLAG - ERROR FLAG.  POSITIVE VALUE INDICATES THAT.
C                 NZSUB ARRAY IS TOO SMALL.
C
C       WORKING PARAMETERS -
C          MRGLNK - A VECTOR OF SIZE NEQNS.  AT THE KTH STEP,
C                   MRGLNK(K), MRGLNK(MRGLNK(K)) , .........
C                   IS A LIST CONTAINING ALL THOSE COLUMNS L(*,J)
C                   WITH J LESS THAN K, SUCH THAT ITS FIRST OFF-
C                   DIAGONAL NONZERO IS L(K,J).  THUS, THE
C                   NONZERO STRUCTURE OF COLUMN L(*,K) CAN BE FOUND
C                   BY MERGING THAT OF SUCH COLUMNS L(*,J) WITH
C                   THE STRUCTURE OF A(*,K).
C          RCHLNK - A VECTOR OF SIZE NEQNS.  IT IS USED TO ACCUMULATE
C                   THE STRUCTURE OF EACH COLUMN L(*,K).  AT THE
C                   END OF THE KTH STEP,
C                       RCHLNK(K), RCHLNK(RCHLNK(K)), ........
C                   IS THE LIST OF POSITIONS OF NONZEROS IN COLUMN K
C                   OF THE FACTOR L.
C          MARKER - AN INTEGER VECTOR OF LENGTH NEQNS. IT IS USED
C                   TO TEST IF MASS SYMBOLIC ELIMINATION CAN BE
C                   PERFORMED.  THAT IS, IT IS USED TO CHECK WHETHER
C                   THE STRUCTURE OF THE CURRENT COLUMN K BEING
C                   PROCESSED IS COMPLETELY DETERMINED BY THE SINGLE
C                   COLUMN MRGLNK(K).
C
C***************************************************************
C
        SUBROUTINE  SMBFCT ( NEQNS, XADJ, ADJNCY, PERM, INVP,
       1                     XLNZ, MAXLNZ, XNZSUB, NZSUB, MAXSUB,
       1                     RCHLNK, MRGLNK, MARKER, FLAG )
C
C***************************************************************
C
        INTEGER ADJNCY(1), INVP(1), MRGLNK(1), NZSUB(1),
       1        PERM(1), RCHLNK(1), MARKER(1)
        INTEGER XADJ(1), XLNZ(1), XNZSUB(1),
       1        FLAG, I, INZ, J, JSTOP, JSTRT, K, KNZ,
       1        KXSUB, MRGK, LMAX, M, MAXLNZ, MAXSUB,
       1        NABOR, NEQNS, NODE, NP1, NZBEG, NZEND,
       1        RCHM, MRKFLG
C
C***************************************************************
C
C       ------------------
```

```
C        INITIALIZATION ...
C        ------------------
         NZBEG = 1
         NZEND = 0
         XLNZ(1) = 1
         DO 100 K = 1, NEQNS
            MRGLNK(K) = 0
            MARKER(K) = 0
  100    CONTINUE
C        -----------------------------------------------------
C        FOR EACH COLUMN ......... .  KNZ COUNTS THE NUMBER
C        OF NONZEROS IN COLUMN K ACCUMULATED IN RCHLNK.
C        -----------------------------------------------------
         NP1 = NEQNS + 1
         DO 1500  K = 1, NEQNS
            KNZ = 0
            MRGK = MRGLNK(K)
            MRKFLG = 0
            MARKER(K) = K
            IF (MRGK .NE. 0 ) MARKER(K) = MARKER(MRGK)
            XNZSUB(K) = NZEND
            NODE = PERM(K)
            JSTRT = XADJ(NODE)
            JSTOP = XADJ(NODE+1) - 1
            IF (JSTRT.GT.JSTOP)  GO TO 1500
C           ------------------------------------------------
C           USE RCHLNK TO LINK THROUGH THE STRUCTURE OF
C           A(*,K) BELOW DIAGONAL
C           ------------------------------------------------
            RCHLNK(K) = NP1
            DO 300 J = JSTRT, JSTOP
               NABOR = ADJNCY(J)
               NABOR = INVP(NABOR)
               IF ( NABOR .LE. K )  GO TO 300
                  RCHM = K
  200             M = RCHM
                  RCHM = RCHLNK(M)
                  IF ( RCHM .LE. NABOR )  GO TO 200
                     KNZ = KNZ+1
                     RCHLNK(M) = NABOR
                     RCHLNK(NABOR) = RCHM
                     IF ( MARKER(NABOR) .NE. MARKER(K) ) MRKFLG = 1
  300       CONTINUE
C           --------------------------------------
C           TEST FOR MASS SYMBOLIC ELIMINATION ...
C           --------------------------------------
            LMAX = 0
            IF ( MRKFLG .NE. 0 .OR. MRGK .EQ. 0 ) GO TO 350
            IF ( MRGLNK(MRGK) .NE. 0 ) GO TO 350
            XNZSUB(K) = XNZSUB(MRGK) + 1
            KNZ = XLNZ(MRGK+1) - (XLNZ(MRGK) + 1)
            GO TO 1400
C           -------------------------------------------------
C           LINK THROUGH EACH COLUMN I THAT AFFECTS L(*,K).
C           -------------------------------------------------
  350       I = K
  400       I = MRGLNK(I)
            IF (I.EQ.0)  GO TO 800
               INZ = XLNZ(I+1) - (XLNZ(I)+1)
               JSTRT = XNZSUB(I) +  1
               JSTOP = XNZSUB(I) + INZ
               IF (INZ.LE.LMAX)  GO TO 500
                  LMAX = INZ
                  XNZSUB(K) = JSTRT
C           -------------------------------------------------
C           MERGE STRUCTURE OF L(*,I) IN NZSUB INTO RCHLNK.
C           -------------------------------------------------
  500          RCHM = K
               DO 700 J = JSTRT, JSTOP
```

```
              NABOR = NZSUB(J)
  600         M = RCHM
              RCHM = RCHLNK(M)
              IF (RCHM.LT.NABOR)  GO TO 600
              IF (RCHM.EQ.NABOR)  GO TO 700
                KNZ = KNZ+1
                RCHLNK(M) = NABOR
                RCHLNK(NABOR) = RCHM
                RCHM = NABOR
  700         CONTINUE
              GO TO 400
C         -----------------------------------------------------
C         CHECK IF SUBSCRIPTS DUPLICATE THOSE OF ANOTHER COLUMN.
C         -----------------------------------------------------
  800     IF (KNZ.EQ.LMAX)  GO TO 1400
C            --------------------------------------------------
C            OR IF TAIL OF K-1ST COLUMN MATCHES HEAD OF KTH.
C            --------------------------------------------------
              IF (NZBEG.GT.NZEND)  GO TO 1200
                I = RCHLNK(K)
                DO 900 JSTRT=NZBEG,NZEND
                  IF (NZSUB(JSTRT)-I)  900, 1000, 1200
  900           CONTINUE
                GO TO 1200
 1000           XNZSUB(K) = JSTRT
                DO 1100 J=JSTRT,NZEND
                  IF (NZSUB(J).NE.I)  GO TO 1200
                  I = RCHLNK(I)
                  IF (I.GT.NEQNS)  GO TO 1400
 1100           CONTINUE
                NZEND = JSTRT - 1
C            -----------------------------------------
C            COPY THE STRUCTURE OF L(*,K) FROM RCHLNK
C            TO THE DATA STRUCTURE (XNZSUB, NZSUB).
C            -----------------------------------------
 1200         NZBEG = NZEND + 1
              NZEND = NZEND + KNZ
              IF (NZEND.GT.MAXSUB)  GO TO 1600
              I = K
              DO 1300 J=NZBEG,NZEND
                I = RCHLNK(I)
                NZSUB(J) = I
                MARKER(I) = K
 1300         CONTINUE
              XNZSUB(K) = NZBEG
              MARKER(K) = K
C         ---------------------------------------------------------------
C         UPDATE THE VECTOR MRGLNK.  NOTE COLUMN L(*,K) JUST FOUND
C         IS REQUIRED TO DETERMINE COLUMN L(*,J), WHERE
C         L(J,K) IS THE FIRST NONZERO IN L(*,K) BELOW DIAGONAL.
C         ---------------------------------------------------------------
 1400     IF (KNZ.LE.1)  GO TO 1500
              KXSUB = XNZSUB(K)
              I = NZSUB(KXSUB)
              MRGLNK(K) = MRGLNK(I)
              MRGLNK(I) = K
 1500     XLNZ(K+1) = XLNZ(K) + KNZ
        MAXLNZ = XLNZ(NEQNS) - 1
        MAXSUB = XNZSUB(NEQNS)
        XNZSUB(NEQNS+1) = XNZSUB(NEQNS)
        FLAG = 0
        RETURN
C     ----------------------------------------------------------------
C     ERROR - INSUFFICIENT STORAGE FOR NONZERO SUBSCRIPTS.
C     ----------------------------------------------------------------
 1600   FLAG = 1
        RETURN
        END
```

Exercises

5.4.1)   Let A be a matrix satisfying $f_i(A) < i$ for $2 \le i \le N$. Show that for each $k < N$, $m_k = k + 1$. Hence or otherwise, show that for $1 < i < N$,

$$Reach\,(x_i, S_{i-1}) = (Adj\,(x_i) \cup Reach\,(x_{i-1}, S_{i-2})) - S_i\,.$$

5.4.2)   Let A be a band matrix with bandwidth β. Assume that the matrix has a full band.

a)   Compare the uncompressed and compressed sparse storage schemes for A.

b)   Compare the two symbolic factorization algorithms as given by Lemma 5.4.1 and Theorem 5.4.3.

5.4.3)   Let R_1 and R_2 be two given sets of integers whose values are less than or equal to N. Assume that a temporary array of size N with all zero entries is provided. Show that the union $R_1 \cup R_2$ can be determined in time proportional to $|R_1| + |R_2|$.

5.5   The Numerical Subroutines for Factorization and Solution

In this section, we describe the subroutines that perform the numerical factorization and solution for linear systems stored using the compressed sparse scheme. The factorization subroutine GSFCT (for general sparse symmetric factorization) uses the inner product form of the factorization algorithm. Since the nonzeros in the lower triangle of A (or the factor L) are stored column by column, the inner product version of the algorithm must be implemented to adapt to this storage mode. The implementation GSFCT is a minor modification of the one in the Yale Sparse Matrix Package.

5.5.1   The Subroutine GSFCT (General sparse Symmetric FaCTorization)

The subroutine GSFCT accepts as input the data structure of the compressed scheme (XLNZ, XNZSUB, NZSUB) and the primary storage vectors DIAG and LNZ. The vectors DIAG and LNZ, on input, contain the nonzeros of the matrix A. On return, the nonzeros of the factor L are overwritten on those of the matrix A. The subroutines use three temporary vectors LINK, FIRST and TEMP, all of size N.

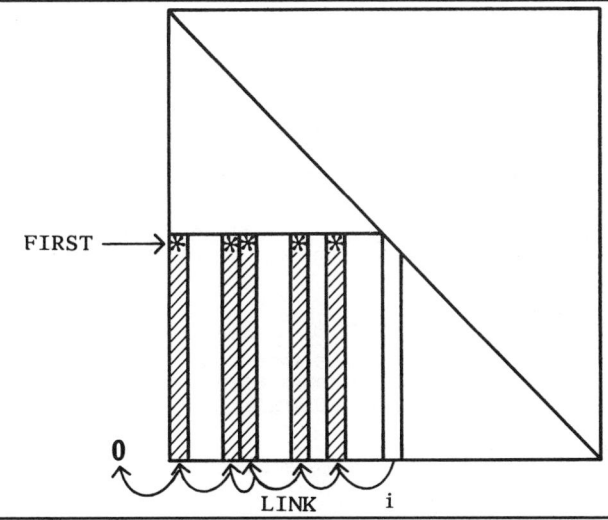

To compute a column L_{*i} of the factor, the columns that are involved in the formation of L_{*i} are exactly those L_{*j} with $\ell_{ij} \neq 0$. The modification can be done one column at a time as follows:

For L_{*j} with $\ell_{ij} \neq 0$,

$$\begin{pmatrix} \ell_{ii} \\ \cdot \\ \cdot \\ \cdot \\ \ell_{Ni} \end{pmatrix} \leftarrow \begin{pmatrix} \ell_{ii} \\ \cdot \\ \cdot \\ \cdot \\ \ell_{Ni} \end{pmatrix} - \ell_{ij} \begin{pmatrix} \ell_{ij} \\ \cdot \\ \cdot \\ \cdot \\ \ell_{Nj} \end{pmatrix}.$$

At step i, all the columns that affect L_{*i} are given by the list

$$\text{LINK}(i), \ \text{LINK}(\text{LINK}(i)), \ \ldots \ .$$

To minimize subscript searching, a work vector FIRST is used so that FIRST(j) points to the location in the storage vector LNZ, where the nonzero ℓ_{ij} resides for $j = \text{LINK}(i), \ \text{LINK}(\text{LINK}(i)), \ \ldots$. In this way, the modification of L_{*i} by L_{*j} can start at the location FIRST(j) in LNZ. The third working vector TEMP is used to accumulate the modifications to the column L_{*i}.

The subroutine GSFCT begins by initializing the working vectors LINK and TEMP. The loop DO 600 J ... processes each column. It accumulates the modifications to the current column in the variable DIAGJ and the vector TEMP. At the same time, it updates the temporary vectors FIRST and LINK. Finally, the modification is applied to the entries in the present column.

```
C***************************************************************
C***************************************************************
C******    GSFCT .....  GENERAL SPARSE SYMMETRIC FACT    ******
C***************************************************************
C***************************************************************
C
C     PURPOSE - THIS SUBROUTINE PERFORMS THE SYMMETRIC
C        FACTORIZATION FOR A GENERAL SPARSE SYSTEM, STORED IN
C        THE COMPRESSED SUBSCRIPT DATA FORMAT.
C
C     INPUT PARAMETERS -
C        NEQNS - NUMBER OF EQUATIONS.
C        XLNZ - INDEX VECTOR FOR LNZ.  XLNZ(I) POINTS TO THE
C               START OF NONZEROS IN COLUMN I OF FACTOR L.
C        (XNZSUB, NZSUB) - THE COMPRESSED SUBSCRIPT DATA
C               STRUCTURE FOR FACTOR L.
C
C     UPDATED PARAMETERS -
C        LNZ - ON INPUT, CONTAINS NONZEROS OF A, AND ON
C               RETURN, THE NONZEROS OF L.
C        DIAG - THE DIAGONAL OF L OVERWRITES THAT OF A.
C        IFLAG - THE ERROR FLAG.  IT IS SET TO 1 IF A ZERO OR
C               NEGATIVE SQUARE ROOT OCCURS DURING THE
C               FACTORIZATION.
C        OPS  - A DOUBLE PRECISION COMMON PARAMETER THAT IS
C               INCREMENTED BY THE NUMBER OF OPERATIONS
C               PERFORMED BY THE SUBROUTINE.
C
C     WORKING PARAMETERS -
C        LINK - AT STEP J, THE LIST IN
C                  LINK(J), LINK(LINK(J)), ...........
C               CONSISTS OF THOSE COLUMNS THAT WILL MODIFY
C               THE COLUMN L(*,J).
C        FIRST - TEMPORARY VECTOR TO POINT TO THE FIRST
C               NONZERO IN EACH COLUMN THAT WILL BE USED
C               NEXT FOR MODIFICATION.
C        TEMP - A TEMPORARY VECTOR TO ACCUMULATE MODIFICATIONS.
C
C***************************************************************
C
      SUBROUTINE  GSFCT ( NEQNS, XLNZ, LNZ, XNZSUB, NZSUB, DIAG,
     1                    LINK, FIRST, TEMP, IFLAG )
C
C***************************************************************
C
         DOUBLE PRECISION COUNT, OPS
         COMMON  /SPKOPS/ OPS
         REAL DIAG(1), LNZ(1), TEMP(1), DIAGJ, LJK
         INTEGER LINK(1), NZSUB(1)
         INTEGER FIRST(1), XLNZ(1), XNZSUB(1),
     1           I, IFLAG, II, ISTOP, ISTRT, ISUB, J,
     1           K, KFIRST, NEQNS, NEWK
C
C***************************************************************
C
C        ---------------------------
C        INITIALIZE WORKING VECTORS ...
C        ---------------------------
         DO 100 I = 1, NEQNS
            LINK(I) = 0
            TEMP(I) = 0.0E0
  100    CONTINUE
C        -----------------------------------------------
C        COMPUTE COLUMN L(*,J) FOR J = 1,...., NEQNS.
```

```
C             -------------------------------------------------
              DO 600 J = 1, NEQNS
C             -------------------------------------------------
C             FOR EACH COLUMN L(*,K) THAT AFFECTS L(*,J).
C             -------------------------------------------------
              DIAGJ = 0.0E0
              NEWK = LINK(J)
     200      K    = NEWK
              IF ( K .EQ. 0 ) GO TO 400
                 NEWK = LINK(K)
C                ----------------------------------------
C                OUTER PRODUCT MODIFICATION OF L(*,J) BY
C                L(*,K) STARTING AT FIRST(K) OF L(*,K).
C                ----------------------------------------
                 KFIRST = FIRST(K)
                 LJK    = LNZ(KFIRST)
                 DIAGJ = DIAGJ + LJK*LJK
                 OPS   = OPS + 1.0D0
                 ISTRT = KFIRST + 1
                 ISTOP = XLNZ(K+1) - 1
                 IF ( ISTOP .LT. ISTRT ) GO TO 200
C                   ------------------------------------------
C                   BEFORE MODIFICATION, UPDATE VECTORS FIRST,
C                   AND LINK FOR FUTURE MODIFICATION STEPS.
C                   ------------------------------------------
                    FIRST(K) = ISTRT
                    I = XNZSUB(K) + (KFIRST-XLNZ(K)) + 1
                    ISUB = NZSUB(I)
                    LINK(K) = LINK(ISUB)
                    LINK(ISUB) = K
C                   ----------------------------------------
C                   THE ACTUAL MOD IS SAVED IN VECTOR TEMP.
C                   ----------------------------------------
                    DO 300 II = ISTRT, ISTOP
                       ISUB = NZSUB(I)
                       TEMP(ISUB) = TEMP(ISUB) + LNZ(II)*LJK
                       I = I + 1
     300            CONTINUE
                    COUNT = ISTOP - ISTRT + 1
                    OPS  = OPS + COUNT
                 GO TO 200
C             -------------------------------------------------
C             APPLY THE MODIFICATIONS ACCUMULATED IN TEMP TO
C             COLUMN L(*,J).
C             -------------------------------------------------
     400      DIAGJ = DIAG(J) - DIAGJ
              IF ( DIAGJ .LE. 0.0E0 ) GO TO 700
              DIAGJ = SQRT(DIAGJ)
              DIAG(J) = DIAGJ
              ISTRT = XLNZ(J)
              ISTOP = XLNZ(J+1) - 1
              IF ( ISTOP .LT. ISTRT ) GO TO 600
                 FIRST(J) = ISTRT
                 I = XNZSUB(J)
                 ISUB = NZSUB(I)
                 LINK(J) = LINK(ISUB)
                 LINK(ISUB) = J
                 DO 500 II = ISTRT, ISTOP
                    ISUB = NZSUB(I)
                    LNZ(II) = ( LNZ(II)-TEMP(ISUB) ) / DIAGJ
                    TEMP(ISUB) = 0.0E0
                    I = I + 1
     500         CONTINUE
                 COUNT = ISTOP - ISTRT + 1
                 OPS  = OPS + COUNT
     600      CONTINUE
              RETURN
C             ----------------------------------------------------------
```

```
C             ERROR - ZERO OR NEGATIVE SQUARE ROOT IN FACTORIZATION.
C             -----------------------------------------------------
  700         IFLAG = 1
              RETURN
        END
```

5.5.2   The Subroutine GSSLV (General sparse Symmetric SoLVe)

The subroutine GSSLV is used to perform the numerical solution of a
factored system, where the matrix is stored in the compressed subscript
sparse format as discussed in Section 5.4.2. It accepts as input the
number of equations NEQNS, together with the data structure and numerical components of the matrix factor. This includes the
compressed subscript structure (XNZSUB, NZSUB), the diagonal
components DIAG of the factor and the off-diagonal nonzeros in the
factor stored in the array pair (XLNZ, LNZ).

Since the nonzeros in the lower triangular factor are stored
column by column, the solution method should be arranged so that
access to the components is made column-wise. The forward substitution uses the "outer-product" form, whereas the backward substitution
loop performs the solution by "inner-products" as discussed in
Section 2.2.1 in Chapter 2.

```
C****************************************************************
C****************************************************************
C******    GSSLV .....  GENERAL SPARSE SYMMETRIC SOLVE    ******
C****************************************************************
C****************************************************************
C
C     PURPOSE - TO PERFORM SOLUTION OF A FACTORED SYSTEM, WHERE
C        THE MATRIX IS STORED IN THE COMPRESSED SUBSCRIPT
C        SPARSE FORMAT.
C
C     INPUT PARAMETERS -
C        NEQNS - NUMBER OF EQUATIONS.
C        (XLNZ, LNZ) - STRUCTURE OF NONZEROS IN L.
C        (XNZSUB, NZSUB) - COMPRESSED SUBSCRIPT STRUCTURE.
C        DIAG - DIAGONAL COMPONENTS OF L.
C
C     UPDATED PARAMETER -
C        RHS - ON INPUT, IT CONTAINS THE RHS VECTOR, AND ON
C              OUTPUT, THE SOLUTION VECTOR.
C
C****************************************************************
C
        SUBROUTINE  GSSLV ( NEQNS, XLNZ, LNZ, XNZSUB, NZSUB,
       1                    DIAG, RHS )
C
C****************************************************************
C
        DOUBLE PRECISION COUNT, OPS
        COMMON /SPKOPS/ OPS
        REAL DIAG(1), LNZ(1), RHS(1), RHSJ, S
        INTEGER NZSUB(1)
        INTEGER XLNZ(1), XNZSUB(1), I, II, ISTOP,
       1        ISTRT, ISUB, J, JJ, NEQNS
```

```
C
C*******************************************************************
C
C             -----------------------
C             FORWARD SUBSTITUTION ...
C             -----------------------
             DO 200 J = 1, NEQNS
                RHSJ = RHS(J) / DIAG(J)
                RHS(J) = RHSJ
                ISTRT = XLNZ(J)
                ISTOP = XLNZ(J+1) - 1
                IF ( ISTOP .LT. ISTRT )  GO TO 200
                   I = XNZSUB(J)
                   DO 100 II = ISTRT, ISTOP
                      ISUB = NZSUB(I)
                      RHS(ISUB) = RHS(ISUB) - LNZ(II)*RHSJ
                      I = I + 1
  100              CONTINUE
  200        CONTINUE
             COUNT = 2*(NEQNS + ISTOP)
             OPS = OPS + COUNT
C             -----------------------
C             BACKWARD SUBSTITUTION ...
C             -----------------------
             J = NEQNS
             DO 500 JJ = 1, NEQNS
                S = RHS(J)
                ISTRT = XLNZ(J)
                ISTOP = XLNZ(J+1) - 1
                IF ( ISTOP .LT. ISTRT )  GO TO 400
                   I = XNZSUB(J)
                   DO 300 II = ISTRT, ISTOP
                      ISUB = NZSUB(I)
                      S = S - LNZ(II)*RHS(ISUB)
                      I = I + 1
  300              CONTINUE
  400           RHS(J) = S / DIAG(J)
                J = J - 1
  500        CONTINUE
             RETURN
          END
```

5.6 Additional Notes

The *element model* (George 1973, Eisenstat 1976) is also used in the study of elimination. It models the factorization process in terms of the clique structure in the elimination graphs. It is motivated by finite element applications, where the clique structure of the matrix graph arises in a natural way. The model is closely related to the quotient graph model studied in Section 5.2.

An implementation of the minimum degree algorithm using the element model can be found in George and McIntyre (1978). In (George 1980) the authors have implemented the minimum degree algorithm using the implicit model via reachable sets on the original graph. Refinements have been included to speed up the execution time.

There are other ordering algorithms that are designed to reduce fill-in. The *minimum deficiency algorithm* (Rose 1972a) numbers a node next if its elimination incurs the least number of fills. It involves substantially more work than the minimum degree algorithm and experience has shown that in practice the ordering produced is rarely much better than the one produced by the minimum degree algorithm.

In (George 1977), a different storage scheme is proposed for general sparse orderings. It makes use of the observation that off-diagonal nonzeros form dense blocks. Only a few items of information are needed to store each non-null block, and standard dense matrix methods can be used to operate on them.

6/ Quotient Tree Methods for Finite Element and Finite Difference Problems

6.0 Introduction

In this and the subsequent two chapters we study methods designed primarily for matrix problems arising in connection with finite difference and finite element methods for solving various problems in structural analysis, fluid flow, elasticity, heat transport and related problems (Zienkiewicz 1977). For our purposes here, the problems we have in mind can be characterized as follows.

Let $\mathcal{M}$ be a planar mesh consisting of the union of triangles and/or quadrilaterals called *elements*, with adjacent elements sharing a common side or a common vertex. There is a node at each vertex of the mesh $\mathcal{M}$, and there may also be nodes lying on element sides and element faces, as shown in the example of Figure 6.0.1. Associated with each node is a variable x_i and for some labelling of the nodes or variables from 1 to N, we define a *finite element system $Ax = b$* associated with $\mathcal{M}$ as one where A is symmetric and positive definite and for which $a_{ij} \neq 0$ implies variables x_i and x_j are associated with nodes of the same element. The graph associated with A will be referred to as the finite element graph associated with $\mathcal{M}$, as shown in Figure 6.0.1.

Figure 6.0.1 An 8 node finite element mesh and its associated finite element graph.

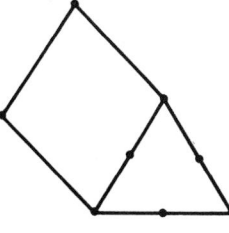

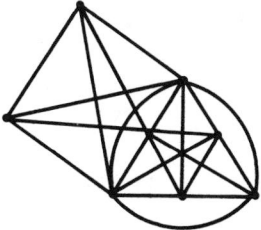

An 8 node finite
element mesh $\mathcal{M}$

The finite element graph
associated with $\mathcal{M}$

In many practical settings, this definition of "finite element system" is not quite general enough, since sometimes more than one variable is associated with some or all of the nodes. However, our definition captures the essential features of such problems and simplifies the presentation of the ideas. Moreover, the extension of the basic ideas to the more general case is immediate, and since our algorithms and programs operate on the associated graph, they work for the general case anyway.

Finite element matrix problems are often solved using the band or profile methods described in Chapter 4, and for relatively small problems these methods are often the most efficient, particularly for one-shot problems where the relatively high cost of finding low fill orderings offsets their lower arithmetic and storage costs. For fairly large problems, and/or in situations where numerous problems having identical structure must be solved, the more sophisticated orderings which attempt to minimize fill, such as the minimum degree ordering of Chapter 5 or the nested dissection orderings of Chapter 8, are attractive.

The methods of Chapters 4 and 5 in a sense represent extremes in the "sophistication spectrum;" the envelope methods do not attempt to exploit much of the structure of A and L, while the methods of Chapter 5 attempt to exploit it all. In this chapter we investigate methods which lie somewhere in between these two extremes, and for certain sizes and types of finite element problems, they turn out to be more efficient than either of the other two strategies. The ordering times and the operation counts are usually comparable with envelope orderings, but the storage requirements are usually substantially lower.

6.1 Solution of Partitioned Systems of Equations

The methods we consider in this chapter rely heavily on the use of partitioned matrices, and some techniques to exploit sparsity in such systems. All the partitionings we consider will be symmetric in the sense that the row and column partitionings will be identical.

6.1.1 *Factorization of a Block Two by Two Matrix*

In order to illustrate most of the important ideas about computations involving sparse partitioned matrices, we consider a block two by two linear system $Ax = b$:

$$\begin{pmatrix} B & V \\ V^T & C \end{pmatrix} \begin{pmatrix} x_1 \\ x_2 \end{pmatrix} = \begin{pmatrix} b_1 \\ b_2 \end{pmatrix}, \qquad (6.1.1)$$

where B and $\overline{C}$ are r by r and s by s submatrices respectively, with $r + s = N$.

The Cholesky factor L of A, correspondingly partitioned, is given by

$$L = \begin{pmatrix} L_B & 0 \\ W^T & L_C \end{pmatrix} , \qquad (6.1.2)$$

where L_B and L_C are the Cholesky factors of the matrices B and $C = \overline{C} - V^T B^{-1} V$ respectively, and $W = L_B^{-1} V$. Here the "modification matrix" subtracted from $\overline{C}$ to obtain C can be written as

$$V^T B^{-1} V = V^T L_B^{-T} L_B^{-1} V = W^T W .$$

The determination of the factor L can be done as described below. For reasons which will be obvious later in this section we refer to it as the *symmetric* block factorization scheme.

Step 1   Factor the matrix B into $L_B L_B^T$.

Step 2   Solve the triangular systems

$$L_B W = V .$$

Step 3   Modify the submatrix remaining to be factored:

$$C = \overline{C} - W^T W .$$

Step 4   Factor the matrix C into $L_C L_C^T$.

This computational sequence is depicted pictorially in Figure 6.1.1.

Does this block-oriented computational scheme have any advantage over the ordinary step by step factorization, in terms of operations? The following result is quoted from George (1974).

Theorem 6.1.1

The number of operations required to compute the factor L of A is the same whether the step by step elimination scheme or the symmetric block factorization scheme is used.

Intuitively, the result holds because the two methods perform exactly the same numerical operations, but in a different order. There is, however, a different way to perform the block factorization, where the arithmetic requirement may decrease or increase. The alternative way depends on the observation that the modification matrix $V^T B^{-1} V$ can be computed in two distinctly different ways, namely as the conventional product

$$(V^T L_B^{-T})(L_B^{-1} V) = W^T W , \qquad (6.1.3)$$

Figure 6.1.1 Diagram indicating the sequence of computations for the symmetric block factorization scheme, and the modes in which the data is processed.

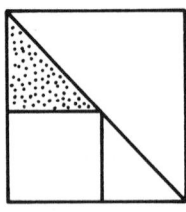

Factorization of B into $L_B L_B{}^T$.

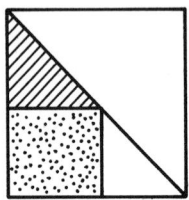

Solution of the system $L_B W = V$.

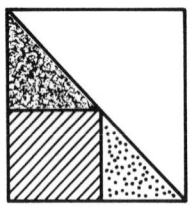

Computation of $C = \overline{C} - W^T W$.

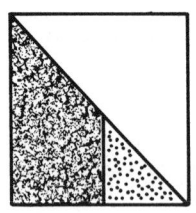

Factorization of C into $L_C L_C^T$.

 accessed only   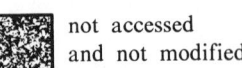 not accessed and not modified   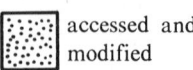 accessed and modified

or as

$$V^T(L_B^{-T}(L_B^{-1}V)) = V^T(L_B^{-T}W) = V^T\tilde{W} . \qquad (6.1.4)$$

We shall refer to this latter way of performing the computation as the *asymmetric block factorization* scheme.

The difference in the two computations hinges on the cost of computing W^TW compared to the cost of solving $L_B^T\tilde{W} = W$, and then computing $V^T\tilde{W}$. As an example illustrating the difference in the arithmetic cost, consider a partitioned matrix A having the structure indicated in Figure 6.1.2.

Figure 6.1.2 Structure of a 2 by 2 partitioned matrix A and its Cholesky factor L.

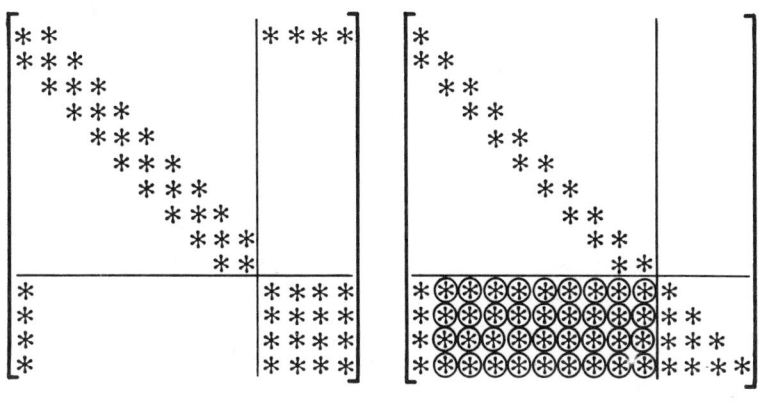

Since the matrix W is full (see Exercise 2.2.3), by Corollary 2.2.2, the cost of solving the equations $L_B^T\tilde{W} = W$ is $4 \times 19 = 76$. The cost of computing $V^T\tilde{W}$ is 10, yielding a total of 86. On the other hand, the cost of computing W^TW is $10 \times 10 = 100$.

In addition to potentially reducing arithmetic operations, this asymmetric scheme may allow us to substantially reduce storage requirements over that for the standard scheme. The key observation is that we do not need W in order to solve for x, provided that V is available. Whenever we need to compute a product such as W^Tz or Wz, we can do so by computing $V^T(L_B^{-T}z)$ or $L_B^{-1}(Vz)$; that is, we solve a triangular system and multiply by a sparse matrix. If V is much sparser than W, as it often is, we save storage and perhaps operations as well. The important point to note in terms of computing the factorization is that if we plan to discard W anyway, computing C in the asymmetric fashion implied by (6.1.4) allows us to avoid ever storing W. We can compute the product $V^T\tilde{W}$ *one column at a time*, discarding each as soon as it has been used to modify a column of $\overline{C}$. Only one temporary vector of length r is required. By comparison, if we compute $V^T B^{-1} V$ as $W^T W$, there appears to be no way to avoid storing all of W at some point, even if we do not intend to retain it for later use.

This asymmetric version of the factorization algorithm can be described as follows.

Step 1   Factor the matrix B into $L_B L_B^T$.

Step 2   For each column $v = V_{*i}$ of V,

2.1)   Solve $L_B w = v$.
2.2)   Solve $L_B^T \tilde{w} = w$.
2.3)   Set $C_{*i} = \overline{C}_{*i} - V^T \tilde{w}$.

Step 3   Factor the matrix C into $L_C L_C^T$.

Of course, the symmetry of C is exploited in forming C_{*i} in Step 2.3. Regardless of which of the above ways we actually employ in calculating the product $V^T B^{-1} V$, there is still some freedom in the order in which we calculate the components. Assuming that we compute the lower triangle, we can compute the elements row by row or column by column, as depicted in Figure 6.1.3; each requires a different order of access to the columns of W or V.

Figure 6.1.3   Diagram showing the access to columns of W or V, when the lower triangle of $V^T B^{-1} V$ is computed column by column and row by row.

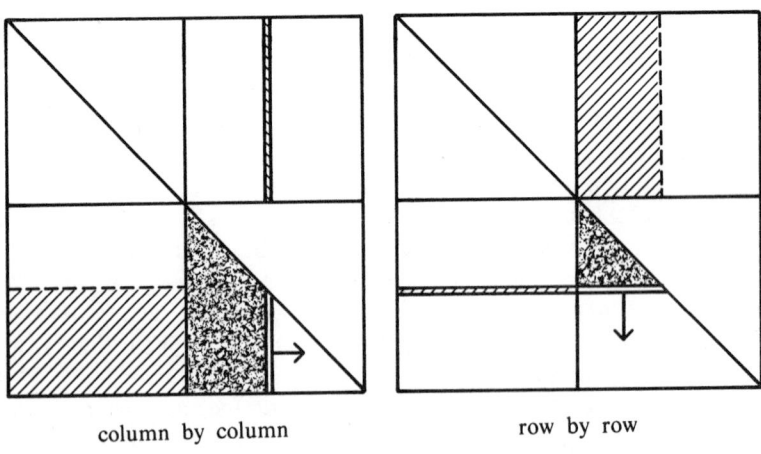

column by column   row by row

6.1.2   Triangular Solution of a Block Two by Two System

With the Cholesky factor of the partitioned matrix available, the solution of the linear system is straightforward, as shown below.

Forward Solve

Solve	$L_B y_1 = b_1$.
Compute	$\tilde{b}_2 = b_2 - W^T y_1$.
Solve	$L_C y_2 = \tilde{b}_2$.

Backward Solve

Solve	$L_C^T x_2 = y_2$.
Compute	$\tilde{y}_1 = y_1 - W x_2$.
Solve	$L_B^T x_1 = \tilde{y}_1$.

This solution method will be referred to as the *standard solution* scheme. However, as we noted in Section 6.1.1, it may be desirable to discard W in favour of storing only V, and use the definition $W = L_B^{-1} V$ whenever we need to operate with the matrix W. If we do this, we obtain the following algorithm, which we refer to as the *implicit solution* scheme. Here t_1 is a temporary vector.

Forward Solve

Solve	$L_B y_1 = b_1$.
Solve	$L_B^T t_1 = y_1$.
Compute	$\tilde{b}_2 = b_2 - V^T t_1$.
Solve	$L_C y_2 = \tilde{b}_2$.

Backward Solve

Solve	$L_C^T x_2 = y_2$.
Solve	$L_B t_1 = V x_2$.
Compute	$\tilde{y}_1 = y_1 - t_1$.
Solve	$L_B^T x_1 = \tilde{y}_1$.

In the implicit scheme, only the submatrices

$$\{L_B, \ L_C, \ V\}$$

are required, compared to

$$\{L_B, \ L_C, \ W\}$$

in the standard scheme. By Corollary 2.2.5,

$$\eta(V) \le \eta(W)$$

and for sparse matrices, V may have *substantially* fewer nonzeros than W. In the matrix example of Figure 6.1.2,

$$\eta(V) = 4$$

while

$$\eta(W) = 40 .$$

Thus, in terms of primary storage requirements, the use of the implicit solution scheme may be quite attractive.

In terms of operations, the relative merits of the two schemes depend on the sparsity of the matrices L_B, V and W. Since the cost of computing Wz is $\eta(W)$ and the cost of computing $L_B^{-1}(Vz)$ is $\eta(V) + \eta(L_B)$, we easily obtain the following result. Here, the sparsity of the vector z is not exploited.

Lemma 6.1.2

The cost of performing implicit solution is no greater than that of doing the standard block solution if and only if $\eta(V) + \eta(L_B) \le \eta(W)$.

In the next section, we shall extend these ideas to block p by p linear systems, for $p > 2$. In sparse partitioned systems, it is typical that the asymmetric version of block factorization and the implicit form of the solution is superior in terms of computation and storage. Thus, we consider only this version in the remainder of this chapter.

Exercises

6.1.1)   Let A be a symmetric positive definite block two by two matrix of the form

$$A = \begin{pmatrix} B & V \\ V^T & \overline{C} \end{pmatrix} ,$$

where both B and $\overline{C}$ are m by m and tridiagonal, and V is diagonal. In your answers to the question below, assume m is large, and ignore low order terms in your operation counts.

a)   Denote the triangular factor of A by

$$L = \begin{pmatrix} L_B & 0 \\ W^T & L_C \end{pmatrix} .$$

Describe the nonzero structures of L_B, L_C and W.

b)   Determine the number of operations (multiplications and divisions) required to compute L using the symmetric and asymmetric factorization algorithms described in Section 6.1.1.

c) Compare the costs of the explicit and implicit solution schemes of Section 6.1.2 for this problem, where you may assume that the right hand side b in the matrix problem $Ax = b$ is full.

d) Answer a), b) and c) above when B and $\overline{C}$ are full, and V is diagonal.

e) Answer a), b) and c) above when B and $\overline{C}$ are full, and V is null except for its first row, which is full.

6.1.2) The asymmetric factorization scheme can be viewed as computing the factorization shown below.

$$A = \begin{pmatrix} B & V \\ V^T & \overline{C} \end{pmatrix} = \begin{pmatrix} L_B L_B^T & 0 \\ V^T & L_C L_C^T \end{pmatrix} \begin{pmatrix} I & \tilde{W} \\ 0 & I \end{pmatrix}$$

where $\tilde{W} = B^{-1}V$, and it is understood that the *factors* of B and C are stored, rather than B and C. Write down explicit and implicit solution procedures analogous to those described in Section 6.1.2, using this factorization. Is there any reduction in the operation counts over those of Section 6.1.2? What about storage requirements if we store the off-diagonal blocks of the factors in each case?

6.1.3) Prove Theorem 6.1.1.

6.2 Quotient Graphs, Trees, and Tree Partitionings

It should be clear that the success of the implicit solution scheme we considered in Section 6.1.2 was due to the very simple form of the off-diagonal block W. For a general p by p partitioned matrix the off-diagonal blocks of its factor will not have such a simple form; to discard them in favor of the original block of A, and then to effectively recompute them when needed, would in general be prohibitively costly in terms of computation. This immediately leads us to ask what characteristics a partitioned matrix should have in order that the off-diagonal blocks of its factor have this simple form. In this section we answer this question, and lay the foundations for an algorithm for finding such partitionings.

6.2.1 Partitioned Matrices and Quotient Graphs

We have already established the connection between symmetric matrices and graphs. In this section we introduce some additional graph theory ideas to allow us to deal with partitioned matrices.

Let A be partitioned into p^2 submatrices A_{ij}, $1 \le i,j \le p$, and suppose we view each block as a single component which is zero if the block is null, and nonzero otherwise. We can then associate a p-node graph with the p by p block matrix A, having edges joining nodes if the corresponding off-diagonal blocks are non-null. Figure 6.2.1 illustrates these ideas. Note that just as in the scalar case, an ordering of this new graph is implied by the matrix to which it corresponds. Also just as before, we are interested in finding partitionings and orderings of *unlabelled* graphs. This motivates the definition of quotient graphs, which we introduced in Chapter 5.

Figure 6.2.1 A partitioned matrix A, the implied partitioning of the node set of its graph, and the graph of its zero-nonzero block structure.

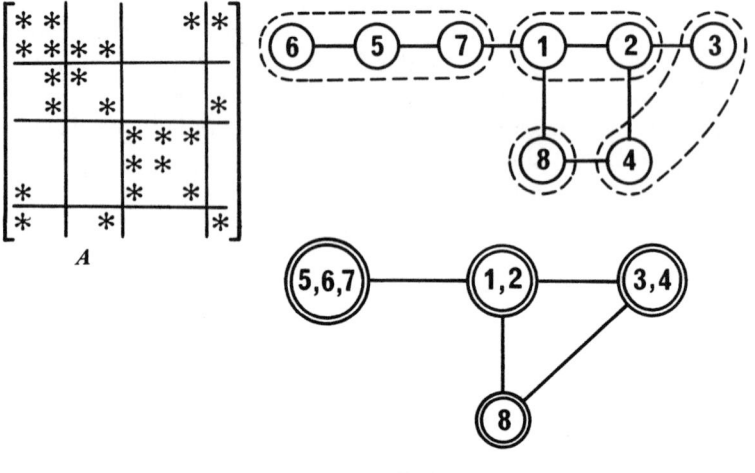

Let $G = (X,E)$ be a given unlabelled graph, and let P be a partition of its node set X:

$$P = \{Y_1, \ Y_2, \ \ldots, \ Y_p\} \ .$$

Recall from Chapter 5 that the *quotient graph* of G with respect to P is the graph $(P,\mathcal{E})$, where $\{Y_i,Y_j\} \in \mathcal{E}$ if and only if $Adj(Y_i) \cap Y_j \ne \emptyset$. We denote this graph by G/P.

Note that our definition of quotient graph is for an unlabelled graph. An ordering α of G is said to be *compatible* with a partitioning

P of G if each member Y_i of P is numbered *consecutively* by α. Clearly orderings and partitionings of graphs corresponding to partitioned matrices *must* have this property. An ordering α which is compatible with P induces or implies an ordering on G/P; conversely, an ordering α_P of G/P induces a *class* of orderings on G which are compatible with P. Figure 6.2.2 illustrates these notions.   Unless we explicitly state otherwise, whenever we refer to an ordering of a partitioned graph, we assume that the ordering is compatible with the partitioning, since our interest is in ordering partitioned matrices.

Figure 6.2.2   An example of induced orderings for the graph example in Figure 6.2.1.

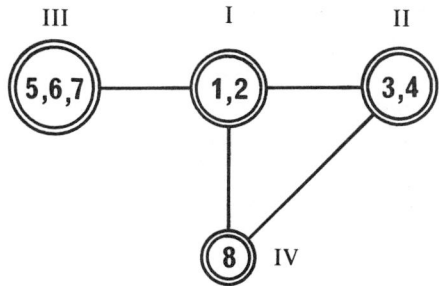

Labelling of G/P induced by the original ordering in Figure 6.2.1

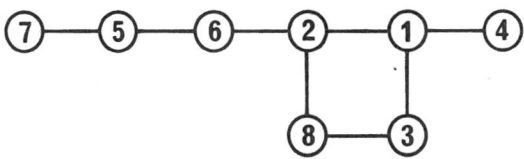

A different ordering compatible with P

In general, when we perform (block) Gaussian elimination on a partitioned matrix, zero blocks may become nonzero. That is, "block fill" may occur just as fill occurs in the scalar case. For example, in the partitioned matrix of Figure 6.2.1, there are two null off-diagonal blocks which will become non-null after factorization. The structure of the triangular factor is given in Figure 6.2.3.

However, for a given partitioning of a symmetric matrix, the way block fill occurs does not necessarily correspond exactly to the scalar situation. In other words, we cannot simply carry out symbolic elimination on the quotient graph, and thereby obtain the block structure of

Figure 6.2.3 Structure of the factor of the matrix in Figure 6.2.1.

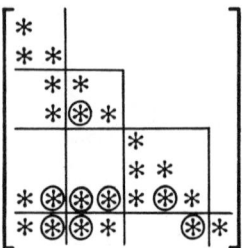

the factor **L**. What this provides is the *worst case fill*, which of course could always happen, since each non-null block could be full. Figure 6.2.4 illustrates this point.

Thus, symbolically factoring the quotient graph of a partitioned matrix may yield a higher block fill than actually occurs. Intuitively, the reason is clear: the elimination model assumes that the product of two nonzero quantities will always be nonzero, which is true for scalars. However, it is quite possible to have two non-null matrices whose product is logically zero.

6.2.2 Trees, Quotient Trees, and Tree Partitionings

A *tree* $T = (X,E)$ is a connected graph with no cycles. It is easy to verify that for a tree T, $|X| = |E| + 1$, and every pair of distinct nodes is connected by exactly one path. A *rooted tree* is an ordered pair (r,T) where r is a distinguished node of T called the *root*. Since every pair of nodes in T is connected by exactly one path, the path from r to any node $x \in X$ is unique. If this path passes through y, then x is a descendant of y, and y is an ancestor of x. If $\{x,y\} \in E$, then x is a *son* of y and y is a *father* of x. If Y consists of a node y and all its descendants, the section graph $T(Y)$ is a *subtree* of T. Note that the ancestor-descendant relationship is only defined for rooted trees. These notions are illustrated in Figure 6.2.5.

A *monotone ordering* α of the rooted tree (r,T) is one for which each node is numbered before its father. Obviously the root must be numbered last. Given an unrooted tree T, an ordering α is monotone if it is monotone for the rooted tree $(\alpha(|X|),T)$. The significance of monotonely ordered trees is that the corresponding matrices *suffer no*

Figure 6.2.4   Example showing that symbolic elimination on G/P may overestimate block fill.

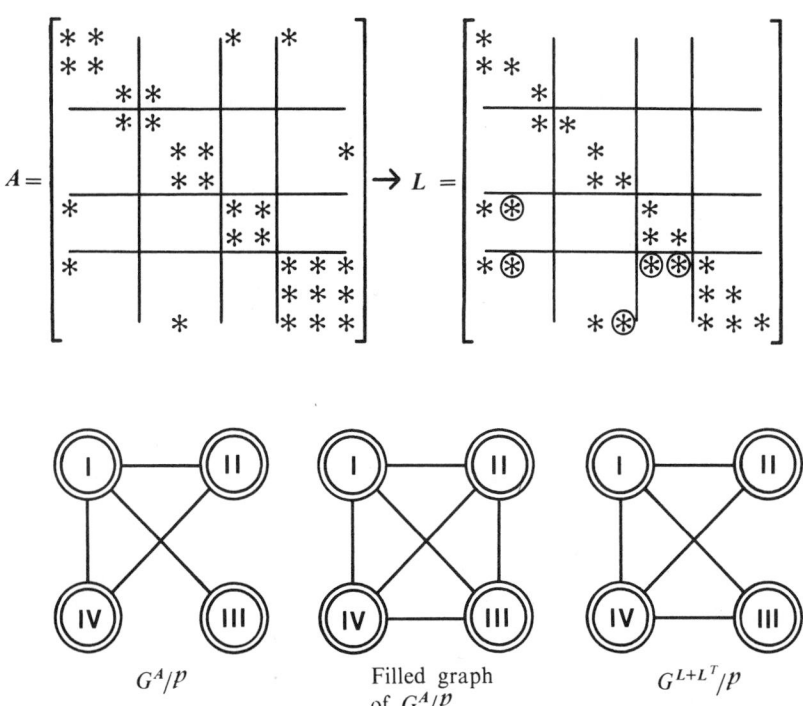

$$G^A/P \qquad \text{Filled graph of } G^A/P \qquad G^{L+L^T}/P$$

Figure 6.2.5   A rooted tree T and a subtree.

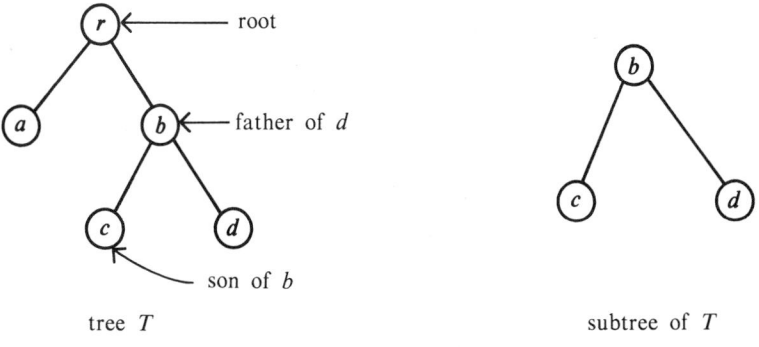

tree T subtree of T

fill during their factorization. The following lemma is due to Parter (1961).

Lemma 6.2.1
Let A be an N by N symmetric matrix whose labelled graph is a monotonely ordered tree. If $A = LL^T$, where L is the Cholesky factor of A, then $a_{ij} = 0$ implies $\ell_{ij} = 0$, $i > j$.

The proof is left as an exercise.

Lemma 6.2.2
Let A and L be as in Lemma 6.2.1. Then $\ell_{ij} = a_{ij}/\ell_{jj}$, $i > j$.

Proof   Recall from Section 2.1.2, that the components of L are given by

$$\ell_{ij} = \left(a_{ij} - \sum_{k=1}^{j-1} \ell_{ik}\ell_{jk} \right)/\ell_{jj}, \quad i > j \ .$$

To prove the result we show that $\sum_{k=1}^{j-1} \ell_{ik}\ell_{jk} = 0$. Suppose for a contradiction that $\ell_{im}\ell_{jm} \neq 0$ for some m satisfying $1 \leq m \leq j-1$. By Lemma 6.2.1, this means that $a_{im}a_{jm} \neq 0$, which implies nodes i and j are both connected to node m in the corresponding tree, with $i > m$ and $j > m$. But this implies that the tree is not monotonely ordered.   □

Lemmas 6.2.1 and 6.2.2 are not as significant as they might at first seem because matrices which arise in applications seldom have graphs which are trees. Their importance lies in the fact that they extend immediately to *partitioned matrices.*

Suppose A is as before, partitioned into p^2 submatrices A_{ij}, $1 \leq i,j \leq p$, and let L_{ij} be the corresponding submatrices of its Cholesky factor L. Suppose further that the labelled quotient graph of the partitioned matrix A is a monotonely ordered tree, as illustrated in Figure 6.2.6. Then it is straightforward to verify the analog of Lemma 6.2.2 for such a partitioned matrix, that is $L_{ij} = A_{ij}L_{jj}^{-T} = (L_{jj}^{-1}A_{ji})^T$ for each non-null submatrix A_{ij} in the lower triangle of A. When a partitioning P of a graph G is such that G/P is a tree, we call P a *tree partitioning* of G.

We have now achieved what we set out to do, namely, to determine the characteristics a partitioned matrix must have in order to apply the ideas developed in Section 6.1. The answer is that we want its labelled quotient graph to be a monotonely ordered tree. If it has this property, we can reasonably discard all the off-diagonal blocks of L, saving only its diagonal blocks and the off-diagonal blocks of the

Figure 6.2.6 Illustration of a partitioned matrix, its graph, and its quotient graph which is a tree.

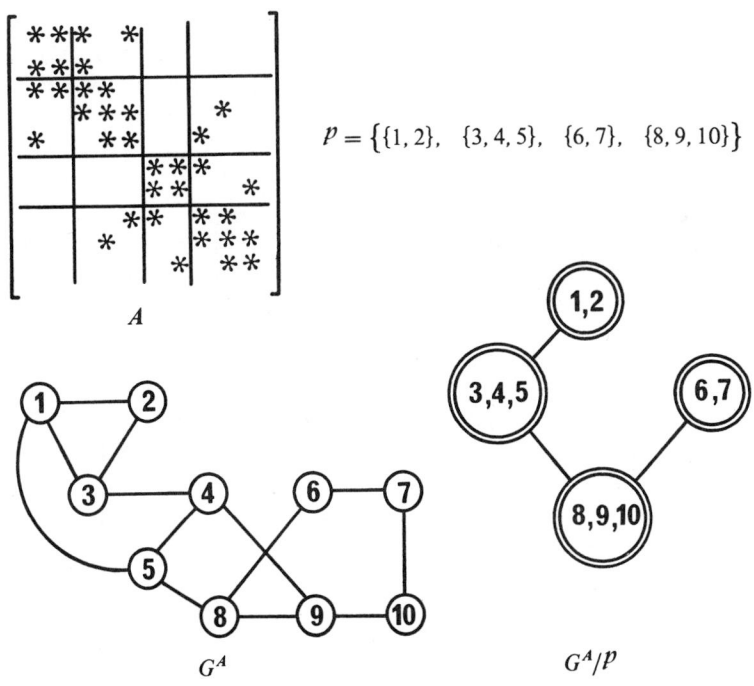

$$p = \{\{1, 2\},\ \{3, 4, 5\},\ \{6, 7\},\ \{8, 9, 10\}\}$$

lower triangle of A.

6.2.3 Asymmetric Block Factorization and Implicit Block Solution of Tree-partitioned Systems

Let A be p by p partitioned with blocks A_{ij}, $1 \leq i,j \leq p$, and L_{ij} be the corresponding blocks of L for $i > j$. If the quotient graph of A is a monotonely ordered tree, there is exactly one non-null block below the diagonal block in A and L (Why?); let this block be $A_{\mu_k, k}$ $1 \leq k \leq p - 1$. The *asymmetric block factorization* algorithm for such problems is as follows.

Step 1   For $k = 1, 2, \ldots, p - 1$ do the following

      1.1)   Factor A_{kk}.

      1.2)   For each column u of A_{k, μ_k}, solve $A_{kk} v = u$, compute $w = A_{k, \mu_k}^T v$ and subtract it from the appropriate column of A_{μ_k,μ_k}.

Step 2   Factor A_{pp}.

Figure 6.2.7   Pictorial illustration of asymmetric block factorization.

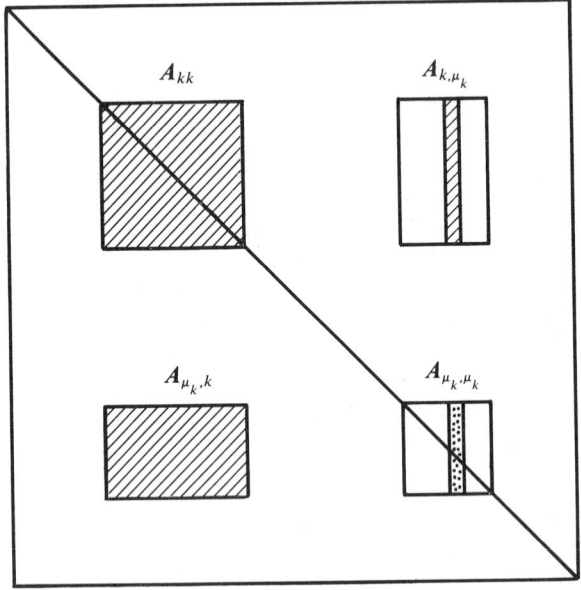

A pictorial illustration of the modification of A_{μ_k,μ_k} is shown in Figure 6.2.7. Note that in the algorithm, temporary storage for the vectors u and w of length equal to the largest block size is all that is needed. (The vector v can overwrite u.) Of course, the symmetry of the diagonal blocks can be exploited.

The implicit solution scheme for such block systems is also straightforward. Here t and $\tilde{t}$ are temporary vectors which can share the same space.

Forward implicit block solve: ($Ly = b$)

Step 1   For $k = 1, \ldots, p - 1$, do the following:

    1.1)   Solve $L_{kk}y_k = b_k$.
    1.2)   Solve $L_{kk}^T t = y_k$.
    1.3)   Compute $b_{\mu_k} \leftarrow b_{\mu_k} - A_{k,\mu_k}^T t$.

Step 2   Solve $L_{pp}y_p = b_p$.

Backward implicit block solve: $(L^T x = y)$

Step 1   Solve $L_{pp}^T x_p = y_p$.
Step 2   For $k = p - 1, \ p - 2, \ldots, 1$ do the following:

2.1)   Compute $t = A_{k, \mu_k} x_{\mu_k}$.
2.2)   Solve $L_{kk} \tilde{t} = t$.
2.3)   Replace $y_k \leftarrow y_k - \tilde{t}$.
2.4)   Solve $L_{kk}^T x_k = y_k$.

Figure 6.2.8 gives the steps on the forward block solve of a block four by four system.

Exercises

6.2.1)   Prove Lemma 6.2.1.

6.2.2)   Let $G^F/P = (P, \mathcal{E}^F)$ be the quotient graph of the filled graph of G with respect to a partitioning P, and let $(G/P)^F = (P, \tilde{\mathcal{E}}^F)$ be the filled graph of G/P. The example in Figure 6.2.4 shows that $\mathcal{E}^F$ may have fewer members than $\tilde{\mathcal{E}}^F$. That is, the block structure of the factor L of a partitioned matrix A may be sparser than the filled graph of the quotient graph G/P would suggest. Show that if the diagonal blocks of L have the propagation property (see Exercise 2.2.3), then the filled graph of G/P will correctly reflect the block structure of L. That is, show that $\mathcal{E}^F = \tilde{\mathcal{E}}^F$ in this case.

6.2.3)   Let $\mathcal{E}^F$ and $\tilde{\mathcal{E}}^F$ be as defined in Exercise 6.2.2, for the labelled graph G, and partitioning $P = \{Y_1, \ Y_2, \ldots, Y_p\}$.

a)   Prove that if the subgraphs $G(Y_i)$, $i = 1, \ 2, \ldots, p$ are connected, then $\mathcal{E}^F = \tilde{\mathcal{E}}^F$.

b)   Give an example to show that the converse need not hold.

c)   Prove   that   if   the   subgraphs $G(\bigcup_{i=1}^{\ell} Y_i)$,
$\ell = 1, \ 2, \ldots, p - 1$ are connected, then $\mathcal{E}^F = \tilde{\mathcal{E}}^F$.

6.2.4)   A tree partitioning $P = \{Y_1, \ Y_2, \ldots, Y_p\}$ of a graph $G = (X, E)$ is *maximal* if there does not exist a tree partitioning $\mathcal{Q} = \{Z_1, \ Z_2, \ldots, Z_t\}$ such that $p < t$, and for each i, $Z_i \subset Y_k$ for some $1 \le k \le p$. In other words, it is maximal if we cannot subdivide one or more of the Y_i's and still

Figure 6.2.8   Forward implicit block solve for a block 4 by 4 system.

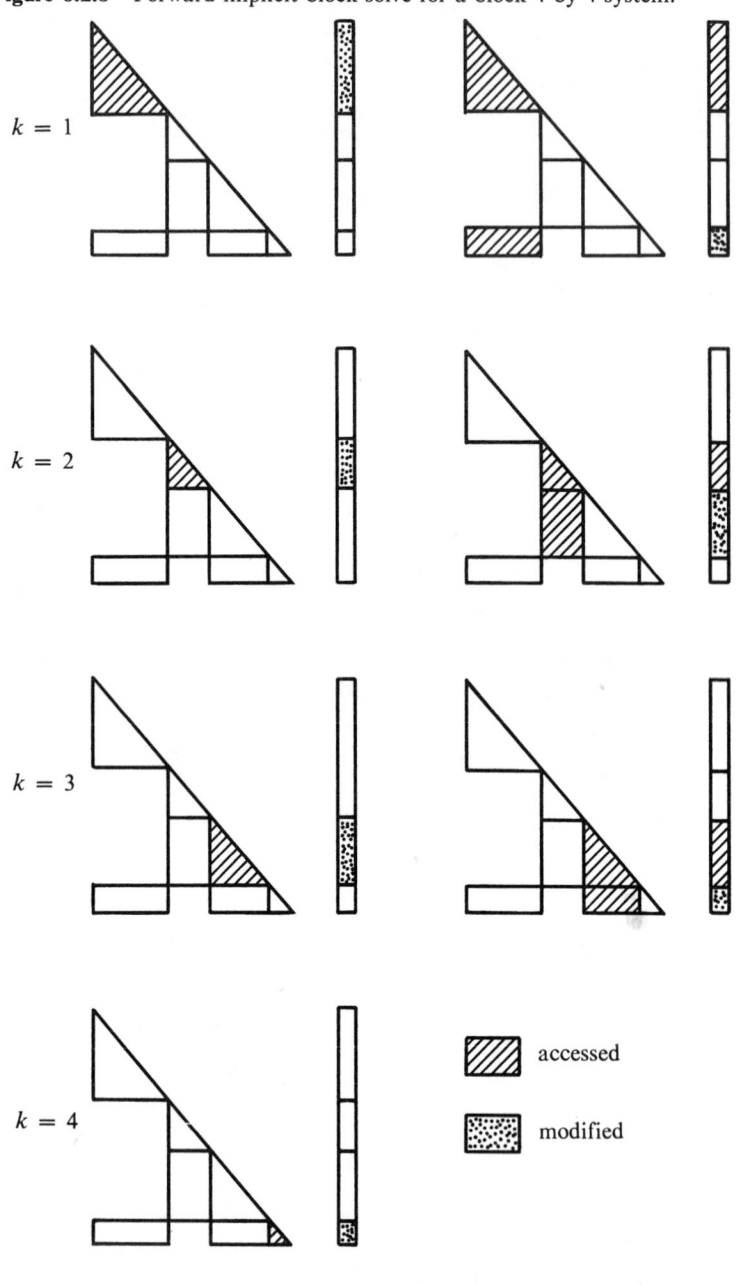

maintain a quotient tree. Suppose that for every pair of distinct nodes in any Y_i, there exist two distinct paths between x and y:

$$x, \; x_1, \; x_2, \ldots , \; x_s, \; y$$

and

$$x, \; y_1, \; y_2, \ldots , \; y_t, \; y$$

such that

$$S = \bigcup \{Z \in \mathcal{P} \mid x_i \in Z, \; 1 \le i \le s\}$$
$$T = \bigcup \{Z \in \mathcal{P} \mid y_i \in Z, \; 1 \le i \le t\}$$

are disjoint. Show that $\mathcal{P}$ is maximal.

6.2.5) Let A be a *block tridiagonal* matrix,

$$A = \begin{pmatrix} A_{11} & V_2 & & & & \\ V_2^T & A_{22} & V_3 & & & \\ & V_3^T & A_{33} & & & \\ & & & \cdot & \cdot & \\ & & & \cdot & \cdot & \cdot \\ & & & & \cdot & \cdot & \cdot \\ & & & & & \cdot & \cdot & V_p \\ & & & & & & V_p^T & A_{pp} \end{pmatrix}$$

where each A_{ii} is an m by m square full matrix and each V_i is a diagonal matrix.

a) What is the arithmetic cost for performing the asymmetric block factorization? How does it compare with that of the symmetric version?

b) What if each submatrix V_i has this sparse form

$$\begin{pmatrix} * & * & \cdots & * & * \\ & & \mathbf{0} & & \end{pmatrix} \; ?$$

Assume m is large, and ignore low order terms in your calculations.

6.3   A Quotient Tree Partitioning Algorithm

6.3.1   A Heuristic Algorithm

The results of Section 6.2 suggest that we would like to find a partitioning $P = \{Y_1, Y_2, \ldots, Y_p\}$ with as many members as possible, consistent with the requirement that G/P be a tree. In this section we provide an algorithm for finding a tree partitioning of a graph.

The algorithm we propose is closely related to level structures (see Section 4.3), so we begin by observing the connection between a level structure on a graph and the partitioning it induces on the corresponding matrix. Let G^A be the unlabelled graph associated with A, and let $\mathcal{L} = \{L_0, L_1, \ldots, L_\ell\}$ be a level structure in G^A. From the definition of level structure, it is clear that the quotient graph $G/\mathcal{L}$ is a chain, so if we number the nodes in each level L_i consecutively, from L_0 to L_ℓ, the levels of $\mathcal{L}$ induce a block tridiagonal structure on the correspondingly ordered matrix. An example appears in Figure 6.3.1.

Figure 6.3.1   Block tridiagonal partitioning induced by a level structure.

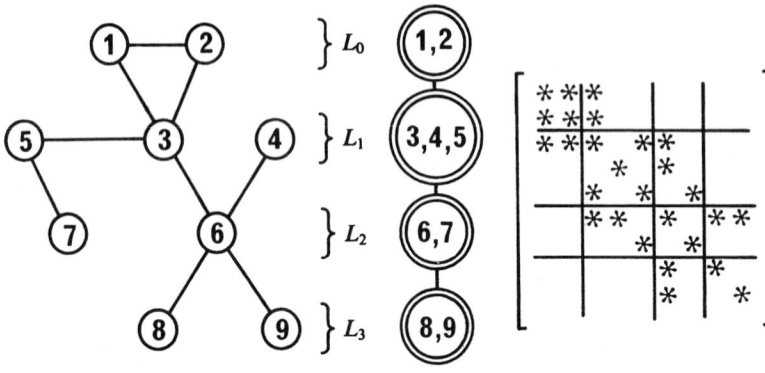

The algorithm we will ultimately present in this section begins with a rooted level structure and then attempts to make the partitioning finer by refining the levels of the level structure. Let $\mathcal{L} = \{L_0, L_1, \ldots, L_\ell\}$ be a rooted level structure and let $P = \{Y_1, Y_2, \ldots, Y_p\}$ be the partitioning obtained by subdividing each L_j as follows. Letting B_j be the section graph prescribed by

$$B_j = G\left(\bigcup_{i=j}^{\ell} L_i\right), \qquad (6.3.1)$$

each L_j is partitioned according to the sets specified by

$$\{Y \mid Y = L_j \cap C, \; G(C) \text{ is a connected component of } B_j\} \; . \quad (6.3.2)$$

Figure 6.3.2 illustrates this refinement for a simple graph.

Consider the refinement on $L_2 = \{d,e\}$. Note that

$$B_2 = G(\{d,e,i,g,f,h\})$$

and it has two connected components with node sets:

$$\{d,i,g\}$$

and

$$\{e,f,h\} \; .$$

Therefore, the level L_2 can be partitioned according to (6.3.2) into $\{d\}$ and $\{e\}$.

Figure 6.3.2   An example showing the refined partitioning P obtained from the level structure $\mathcal{L}$.

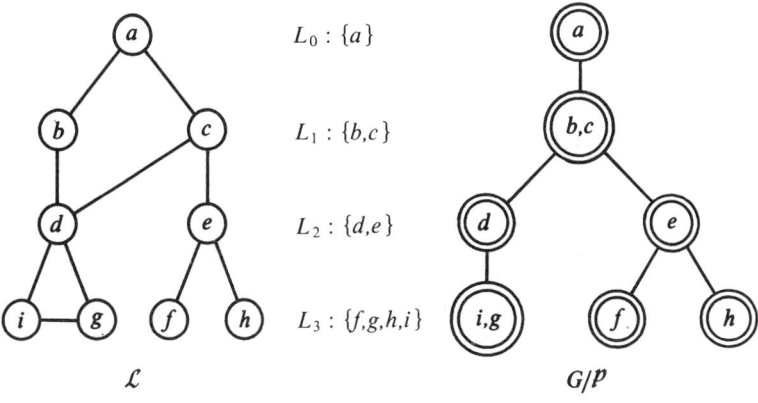

$L_0 : \{a\}$

$L_1 : \{b,c\}$

$L_2 : \{d,e\}$

$L_3 : \{f,g,h,i\}$

$\mathcal{L}$

G/P

We are now ready to describe the algorithm for finding a tree partitioning. Our description makes use of the definition $SPAN(Y)$, which is defined for a subset Y of X by

$$SPAN(Y) = \{x \in X \mid \text{there exists a path from } y$$

$$\text{to } x, \text{ for some } y \in Y\} \; . \quad (6.3.3)$$

When Y is a single node y, $SPAN(Y)$ is simply the connected component containing y.

The algorithm we now describe makes use of a stack, and thereby avoids explicitly finding the connected components of the B_j which appeared in the description of the level refinement (6.3.2). We assume that we are given a root r for the rooted level structure to be refined.

We discuss the choice of r later in this section.

Step 0   (Initialization): Empty the stack. Generate the level structure $\mathcal{L}(r) = \{L_0, L_1, L_2, \ldots, L_{\ell(r)}\}$ rooted at r, and choose any node $y \in L_{\ell(r)}$. Set $\ell = \ell(r)$ and $S = \{y\}$.

Step 1   (Pop stack): If the node set T on the top of the stack belongs to L_ℓ, pop T and set $S \leftarrow S \cup T$.

Step 2   (Form possible partition member): Determine the set $Y = SPAN(S)$ in the subgraph $G(L_\ell)$. If $\ell < \ell(r)$ and some node in $Adj(Y) \cap L_{\ell+1}$ has not yet been placed in a partition member, go to Step 5.

Step 3   (New partition member): Put Y in $\mathcal{P}$.

Step 4   (Next level): Determine the set $S = Adj(Y) \cap L_{\ell-1}$, and set $\ell \leftarrow \ell - 1$. If $\ell \geq 0$, go to Step 1, otherwise stop.

Step 5   (Partially formed partition member): Push S onto the stack. Pick $y_{\ell+1} \in Adj(Y) \cap L_{\ell+1}$ and trace a path $y_{\ell+1}$, $y_{\ell+2}, \ldots, y_{\ell+t}$, where $y_{\ell+i} \in L_{\ell+i}$ and $Adj(y_{\ell+t}) \cap L_{\ell+t+1} = \emptyset$. Set $S = \{y_{\ell+t}\}$ and $\ell \leftarrow \ell + t$, and then go to Step 2.

The example in Figure 6.3.4, taken from George and Liu (1978c), illustrates how the algorithm operates. The level structure rooted at node 1 is refined to obtain a quotient tree having 10 nodes. In the example, $Y_1 = \{20\}$, $Y_2 = \{18, 19\}$, $Y_3 = \{16\}$, $Y_4 = \{10, 15\}$, $Y_5 = \{9, 14, 17\}$, and $Y_6 = \{5, 11\}$, with $L_4 = Y_5 \cup Y_6$, $L_5 = Y_2 \cup Y_4$, and $L_6 = Y_1 \cup Y_3$.

In order to complete the description of the tree partitioning algorithm we must specify how to find the root node r for the level structure. We obtain r and $\mathcal{L}(r)$ by using the subroutine FNROOT, described in Section 4.3.3. Since we want a partitioning with as many members as possible, this seems to be a sensible choice since it will tend to provide a level structure having relatively many levels.

6.3.2 Subroutines for Finding a Quotient Tree Partitioning

In this section, a set of subroutines which implements the quotient tree algorithm is discussed. The parameters NEQNS, XADJ and ADJNCY, as before, are used to store the adjacency structure of the given graph. The vector PERM returns the computed quotient tree ordering. In addition to the ordering, the partitioning information is returned in the variable NBLKS and the vector XBLK. The number of partition blocks is given in NBLKS, while the node numbers of a particular block, say block k, are given by

$$\{\text{PERM}(j) \mid \text{XBLK}(k) \leq j < \text{XBLK}(k+1)\}$$

Figure 6.3.3 A graph, rooted level structure, and refined quotient tree.

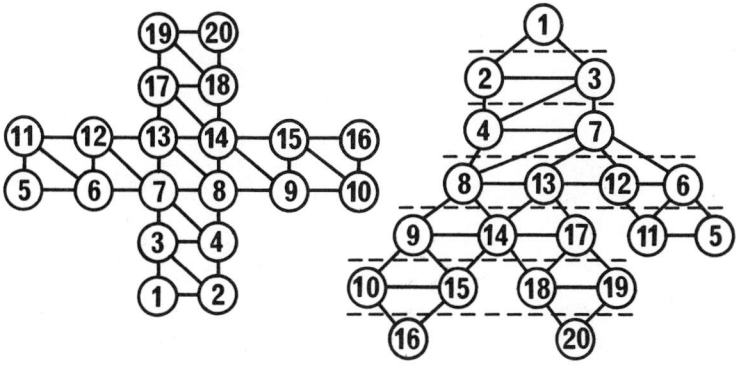

A '+' shaped graph   Rooted level structure $\mathcal{L}(1)$

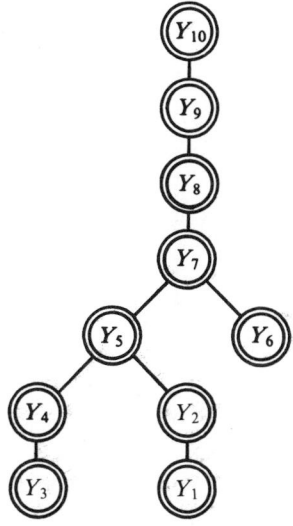

Figure 6.3.5 contains the representation of the quotient tree ordering for the example in Figure 6.3.3.

As we see from the example, the vector XBLK has size NBLKS+1. The last extra pointer is included so that blocks can be retrieved in a uniform manner. In the example, to obtain the fifth block, we note that

$$XBLK(5) = 7 ,$$

Figure 6.3.4 Numbering in the refined quotient tree algorithm.

k	Partitioning	Y_k	Level	Adjacent Set	Stack
1		$\{20\}$	6	$\{18, 19\}$	$\emptyset$
2		$\{18, 19\}$	5	$\{14, 17\}$	$\emptyset$
3	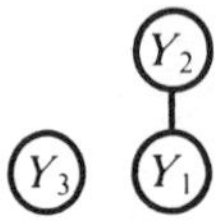	$\{16\}$	6	$\{10, 15\}$	$\{14, 17\}$
4	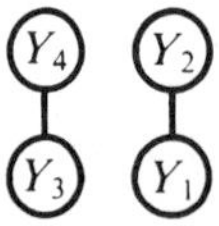	$\{10, 15\}$	5	$\{9, 14\}$	$\{14, 17\}$
5	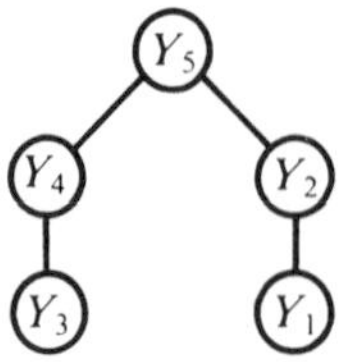	$\{9, 14, 17\}$	4	$\{8, 13\}$	$\emptyset$
6	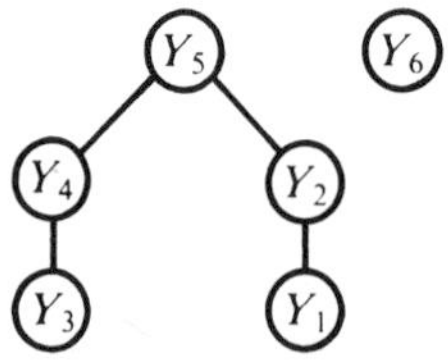	$\{5, 11\}$	4	$\{6, 12\}$	$\{8, 13\}$
7	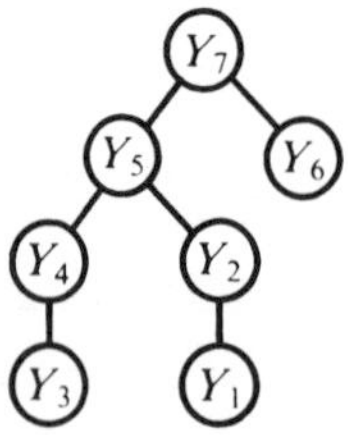	$\{8, 13, 12, 6\}$	3	$\{4, 7\}$	$\emptyset$

Figure 6.3.4 (continued)

k	Partitioning	Y_k	Level	Adjacent Set	Stack
8	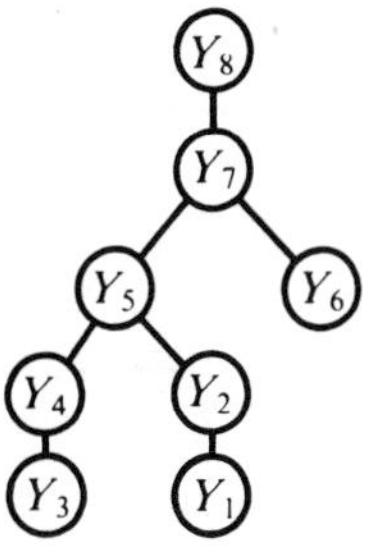	{4, 7}	2	{2, 3}	∅
9	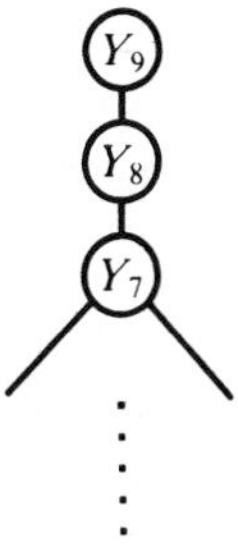 etc.	{2, 3}	1	{1}	∅
10	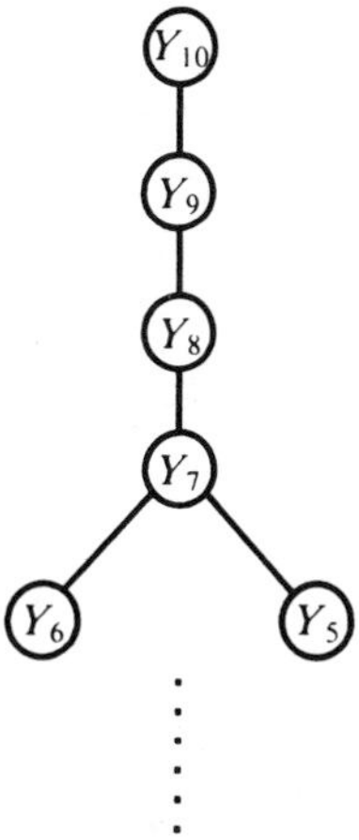 etc.	{1}	0	∅	∅

Figure 6.3.5 An example of the data structure for a partitioning.

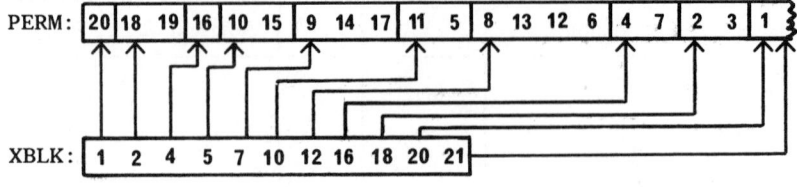

NBLKS : **10**

$$XBLK(6) = 10 \ .$$

Thus, the nodes in this block are given by PERM(7), PERM(8) and PERM(9).

There are seven subroutines in this set, two of which have been considered in detail in Chapter 4. We first consider their control relationship as shown in Figure 6.3.6.

Figure 6.3.6 Control relation of subroutines for the refined quotient tree algorithm.

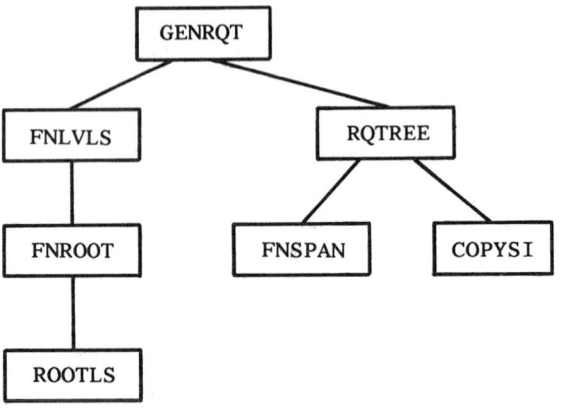

The subroutines FNROOT and ROOTLS are used to determine a pseudo-peripheral node of a connected component of a given graph. For details of these two subroutines, readers are referred back to Section 4.3.3. The subroutine COPYSI is a simple utility program that copies an integer array into another one. (A listing of the subroutine appears after that of RQTREE.) We now describe in detail the remaining subroutines in this group.

GENRQT (GENeral Refined Quotient Tree)

This subroutine is the driver subroutine for finding the quotient tree ordering of a general disconnected graph. It goes through the graph and calls the subroutine RQTREE to number each connected component in the graph. It requires three working arrays XLS, LS and NODLVL. The array pair (XLS, LS) is used by FNLVLS to obtain a level structure rooted at a pseudo-peripheral node, while the vector NODLVL is used to store the level number of nodes in the level structure.

The subroutine begins by initializing the vector NODLVL and the variable NBLKS. It then goes through the graph until it finds a node i not yet numbered. Note that numbered nodes have their NODLVL values set to zero. This node i together with the array NODLVL defines a connected subgraph of the original graph. The subroutines FNLVLS and RQTREE are then called to order the nodes of this subgraph. The subroutine returns after it has processed all the components of the graph.

```
C***************************************************************
C***************************************************************
C******      GENRQT ..... GENERAL REFINED QUOTIENT TREE   *****
C***************************************************************
C***************************************************************
C
C       PURPOSE - THIS ROUTINE IS A DRIVER FOR DETERMINING A
C          PARTITIONED ORDERING FOR A POSSIBLY DISCONNECTED
C          GRAPH USING THE REFINED QUOTIENT TREE ALGORITHM.
C
C       INPUT PARAMETERS -
C          NEQNS - NUMBER OF VARIABLES.
C          (XADJ, ADJNCY) - THE ADJACENCY STRUCTURE.
C
C       OUTPUT PARAMETERS -
C          (NBLKS, XBLK) - THE QUOTIENT TREE PARTITIONING.
C          PERM - THE PERMUTATION VECTOR.
C
C       WORKING PARAMETERS -
C          (XLS, LS) - THIS LEVEL STRUCTURE PAIR IS USED BY
C             FNROOT TO FIND A PSEUDO-PERIPHERAL NODE.
C          NODLVL - A TEMPORARY VECTOR TO STORE THE LEVEL
C             NUMBER OF EACH NODE IN A LEVEL STRUCTURE.
C
C       PROGRAM SUBROUTINES -
C          FNLVLS, RQTREE.
C
C***************************************************************
C
        SUBROUTINE  GENRQT ( NEQNS, XADJ, ADJNCY, NBLKS, XBLK,
     1                       PERM, XLS, LS, NODLVL )
C
C***************************************************************
C
        INTEGER ADJNCY(1), LS(1), NODLVL(1), PERM(1),
     1          XBLK(1), XLS(1)
        INTEGER XADJ(1), I, IXLS, LEAF, NBLKS, NEQNS, NLVL,
     1          ROOT
C
C***************************************************************
C
```

```
C          ------------------
C          INITIALIZATION ...
C          ------------------
           DO 100 I = 1, NEQNS
              NODLVL(I) = 1
   100     CONTINUE
           NBLKS = 0
           XBLK(1) = 1
C          -----------------------------------------------------
C          FOR EACH CONNECTED COMPONENT, FIND A ROOTED LEVEL
C          STRUCTURE, AND THEN CALL RQTREE FOR ITS BLOCK ORDER.
C          -----------------------------------------------------
           DO 200 I = 1, NEQNS
              IF (NODLVL(I) .LE. 0) GO TO 200
              ROOT = I
              CALL   FNLVLS ( ROOT, XADJ, ADJNCY, NODLVL,
     1                        NLVL, XLS, LS )
              IXLS = XLS(NLVL)
              LEAF = LS(IXLS)
              CALL   RQTREE ( LEAF, XADJ, ADJNCY, PERM,
     1                        NBLKS, XBLK, NODLVL, XLS, LS )
   200     CONTINUE
           RETURN
        END
```

FNLVLS (FiNd LeVeL Structure)

This subroutine FNLVLS generates a rooted level structure for a component, specified by NODLVL and rooted at a pseudo-peripheral node. In addition, it also records the level number of the nodes in the level structure.

The connected component is specified by the input parameters ROOT, XADJ, ADJNCY and NODLVL. The subroutine first calls the subroutine FNROOT to obtain the required rooted level structure, given by (NLVL, XLS, LS). It then loops through the level structure to determine the level numbers and puts them into NODLVL (loop DO 200 LVL = ...).

```
C*****************************************************************
C*****************************************************************
C******    FNLVLS ..... FIND LEVEL STRUCTURE      *******
C*****************************************************************
C*****************************************************************
C
C    PURPOSE - FNLVLS GENERATES A ROOTED LEVEL STRUCTURE FOR
C       A MASKED CONNECTED SUBGRAPH, ROOTED AT A PSEUDO-
C       PERIPHERAL NODE.   THE LEVEL NUMBERS ARE RECORDED.
C
C    INPUT PARAMETERS -
C       (XADJ, ADJNCY) - THE ADJACENCY STRUCTURE.
C
C    OUTPUT PARAMETERS -
C       NLVL - NUMBER OF LEVELS IN THE LEVEL STRUCTURE FOUND.
C       (XLS, LS) - THE LEVEL STRUCTURE RETURNED.
C
C    UPDATED PARAMETERS -
C       ROOT - ON INPUT, WITH THE ARRAY NODLVL, SPECIFIES
C              THE COMPONENT WHOSE PSEUDO-PERIPHERAL NODE IS
C              TO BE FOUND. ON OUTPUT, IT CONTAINS THAT NODE.
C       NODLVL - ON INPUT, IT SPECIFIES A SECTION SUBGRAPH.
```

```
C                   ON RETURN, IT CONTAINS THE NODE LEVEL NUMBERS.
C
C       PROGRAM SUBROUTINES -
C          FNROOT.
C
C*********************************************************************
C
        SUBROUTINE  FNLVLS ( ROOT, XADJ, ADJNCY, NODLVL,
     1                       NLVL, XLS, LS )
C
C*********************************************************************
C
        INTEGER ADJNCY(1), LS(1), NODLVL(1), XLS(1)
        INTEGER XADJ(1), J, LBEGIN, LVL, LVLEND, NLVL,
     1          NODE, ROOT
C
C*********************************************************************
C
        CALL  FNROOT ( ROOT, XADJ, ADJNCY, NODLVL,
     1                 NLVL, XLS, LS )
        DO 200 LVL = 1, NLVL
           LBEGIN = XLS(LVL)
           LVLEND = XLS(LVL + 1) - 1
           DO 100 J = LBEGIN, LVLEND
              NODE = LS(J)
              NODLVL(NODE) = LVL
100        CONTINUE
200     CONTINUE
        RETURN
        END
```

RQTREE (Refined Quotient TREE)

This is the subroutine that actually applies the quotient tree algorithm as described in Section 6.3.1. Throughout the procedure, it maintains a stack of node subsets. Before going into the details of the subroutine, we first consider the organization of the stack.

For each node subset in the stack, we need to store its size and the level number of its nodes. In the storage array called STACK, we store the nodes in the subset in contiguous locations, and then the subset size and the level number in the next two locations. We also keep a variable TOPSTK that stores the current number of locations used in STACK. Figure 6.3.7 contains an illustration of the organization of the vector STACK.

To push a subset S of level i into the stack, we simply copy the nodes in S into the vector STACK starting at location TOPSTK+1. We then enter the size $|S|$ and the level number i and finally update the value of TOPSTK. On the other hand, to pop a node subset from the stack, we first obtain the size of the subset from STACK(TOPSTK-1) and then the subset can be retrieved from STACK starting at TOPSTK-size-1. The value of TOPSTK is also updated to reflect the current status of the stack.

We now consider the details of the subroutine RQTREE. It operates on a connected subgraph as specified by LEAF, XADJ, ADJNCY and NODLVL. It implicitly assumes that a level structure has been formed on this component, where NODLVL contains the level numbers for its nodes and LEAF is a *leaf node* in the level structure. In a level structure $\mathcal{L} = \{L_0, L_1, \ldots, L_\ell\}$, a node x is said to be a *leaf* in $\mathcal{L}$ if $Adj(x) \cap L_{i+1} = \emptyset$ where $x \in L_i$.

Figure 6.3.7   Organization of the stack in the subroutine RQTREE.

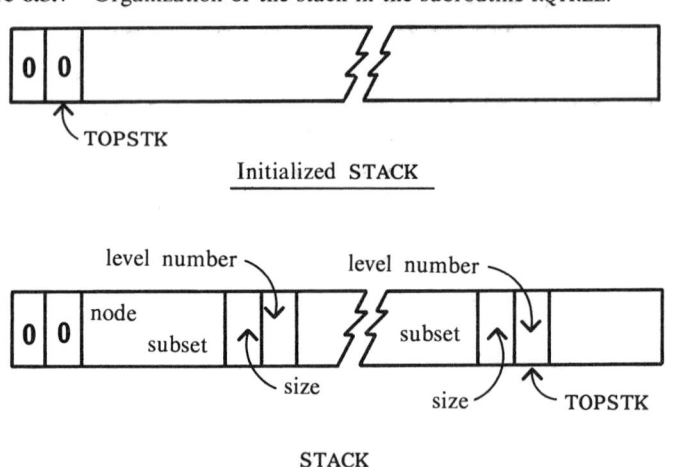

In addition to STACK, the subroutine uses a second working vector ADJS. This vector is used to store the adjacent set of the current block in the lower level, and it is a potential subset for the next block.

The subroutine starts by initializing the STACK vector, its pointer TOPSTK and the variable TOPLVL. The variable TOPLVL is local to the subroutine and it stores the level number of the top subset in the stack. A leaf block is then determined by calling the subroutine FNSPAN on the node LEAF. (A *leaf block* is a subset Y such that $Adj(Y) \cap L_{i+1} = \emptyset$ where $Y \subset L_i$.) It is numbered as the next block (statement labelled 300).

We then march onto the next lower level (LEVEL = LEVEL − 1 and following). The adjacent set of the previous block in this level is used to start building up the next potential block. If the node subset at the top of the STACK vector belongs to the same level, it is popped from the stack and included into the potential block. Then, the subroutine FNSPAN is called (statement labelled 400) to obtain the span of this subset. If the span does not have any unnumbered neighbors in the higher level, it becomes the next block to be numbered. Otherwise, the span is pushed into the stack and instead a leaf block is determined as the next one to be numbered.

The subroutine goes through all the levels until it comes to the first one.   By this time, all the nodes in the component should have been numbered and the subroutine returns.

```
C**************************************************************
C**************************************************************
C*********·    RQTREE ..... REFINED QUOTIENT TREE    ******
C**************************************************************
C**************************************************************
C
C     PURPOSE - THIS SUBROUTINE FINDS A QUOTIENT TREE ORDERING
C        FOR THE COMPONENT SPECIFIED BY LEAF AND NODLVL.
C
C     INPUT PARAMETERS -
C        (XADJ, ADJNCY) - THE ADJACENCY STRUCTURE.
C        LEAF - THE INPUT NODE THAT DEFINES THE CONNECTED
C               COMPONENT.  IT IS ALSO A LEAF NODE IN THE
C               ROOTED LEVEL STRUCTURE PASSED TO RQTREE.
C               I.E. IT HAS NO NEIGHBOR IN THE NEXT LEVEL.
C
C     OUTPUT PARAMETERS -
C        PERM - THE PERMUTATION VECTOR CONTAINING THE ORDERING.
C        (NBLKS, XBLK) - THE QUOTIENT TREE PARTITIONING.
C
C     UPDATED PARAMETERS -
C        NODLVL - THE NODE LEVEL NUMBER VECTOR.  NODES IN THE
C               COMPONENT HAVE THEIR NODLVL SET TO ZERO AS
C               AS THEY ARE NUMBERED.
C
C     WORKING PARAMETERS -
C        ADJS - TEMPORARY VECTOR TO STORE THE ADJACENT SET
C               OF NODES IN A PARTICULAR LEVEL.
C        STACK - TEMPORARY VECTOR USED TO MAINTAIN THE STACK
C               OF NODE SUBSETS.  IT IS ORGANISED AS -
C               ( SUBSET NODES, SUBSET SIZE, SUBSET LEVEL )
C
C     PROGRAM SUBROUTINES -
C        FNSPAN, COPYSI.
C
C**************************************************************
C
      SUBROUTINE  RQTREE ( LEAF, XADJ, ADJNCY, PERM,
     1                     NBLKS, XBLK, NODLVL, ADJS, STACK )
C
C**************************************************************
C
      INTEGER ADJNCY(1), ADJS(1), NODLVL(1), PERM(1),
     1        STACK(1), XBLK(1)
      INTEGER XADJ(1), BLKSZE, IP, J, JP, LEAF, LEVEL,
     1        NADJS, NBLKS, NODE, NPOP, NULEAF,
     1        NUM, TOPLVL, TOPSTK
C
C**************************************************************
C
C        ---------------------------------------------
C        INITIALIZE THE STACK VECTOR AND ITS POINTERS.
C        ---------------------------------------------
      STACK(1) = 0
      STACK(2) = 0
      TOPSTK = 2
      TOPLVL = 0
      NUM = XBLK(NBLKS+1) - 1
C        ---------------------------------------------
C        FORM A LEAF BLOCK, THAT IS, ONE WITH NO NEIGHBORS
C        IN ITS NEXT HIGHER LEVEL.
```

```
C            ------------------------------------------------------
   100       LEVEL = NODLVL(LEAF)
             NODLVL(LEAF) = 0
             PERM(NUM+1) = LEAF
             BLKSZE = 1
             CALL   FNSPAN ( XADJ, ADJNCY, NODLVL, BLKSZE, PERM(NUM+1),
        1                    LEVEL, NADJS, ADJS, NULEAF )
             IF ( NULEAF .LE. 0 )  GO TO 300
                JP = NUM
                DO 200 J = 1, BLKSZE
                   JP = JP + 1
                   NODE = PERM(JP)
                   NODLVL(NODE) = LEVEL
   200          CONTINUE
             LEAF = NULEAF
             GO TO 100
C            ------------------------------
C            A NEW BLOCK HAS BEEN FOUND ...
C            ------------------------------
   300       NBLKS = NBLKS + 1
             XBLK(NBLKS) = NUM + 1
             NUM = NUM + BLKSZE
C            ------------------------------------------------------
C            FIND THE NEXT POSSIBLE BLOCK BY USING THE ADJACENT
C            SET IN THE LOWER LEVEL AND THE TOP NODE SUBSET (IF
C            APPROPRIATE) IN THE STACK.
C            ------------------------------------------------------
             LEVEL = LEVEL - 1
             IF ( LEVEL .LE. 0 )  GO TO 500
             CALL   COPYSI ( NADJS, ADJS, PERM(NUM+1) )
             BLKSZE = NADJS
             IF ( LEVEL .NE. TOPLVL )  GO TO 400
C               ---------------------------------------------------
C               THE LEVEL OF THE NODE SUBSET AT THE TOP OF THE
C               STACK IS THE SAME AS THAT OF THE ADJACENT SET.
C               POP THE NODE SUBSET FROM THE STACK.
C               ---------------------------------------------------
                NPOP = STACK(TOPSTK-1)
                TOPSTK = TOPSTK - NPOP - 2
                IP = NUM + BLKSZE + 1
                CALL   COPYSI ( NPOP, STACK(TOPSTK+1), PERM(IP) )
                BLKSZE = BLKSZE + NPOP
                TOPLVL = STACK(TOPSTK)
   400          CALL   FNSPAN ( XADJ, ADJNCY, NODLVL, BLKSZE,
        1                       PERM(NUM+1), LEVEL, NADJS, ADJS, NULEAF )
             IF ( NULEAF .LE. 0 )  GO TO 300
C            -----------------------------------------
C            PUSH THE CURRENT NODE SET INTO THE STACK.
C            -----------------------------------------
             CALL   COPYSI ( BLKSZE, PERM(NUM+1), STACK(TOPSTK+1) )
             TOPSTK = TOPSTK + BLKSZE + 2
             STACK(TOPSTK-1) = BLKSZE
             STACK(TOPSTK) = LEVEL
             TOPLVL = LEVEL
             LEAF   = NULEAF
             GO TO 100
C            ---------------
C            BEFORE EXIT ...
C            ---------------
   500       XBLK(NBLKS+1) = NUM + 1
             RETURN
          END


C******************************************************************
C******************************************************************
C**********       COPYSI ..... COPY INTEGER VECTOR       *********
C******************************************************************
```

```
C**************************************************************
C
C      PURPOSE - THIS ROUTINE COPIES THE N INTEGER ELEMENTS FROM
C         THE VECTOR A TO B. (ARRAYS OF SHORT INTEGERS)
C
C      INPUT PARAMETERS -
C         N - SIZE OF VECTOR A.
C         A - THE INTEGER VECTOR.
C
C      OUTPUT PARAMETER -
C         B - THE OUTPUT INTEGER VECTOR.
C
C**************************************************************
C
       SUBROUTINE  COPYSI ( N, A, B )
C
C**************************************************************
C
       INTEGER A(1), B(1)
       INTEGER I, N
C
C**************************************************************
C
       IF ( N .LE. 0 ) RETURN
       DO 100 I = 1, N
          B(I) = A(I)
  100     CONTINUE
       RETURN
       END
```

FNSPAN (FiNd SPAN)

This subroutine is used by the subroutine RQTREE and has several functions, one of which is to find the span of a set. Let $\mathcal{L} = \{L_0, L_1, \ldots, L_\ell\}$ be a given level structure and let S be a subset in level L_i. This subroutine determines the span of S in the subgraph $G(L_i)$ and finds the adjacent set of S in level L_{i-1}. Moreover, if the span of S has some unnumbered neighbors in level L_{i+1}, the subroutine returns an unnumbered leaf node and in that case, the span of S may only be partially formed.

Inputs to this subroutine are the graph structure in the array pair (XADJ, ADJNCY), the level structure stored implicitly in the vector NODLVL, and the subset (NSPAN, SET) in level LEVEL of the level structure. On return, the vector SET is expanded to accommodate the span of this given set. The variable NSPAN will be increased to the size of the span set.

After initialization, the subroutine goes through each node in the partially spanned set. Here, the variable SETPTR points to the current node in the span set under consideration. The loop DO 500 J = ... is then executed to inspect the level numbers of its neighbors. Depending on the level number, the neighbor is either bypassed or included in the span set or included in the adjacent set. A final possibility is when the neighbor belongs to a higher level. In this case, a path through unnumbered nodes is traced down the level structure until we hit a leaf node. The subroutine returns after recovering the nodes in the partially formed adjacent set (loop DO 900 I = ...).

A normal return from FNSPAN will have the span set in (NSPAN, SET) and the adjacent set in (NADJS, ADJS) completely formed, and have zero in the variable LEAF.

```
C************************************************************
C************************************************************
C*********       FNSPAN ..... FIND SPAN SET      **********
C************************************************************
C************************************************************
C
C     PURPOSE - THIS SUBROUTINE IS ONLY USED BY RQTREE.  ITS
C        MAIN PURPOSE IS TO FIND THE SPAN OF A GIVEN SUBSET
C        IN A GIVEN LEVEL SUBGRAPH IN A LEVEL STRUCTURE.
C        THE ADJACENT SET OF THE SPAN IN THE LOWER LEVEL IS
C        ALSO DETERMINED.  IF THE SPAN HAS AN UNNUMBERED NODE
C        IN THE HIGHER LEVEL, AN UNNUMBERED LEAF NODE (I.E. ONE
C        WITH NO NEIGHBOR IN NEXT LEVEL) WILL BE RETURNED.
C
C     INPUT PARAMETERS -
C        (XADJ, ADJNCY) - THE ADJACENT STRUCTURE.
C        LEVEL - LEVEL NUMBER OF THE CURRENT SET.
C
C     UPDATED PARAMETERS -
C        (NSPAN, SET) - THE INPUT SET.  ON RETURN, IT CONTAINS
C              THE RESULTING SPAN SET.
C        NODLVL - THE LEVEL NUMBER VECTOR.  NODES CONSIDERED
C              WILL HAVE THEIR NODLVL CHANGED TO ZERO.
C
C     OUTPUT PARAMETERS -
C        (NADJS, ADJS) - THE ADJACENT SET OF THE SPAN IN THE
C              LOWER LEVEL.
C        LEAF - IF THE SPAN HAS AN UNNUMBERED HIGHER LEVEL NODE,
C              LEAF RETURNS AN UNNUMBERED LEAF NODE IN THE LEVEL
C              STRUCTURE, OTHERWISE, LEAF IS ZERO.
C
C
C************************************************************
C
        SUBROUTINE  FNSPAN ( XADJ, ADJNCY, NODLVL, NSPAN, SET,
     1                       LEVEL, NADJS, ADJS, LEAF )
C
C************************************************************
C
        INTEGER ADJNCY(1), ADJS(1), NODLVL(1), SET(1)
        INTEGER XADJ(1), I, J, JSTOP, JSTRT, LEAF, LEVEL,
     1          LVL, LVLM1, NADJS, NBR, NBRLVL, NODE,
     1          NSPAN, SETPTR
C
C************************************************************
C
C        -----------------
C        INITIALIZATION ...
C        -----------------
         LEAF = 0
         NADJS = 0
         SETPTR = 0
  100    SETPTR = SETPTR + 1
         IF ( SETPTR .GT. NSPAN ) RETURN
C        -----------------------------------------------
C        FOR EACH NODE IN THE PARTIALLY SPANNED SET ...
C        -----------------------------------------------
         NODE = SET(SETPTR)
         JSTRT = XADJ(NODE)
```

```
                JSTOP = XADJ(NODE + 1) - 1
                IF ( JSTOP .LT. JSTRT ) GO TO 100
C               --------------------------------------------------------
C               FOR EACH NEIGHBOR OF NODE, TEST ITS NODLVL VALUE
C               --------------------------------------------------------
                DO 500 J = JSTRT, JSTOP
                    NBR = ADJNCY(J)
                    NBRLVL = NODLVL(NBR)
                    IF (NBRLVL .LE. 0) GO TO 500
                    IF (NBRLVL - LEVEL) 200, 300, 600
C               --------------------------------------
C               NBR IS IN LEVEL-1, ADD IT TO ADJS.
C               --------------------------------------
  200               NADJS = NADJS + 1
                    ADJS(NADJS) = NBR
                    GO TO 400
C               ------------------------------------------
C               NBR IS IN LEVEL, ADD IT TO THE SPAN SET.
C               ------------------------------------------
  300               NSPAN = NSPAN + 1
                    SET(NSPAN) = NBR
  400               NODLVL(NBR) = 0
  500           CONTINUE
                GO TO 100
C               --------------------------------------------------------
C               NBR IS IN LEVEL+1. FIND AN UNNUMBERED LEAF NODE BY
C               TRACING A PATH UP THE LEVEL STRUCTURE.    THEN
C               RESET THE NODLVL VALUES OF NODES IN ADJS.
C               --------------------------------------------------------
  600           LEAF = NBR
                LVL  = LEVEL + 1
  700           JSTRT = XADJ(LEAF)
                JSTOP = XADJ(LEAF+1) - 1
                DO 800 J = JSTRT, JSTOP
                    NBR = ADJNCY(J)
                    IF ( NODLVL(NBR) .LE. LVL ) GO TO 800
                    LEAF = NBR
                    LVL  = LVL + 1
                    GO TO 700
  800           CONTINUE
                IF (NADJS .LE. 0) RETURN
                LVLM1 = LEVEL - 1
                DO 900 I = 1, NADJS
                    NODE = ADJS(I)
                    NODLVL(NODE) = LVLM1
  900           CONTINUE
                RETURN
            END
```

Exercises

6.3.1) Let $P = \{Y_1, Y_2, \ldots, Y_p\}$ be the partitioning of G generated by the algorithm described in Section 6.3.1.

   a) Show that G/P is a quotient tree.

   b) Prove that the quotient tree generated by the algorithm of Section 6.3.1 is maximal, as defined in Exercise 6.2.4. (Hint: Let $Y \in P$ and $Y \subset L_j(r)$, where $L_j(r)$ is defined in Section 6.3.1. Show that for any two nodes x and y

in Y, there exists a path joining them in $G(\bigcup_{i=0}^{j-1} L_i(r))$
and one in $G(\bigcup_{i=j}^{\ell} L_i(r))$. Then use the result of Exercise 6.2.4.)

6.4   A Storage Scheme and Storage Allocation Procedure

In this section we describe a storage scheme which is specially designed for solving partitioned matrix problems whose quotient graphs are monotonely ordered trees. The assumption is that all the off-diagonal blocks of the triangular factor L are to be discarded in favor of the blocks of the original matrix A. In other words, the implicit solution scheme described at the end of Section 6.2.3 is to be used.

6.4.1   The Storage Scheme

For illustrative purposes we again assume A is partitioned into p^2 submatrices A_{ij}, $1 \leq i,j \leq p$, and let L_{ij} be the corresponding submatrices of L, where $A = LL^T$. Since we do not know whether A will have the form

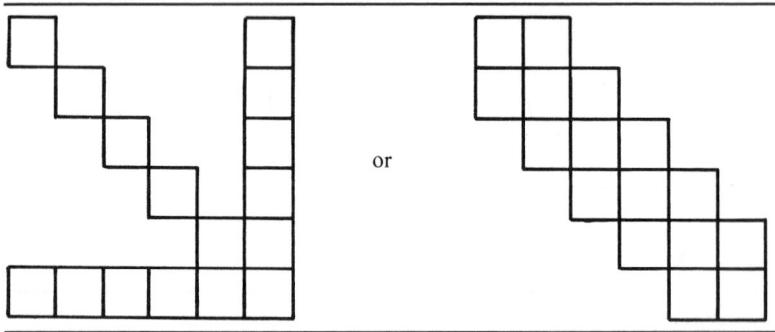

or

(which correspond to quite different quotient trees), or something in between, our storage scheme must be quite flexible. Define the matrices

$$V_k = \begin{pmatrix} A_{1k} \\ A_{2k} \\ \cdot \\ \cdot \\ \cdot \\ A_{k-1,k} \end{pmatrix}, \ 2 \le k \le p \ . \qquad (6.4.1)$$

Thus, A can be viewed as follows, where p is chosen to be 5.

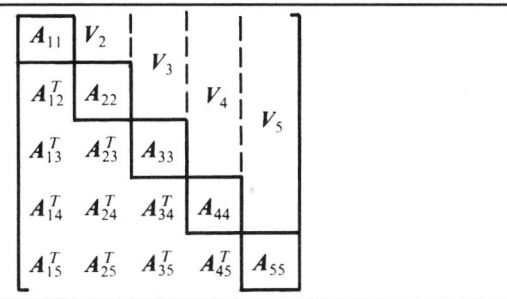

Now our computational scheme requires that we store the diagonal blocks L_{kk}, $1 \le k \le p$, and the non-null off-diagonal blocks of A. The storage scheme we use is illustrated in Figure 6.4.1. The diagonal blocks of L are viewed as forming a single block diagonal matrix which is stored using the envelope storage scheme already described in Section 4.4.1. That is, the diagonal is stored in the array DIAG, and the rows of the lower envelope are stored using the array pair (XENV, ENV). In addition, an array XBLK of length $p+1$ is used to record the partitioning P: XBLK(k) is the number of the first row of the k-th diagonal block, and for convenience we set XBLK($p + 1$) $= N + 1$.

The nonzero components of the V_k, $1 < k \le p$ are stored in a single one dimensional array NONZ, column by column, beginning with those of V_2. A parallel integer array NZSUBS is used to store the row subscripts of the numbers in NONZ, and a vector XNONZ of length $N + 1$ contains the positions in NONZ where each column resides. For programming convenience, we set XNONZ($N + 1$) $= \eta + 1$, where η denotes the number of components in NONZ. Note that XNONZ($i + 1$) $=$ XNONZ(i) implies that the corresponding column of V_k is null.

Suppose XBLK(k) $\le i <$ XBLK($k + 1$), and we wish to print the $i -$ XBLK(k) $+ 1$-st column of A_{jk}, where $j < k$. The following code segment illustrates how this could be done. The elements of each row in NONZ are assumed to be stored in order of increasing row subscript.

Figure 6.4.1 Example showing the arrays used in the quotient tree storage scheme.

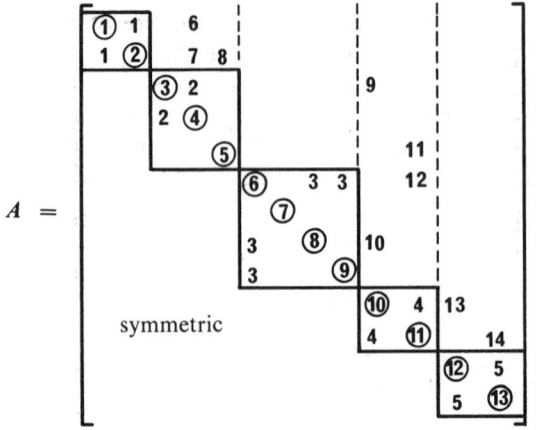

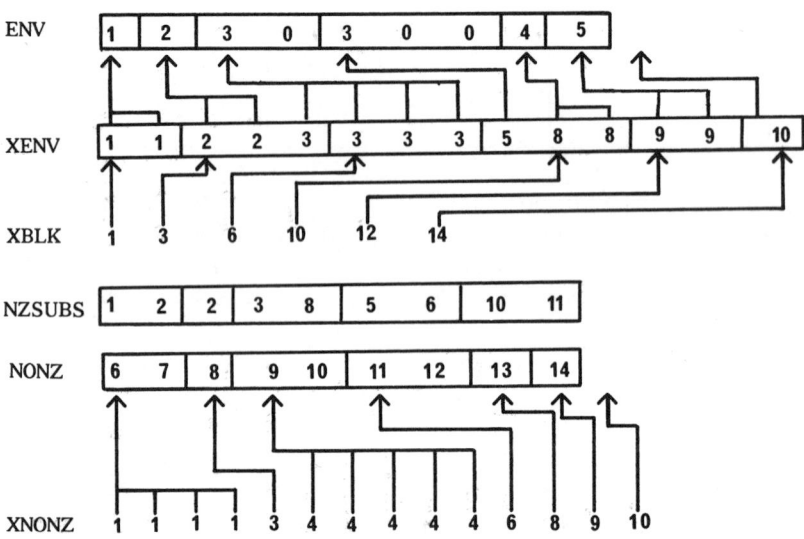

```
          MSTRT = XNONZ(I)
          MSTOP = XNONZ(I+1)-1
          IF (MSTOP.LT.MSTRT) GO TO 200
          DO 100 M = MSTRT, MSTOP
               ROW = NZSUBS(M)
               IF(ROW.LT.XBLK(J)) GO TO 100
               IF(ROW.GT.XBLK(J+1)) GO TO 200
               VALUE = NONZ(M)
               WRITE (6,3000) ROW, VALUE
3000           FORMAT (1X,15H ROW SUBSCRIPT=,I3,7H VALUE=,F12.6)
100       CONTINUE
200       CONTINUE
           ⋮
```

The storage required for the vectors XENV, XBLK, XNONZ and NZSUBS should be regarded as overhead storage, (recall our remarks in Section 2.3.1) since it is not used for actual data. In addition we will need some temporary storage to implement the factorization and solution procedures. We discuss this aspect in Section 6.5, where we deal with the numerical computation subroutines TSFCT (Tree Symmetric FaCTorization) and TSSLV (Tree Symmetric SoLVe).

6.4.2 Internal Renumbering of the Blocks

The ordering algorithm described in Section 6.3 determines a tree partitioning for a connected graph. So far, we have assumed that nodes *within* a block (or a partition member) are labelled arbitrarily. This certainly does not affect the number of off-block-diagonal nonzeros in the original matrix.

However, since the storage scheme stores the diagonal blocks using the envelope structure, the way nodes are arranged within a block can affect the primary storage for the diagonal envelope. It is the purpose of this section to discuss an internal numbering strategy and describe its implementation.

The strategy should use some envelope/profile reduction scheme on each block, and the reverse Cuthill-McKee algorithm, which is simple and quite effective (see Section 4.3.1), seems to be suitable for this purpose. The method is described below. Let $P = \{Y_1, Y_2, \ldots, Y_p\}$ be a given monotonely ordered quotient tree partitioning.

For each block Y_k in P, do the following:

Step 1   Determine the subset

$$U = \{y \in Y_k \mid Adj(y) \cap (Y_1 \cup \cdots \cup Y_{k-1}) = \emptyset\} .$$

Step 2   Reorder the nodes in $G(U)$ by the reverse Cuthill-McKee algorithm.

Step 3   Number the nodes in $Y_k - U$ after U in arbitrary order.

The example in Figure 6.4.2 serves to demonstrate the effect of this renumbering step. The envelope of the diagonal blocks for the ordering α_1 has size 24, whereas the diagonal blocks for α_2 have only a total of 11 entries in their envelopes. Indeed, the relabelling can yield a significant reduction of the storage requirement.

The implementation of this internal-block renumbering scheme is quite straightforward. It consists of two new subroutines BSHUFL and SUBRCM, along with the use of three others which have already been discussed in previous chapters.

They are discussed in detail below.

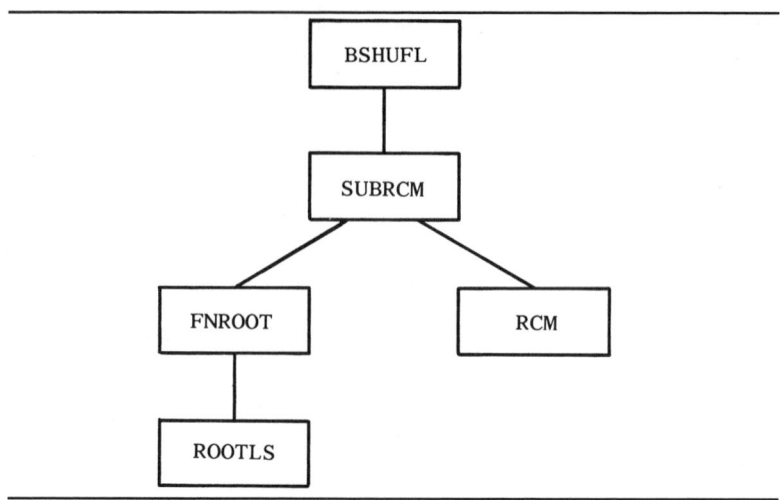

BSHUFL (Block SHUFfLe)

Inputs to this subroutine are the graph structure in (XADJ, ADJNCY), the quotient tree partitioning in (NBLKS, XBLK) and PERM. The subroutine will shuffle the permutation vector PERM according to the scheme described earlier in this section. It needs four working vectors: BNUM for storing the block number of each node, SUBG for accumulating nodes in a subgraph, and MASK and XLS for the execution of the subroutine SUBRCM.

The subroutine begins by initializing the working vectors BNUM and MASK (loop DO 200 K = ...). The loop DO 500 K = ... goes through each block in the partitioning. For each block, all those nodes with no neighbors in the previous blocks are accumulated in the vector SUBG (loop DO 400 ...). The variable NSUBG keeps the number of nodes in the subgraph. The subroutine SUBRCM is then called to renumber this subgraph using the RCM algorithm. The program returns after all the blocks have been processed.

Figure 6.4.2 Example to show the effect of within-block relabelling.

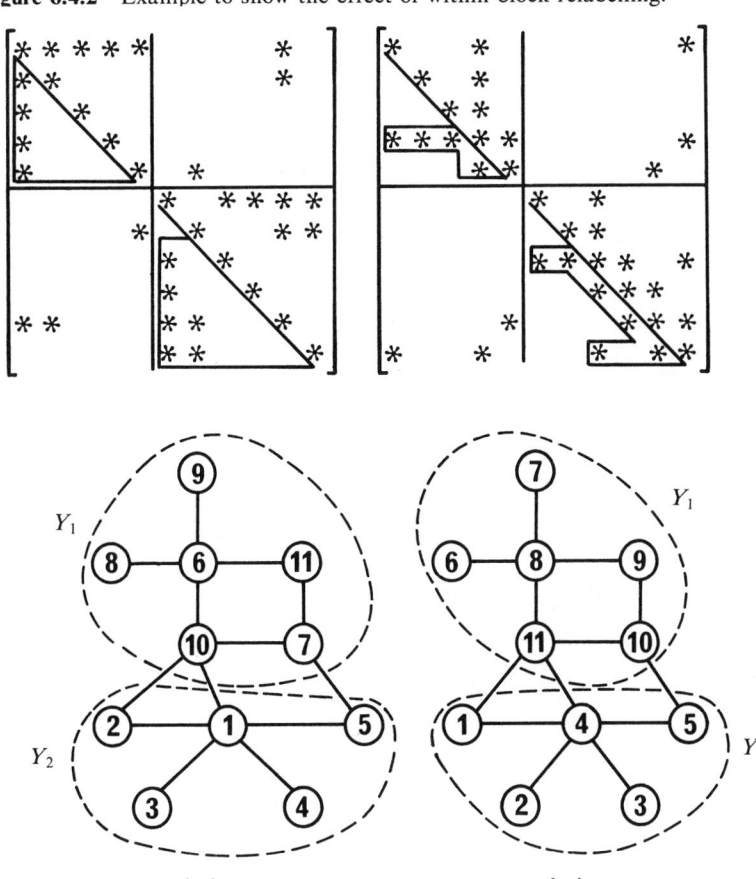

ordering α_1 ordering α_2

```
C********************************************************************
C********************************************************************
C**********      BSHUFL .....  INTERNAL BLOCK SHUFFLE     *******
C********************************************************************
C********************************************************************
C
C      PURPOSE - TO  RENUMBER THE NODES OF EACH BLOCK
C          SO AS TO REDUCE ITS ENVELOPE.
C          NODES IN A BLOCK WITH NO NEIGHBORS IN PREVIOUS
C          BLOCKS ARE RENUMBERED BY SUBRCM BEFORE THE OTHERS.
C
C      INPUT PARAMETERS -
C          (XADJ, ADJNCY) - THE GRAPH ADJACENCY STRUCTURE.
C          (NBLKS, XBLK ) - THE TREE PARTITIONING.
C
C      UPDATED PARAMETER -
C          PERM - THE PERMUTATION VECTOR. ON RETURN, IT CONTAINS
C                 THE NEW PERMUTATION.
C
```

```
C       WORKING VECTORS -
C            BNUM - STORES THE BLOCK NUMBER OF EACH VARIABLE.
C            MASK - MASK VECTOR USED TO PRESCRIBE A SUBGRAPH.
C            SUBG - VECTOR USED TO CONTAIN A SUBGRAPH.
C            XLS - INDEX VECTOR TO A LEVEL STRUCTURE.
C
C       PROGRAM SUBROUTINE -
C            SUBRCM.
C
C*************************************************************
C
        SUBROUTINE   BSHUFL ( XADJ, ADJNCY, PERM, NBLKS, XBLK,
     1                        BNUM, MASK, SUBG, XLS )
C
C*************************************************************
C
        INTEGER ADJNCY(1), BNUM(1), MASK(1), PERM(1),
     1          SUBG(1), XBLK(1), XLS(1)
        INTEGER XADJ(1), I, IP, ISTOP, ISTRT, J,
     1          JSTRT, JSTOP, K, NABOR, NBLKS, NBRBLK,
     1          NODE, NSUBG
C
C*************************************************************
C
        IF ( NBLKS .LE. 0 ) RETURN
C       ----------------------------------------------------
C       INITIALIZATION ..... FIND THE BLOCK NUMBER FOR EACH
C       VARIABLE AND INITIALIZE THE VECTOR MASK.
C       ----------------------------------------------------
        DO 200 K = 1, NBLKS
           ISTRT = XBLK(K)
           ISTOP = XBLK(K+1) - 1
           DO 100 I = ISTRT,ISTOP
              NODE = PERM(I)
              BNUM(NODE) = K
              MASK(NODE) = 0
100        CONTINUE
200     CONTINUE
C       ----------------------------------------------------
C       FOR EACH BLOCK, FIND THOSE NODES WITH NO NEIGHBORS
C       IN PREVIOUS BLOCKS AND ACCUMULATE THEM IN SUBG.
C       THEY WILL BE RENUMBERED BEFORE OTHERS IN THE BLOCK.
C       ----------------------------------------------------
        DO 500 K = 1,NBLKS
           ISTRT = XBLK(K)
           ISTOP = XBLK(K+1) - 1
           NSUBG = 0
           DO 400 I = ISTRT, ISTOP
              NODE = PERM(I)
              JSTRT = XADJ(NODE)
              JSTOP = XADJ(NODE+1) - 1
              IF (JSTOP .LT. JSTRT) GO TO 400
                 DO 300 J = JSTRT, JSTOP
                    NABOR = ADJNCY(J)
                    NBRBLK = BNUM(NABOR)
                    IF (NBRBLK .LT. K) GO TO 400
300              CONTINUE
                 NSUBG = NSUBG + 1
                 SUBG(NSUBG) = NODE
                 IP = ISTRT + NSUBG - 1
                 PERM(I) = PERM(IP)
400        CONTINUE
C          -------------------------------------------------
C          CALL SUBRCM TO RENUMBER THE SUBGRAPH STORED
C          IN (NSUBG, SUBG).
C          -------------------------------------------------
           IF ( NSUBG .GT. 0 )
     1        CALL  SUBRCM ( XADJ, ADJNCY, MASK, NSUBG,
```

```
        1                          SUBG, PERM(ISTRT), XLS )
  500      CONTINUE
           RETURN
        END
```

SUBRCM (SUBgraph RCM)

This subroutine is similar to GENRCM except that it operates on a subgraph. The subgraph, which may be disconnected, is given in the pair (NSUBG, SUBG). The arrays MASK and XLS are working vectors used by the subroutines FNROOT and RCM (see Sections 4.3.3 and 4.3.4).

```
C****************************************************************
C****************************************************************
C********     SUBRCM .....  REVERSE CM ON SUBGRAPH    ********
C****************************************************************
C****************************************************************
C
C      PURPOSE - THIS ROUTINE FINDS THE RCM ORDERING FOR A
C          GIVEN SUBGRAPH (POSSIBLY DISCONNECTED).
C
C      INPUT PARAMETERS -
C          (XADJ, ADJNCY) - ADJACENCY STRUCTURE PAIR FOR THE GRAPH.
C          (NSUBG, SUBG) - THE GIVEN SUBGRAPH.  NSUBG IS THE
C                   THE SIZE OF THE SUBGRAPH, AND SUBG CONTAINS
C                   THE NODES IN IT.
C
C      OUTPUT PARAMETER -
C          PERM - THE PERMUTATION VECTOR. IT IS ALSO USED
C                   TEMPORARILY TO STORE A LEVEL STRUCTURE.
C
C      WORKING PARAMETERS -
C          MASK - MASK VECTOR WITH ALL ZEROS.  IT IS USED TO
C                   SPECIFY NODES IN THE SUBGRAPH.
C          XLS - INDEX TO A LEVEL STRUCTURE.  NOTE THAT THE LEVEL
C                   STRUCTURE IS STORED IN PART OF PERM.
C
C      PROGRAM SUBROUTINES -
C          FNROOT, RCM.
C
C****************************************************************
C
        SUBROUTINE  SUBRCM ( XADJ, ADJNCY, MASK, NSUBG, SUBG,
       1                     PERM, XLS )
C
C****************************************************************
C
        INTEGER ADJNCY(1), MASK(1), PERM(1), SUBG(1),
       1         XLS(1)
        INTEGER XADJ(1), CCSIZE, I, NLVL, NODE, NSUBG, NUM
C
C****************************************************************
C
        DO 100 I = 1, NSUBG
           NODE = SUBG(I)
           MASK(NODE) = 1
  100   CONTINUE
        NUM = 0
        DO 200 I = 1, NSUBG
           NODE = SUBG(I)
           IF ( MASK(NODE) .LE. 0 ) GO TO 200
C          -----------------------------------------------
C              FOR EACH CONNECTED COMPONENT IN THE SUBGRAPH,
```

```
C                    CALL FNROOT AND RCM FOR THE ORDERING.
C                    --------------------------------------------
                     CALL  FNROOT ( NODE, XADJ, ADJNCY, MASK,
      1                             NLVL, XLS, PERM(NUM+1) )
                     CALL    RCM ( NODE, XADJ, ADJNCY, MASK,
      1                            PERM(NUM+1), CCSIZE, XLS )
                     NUM = NUM + CCSIZE
                     IF ( NUM .GE. NSUBG )  RETURN
      200   CONTINUE
            RETURN
          END
```

6.4.3   Storage Allocation and the Subroutines FNTENV, FNOFNZ, and FNTADJ

We now describe two subroutines FNTENV (FiNd Tree ENVelope) and FNOFNZ (FiNd OFf-diagonal NonZeros) which are designed to accept as input a graph G, an ordering α, and a partitioning P, and set up the data structure we described in Section 6.4.1. In addition, in order to obtain an efficient implementation of the numerical factorization procedure, it is necessary to construct a vector containing the adjacency structure of the associated quotient tree G/P. This is the function of the third subroutine FNTADJ (FiNd Tree ADJacency) which we also describe in this section.

FNTENV (FiNd Tree ENVelope)

This subroutine finds the envelope structure of the diagonal blocks in a partitioned matrix. It accepts as input the adjacency structure (XADJ, ADJNCY), the ordering (PERM, INVP) and the quotient tree partitioning (NBLKS, XBLK). The structure in XENV produced by FNTENV may not be exactly the envelope structure of the diagonal blocks, although it always contains the actual envelope structure. For the sake of simplicity and efficiency, it uses the following observation in the construction of the envelope structure.

Let $P = \{Y_1, \ldots, Y_p\}$ be the given tree partitioning, and $x_i, x_j \in Y_k$. If

$$Adj(x_i) \cap \{Y_1, \ldots, Y_{k-1}\} \neq \varnothing$$

and

$$Adj(x_j) \cap \{Y_1, \ldots, Y_{k-1}\} \neq \varnothing,$$

the subroutine will include $\{x_i, x_j\}$ in the envelope structure of the diagonal blocks.

Although this algorithm can yield an unnecessarily large envelope for the diagonal block, (Why? Give an example.) for orderings generated by the RQT algorithm, it usually comes very close to obtaining the exact envelope. Because it works so well, we use it rather than a more sophisticated (and more expensive) scheme which would find the exact envelope. For other quotient tree ordering algorithms, such as the one-way dissection algorithm described in Chapter 7, a more sophisticated scheme is required. (See Section 7.3.3.)

```
C***********************************************************
C***********************************************************
C********     FNTENV ..... FIND TREE DIAGONAL ENVELOPE ********
C***********************************************************
C***********************************************************
C
C     PURPOSE - THIS SUBROUTINE DETERMINES THE ENVELOPE INDEX
C        VECTOR FOR THE ENVELOPE OF THE DIAGONAL BLOCKS OF A
C        TREE PARTITIONED SYSTEM.
C
C     INPUT PARAMETERS -
C        (XADJ, ADJNCY) - ADJACENCY STRUCTURE PAIR FOR THE GRAPH.
C        (PERM, INVP) - THE PERMUTATION VECTORS.
C        (NBLKS, XBLK) - THE TREE PARTITIONING.
C
C     OUTPUT PARAMETERS -
C        XENV - THE ENVELOPE INDEX VECTOR.
C        ENVSZE - THE SIZE OF THE ENVELOPE FOUND.
C
C***********************************************************
C
        SUBROUTINE   FNTENV ( XADJ, ADJNCY, PERM, INVP,
      1                        NBLKS, XBLK, XENV, ENVSZE )
C
C***********************************************************
C
        INTEGER ADJNCY(1), INVP(1), PERM(1), XBLK(1)
        INTEGER XADJ(1), XENV(1), BLKBEG, BLKEND,
      1          I, IFIRST, J, JSTOP, JSTRT, K, KFIRST,
      1          ENVSZE, NBLKS, NBR, NODE
C
C***********************************************************
C
        ENVSZE = 1
C        ----------------------------------------------
C        LOOP THROUGH EACH BLOCK IN THE PARTITIONING ...
C        ----------------------------------------------
        DO 400 K = 1, NBLKS
          BLKBEG = XBLK(K)
          BLKEND = XBLK(K+1) - 1
C          --------------------------------------------
C          KFIRST STORES THE FIRST NODE IN THE K-TH BLOCK
C          THAT HAS A NEIGHBOUR IN THE PREVIOUS BLOCKS.
C          --------------------------------------------
          KFIRST = BLKEND
          DO 300 I = BLKBEG, BLKEND
            XENV(I) = ENVSZE
            NODE = PERM(I)
            JSTRT = XADJ(NODE)
            JSTOP = XADJ(NODE+1) - 1
            IF ( JSTOP .LT. JSTRT ) GO TO 300
C            ------------------------------------------
C            IFIRST STORES THE FIRST NONZERO IN THE
C            I-TH ROW WITHIN THE K-TH BLOCK.
C            ------------------------------------------
            IFIRST = I
            DO 200 J = JSTRT, JSTOP
              NBR = ADJNCY(J)
              NBR = INVP(NBR)
```

```
                    IF ( NBR .LT. BLKBEG )  GO TO 100
                       IF ( NBR .LT. IFIRST ) IFIRST = NBR
                       GO TO 200
   100               IF ( KFIRST .LT. IFIRST )  IFIRST = KFIRST
                       IF ( I .LT. KFIRST )  KFIRST = I
   200             CONTINUE
                   ENVSZE = ENVSZE + I - IFIRST
   300         CONTINUE
   400     CONTINUE
           XENV(BLKEND+1) = ENVSZE
           ENVSZE = ENVSZE - 1
           RETURN
       END
```

FNOFNZ (FiNd OFf-diagonal NonZeros)

The subroutine FNOFNZ is used to determine the structure of the off-block-diagonal nonzeros in a given partitioned matrix. With respect to the storage scheme in Section 6.4.1, this subroutine finds the subscript vector NZSUBS and the subscript or nonzero index vector XNONZ. It also returns a number in MAXNZ which is the number of off-block-diagonal nonzeros in the matrix.

Input to the subroutine is the adjacency structure of the graph (XADJ, ADJNCY), the quotient tree ordering (PERM, INVP), the quotient tree partitioning (NBLKS, XBLK), and the size of the array NZSUBS, contained in MAXNZ. The subroutine loops through the blocks in the partitioning. Within each block, the loop DO 200 J = ... is executed to consider each node in the block. Each neighbor belonging to an earlier block corresponds to an off-diagonal nonzero and it is added to the data structure. After the subscripts in a row have been determined, they are sorted (using SORTS1) into ascending sequence. The subroutine SORTS1 is straightforward and needs no explanation. A listing of it follows that of FNOFNZ.

Note that if the user does not provide a large enough subscript vector, the subroutine will detect this from the input parameter MAXNZ. It will continue to count the nonzeros, but will not store their column subscripts. Before returning, MAXNZ is set to the number of nonzeros found. Thus, the user should check that the value of MAXNZ has not been *increased* by the subroutine, as this indicates that not enough space in NZSUBS was provided.

```
C***************************************************************
C***************************************************************
C****        FNOFNZ ..... FIND OFF-BLOCK-DIAGONAL NONZEROS  ****
C***************************************************************
C***************************************************************
C
C      PURPOSE - THIS SUBROUTINE FINDS THE COLUMN SUBSCRIPTS OF
C         THE OFF-BLOCK-DIAGONAL NONZEROS IN THE LOWER TRIANGLE
C         OF A PARTITIONED MATRIX.
C
C      INPUT PARAMETERS -
C         (XADJ, ADJNCY) - ADJACENCY STRUCTURE PAIR FOR THE GRAPH.
C         (PERM, INVP) - THE PERMUTATION VECTORS.
C         (NBLKS, XBLK) - THE BLOCK PARTITIONING.
C
C      OUTPUT PARAMETERS -
C         (XNONZ, NZSUBS) - THE COLUMN SUBSCRIPTS OF THE NONZEROS
C                  OF A TO THE LEFT OF THE DIAGONAL BLOCKS ARE
C                  STORED ROW BY ROW IN CONTINGUOUS LOCATIONS IN THE
C                  ARRAY NZSUBS.  XNONZ IS THE INDEX VECTOR TO IT.
C
C      UPDATED PARAMETER -
C         MAXNZ - ON INPUT, IT CONTAINS THE SIZE OF THE VECTOR
C                  NZSUBS; AND ON OUTPUT, THE NUMBER OF NONZEROS
C                  FOUND.
C
C***************************************************************
C
        SUBROUTINE  FNOFNZ ( XADJ, ADJNCY, PERM, INVP,
     1                       NBLKS, XBLK, XNONZ, NZSUBS, MAXNZ )
C
C***************************************************************
C
        INTEGER ADJNCY(1), INVP(1), NZSUBS(1), PERM(1),
     1          XBLK(1)
        INTEGER XADJ(1), XNONZ(1), BLKBEG, BLKEND, I, J,
     1          JPERM, JXNONZ, K, KSTOP, KSTRT, MAXNZ,
     1          NABOR, NBLKS, NZCNT
C
C***************************************************************
C
        NZCNT = 1
        IF ( NBLKS .LE. 0 )  GO TO 400
C       ------------------------
C       LOOP OVER THE BLOCKS ....
C       ------------------------
        DO 300 I = 1, NBLKS
           BLKBEG = XBLK(I)
           BLKEND = XBLK(I+1) - 1
C          ------------------------------------------
C          LOOP OVER THE ROWS OF THE I-TH BLOCK ...
C          ------------------------------------------
           DO 200 J = BLKBEG, BLKEND
              XNONZ(J) = NZCNT
              JPERM = PERM(J)
              KSTRT = XADJ(JPERM)
              KSTOP = XADJ(JPERM+1) - 1
              IF ( KSTRT .GT. KSTOP )  GO TO 200
C             -----------------------------------
C             LOOP OVER THE NONZEROS OF ROW J ...
C             -----------------------------------
              DO 100 K = KSTRT, KSTOP
                 NABOR = ADJNCY(K)
                 NABOR = INVP(NABOR)
C                -------------------------------------------
C                CHECK TO SEE IF IT IS TO THE LEFT OF THE
C                I-TH DIAGONAL BLOCK.
```

```
C                      -------------------------------------------
                       IF ( NABOR .GE. BLKBEG ) GO TO 100
                       IF ( NZCNT .LE. MAXNZ ) NZSUBS(NZCNT) = NABOR
                       NZCNT = NZCNT + 1
      100              CONTINUE
C                      --------------------------
C                      SORT THE SUBSCRIPTS OF ROW J
C                      --------------------------
                       JXNONZ = XNONZ(J)
                       IF ( NZCNT - 1 .LE. MAXNZ )
     1                    CALL SORTS1 (NZCNT - JXNONZ, NZSUBS(JXNONZ))
      200         CONTINUE
      300         CONTINUE
                 XNONZ(BLKEND+1) = NZCNT
      400         MAXNZ = NZCNT - 1
                 RETURN
           END


C****************************************************************
C****************************************************************
C***********    SORTS1 ..... LINEAR INSERTION SORT    ******
C****************************************************************
C****************************************************************
C
C    PURPOSE - SORTS1 USES LINEAR INSERTION TO SORT THE
C       GIVEN ARRAY OF SHORT INTEGERS INTO INCREASING ORDER.
C
C    INPUT PARAMETER -
C       NA - THE SIZE OF INTEGER ARRAY.
C
C    UPDATED PARAMETER -
C       ARRAY - THE INTEGER VECTOR, WHICH ON OUTPUT WILL BE
C               IN INCREASING ORDER.
C
C****************************************************************
C
           SUBROUTINE  SORTS1 ( NA, ARRAY )
C
C****************************************************************
C
           INTEGER ARRAY(1)
           INTEGER K, L, NA, NODE
C
C****************************************************************
C
           IF (NA .LE. 1) RETURN
           DO 300 K = 2, NA
              NODE = ARRAY(K)
              L = K - 1
      100     IF (L .LT. 1) GO TO 200
                 IF ( ARRAY(L)  .LE.  NODE )   GO TO 200
                 ARRAY(L+1) = ARRAY(L)
                 L = L - 1
                 GO TO 100
      200        ARRAY(L+1) = NODE
      300     CONTINUE
           RETURN
           END
```

FNTADJ (FiNd Tree ADJacency)

The purpose of this subroutine is to determine the adjacency structure of a given monotonely-ordered quotient tree. Recall from Section 6.2.2 that the structure of a monotonely ordered rooted tree is completely characterized by the *Father* function, where for a node x, $Father(x) = y$ means that $y \in Adj(x)$ and that the (unique) path from the root to x goes through y. Our representation of the structure of our quotient tree is in a vector, called FATHER, of size p, where p is the number of blocks. Figure 6.4.3 contains the FATHER vector for the quotient tree ordering in Figure 6.4.2. Note that FATHER(p) is always set to zero.

Figure 6.4.3   An example of the FATHER vector.

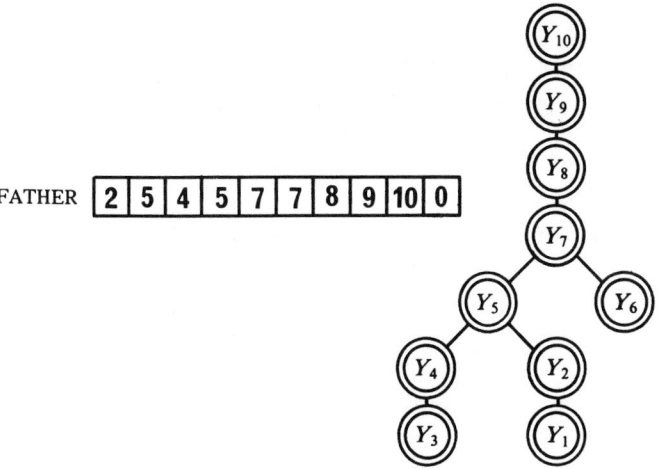

FATHER

The subroutine FNTADJ accepts as input the adjacency structure of the graph (XADJ, ADJNCY), the quotient tree ordering (PERM, INVP), and the quotient tree partitioning (NBLKS, XBLK). It uses a working vector BNUM of size N to store the block numbers of the nodes in the partitioning.

The subroutine begins by setting up the BNUM vector for each node (loop DO 200 K = ...). It then loops through each block in the partitioning in the loop DO 600 K = ... to obtain its father block number. If it does not have any father block, the corresponding FATHER value is set to 0.

```
C*********************************************************
C*********************************************************
C**********     FNTADJ .....  FIND TREE ADJACENCY    *********
C*********************************************************
C*********************************************************
C
C       PURPOSE - TO DETERMINE THE QUOTIENT TREE
C       ADJACENCY STRUCTURE OF A GRAPH.  THE STRUCTURE IS
C       REPRESENTED BY THE FATHER VECTOR.
C
C       INPUT PARAMETERS -
C       (XADJ, ADJNCY) - ADJACENCY STRUCTURE PAIR FOR THE GRAPH.
C       (PERM, INVP) - THE PERMUTATION VECTORS.
C       (NBLKS, XBLK) - THE TREE PARTITIONING.
C
C       OUTPUT PARAMETERS -
C       FATHER - THE FATHER VECTOR OF THE QUOTIENT TREE.
C
C       WORKING PARAMETERS -
C       BNUM - TEMPORARY VECTOR TO STORE THE BLOCK NUMBER OF
C               OF EACH VARIABLE.
C
C*********************************************************
C
        SUBROUTINE  FNTADJ ( XADJ, ADJNCY, PERM, INVP,
     1                       NBLKS, XBLK, FATHER, BNUM )
C
C*********************************************************
C
        INTEGER ADJNCY(1), BNUM(1), FATHER(1), INVP(1),
     1          PERM(1), XBLK(1)
        INTEGER XADJ(1), I, ISTOP, ISTRT, J, JSTOP, JSTRT,
     1          K, NABOR, NBLKS, NBM1, NBRBLK, NODE
C
C*********************************************************
C
C       -----------------------------------
C       INITIALIZE THE BLOCK NUMBER VECTOR.
C       -----------------------------------
        DO 200 K = 1, NBLKS
           ISTRT = XBLK(K)
           ISTOP = XBLK(K+1) - 1
           DO 100 I = ISTRT, ISTOP
              NODE = PERM(I)
              BNUM(NODE) = K
  100      CONTINUE
  200   CONTINUE
C       ------------------
C       FOR EACH BLOCK ...
C       ------------------
        FATHER(NBLKS) = 0
        NBM1 = NBLKS - 1
        IF ( NBM1 .LE. 0 )  RETURN
        DO 600 K = 1, NBM1
           ISTRT = XBLK(K)
           ISTOP = XBLK(K+1) - 1
C          ------------------------------------------------
C          FIND ITS FATHER BLOCK IN THE TREE STRUCTURE.
C          ------------------------------------------------
           DO 400 I = ISTRT, ISTOP
              NODE = PERM(I)
              JSTRT = XADJ(NODE)
              JSTOP = XADJ(NODE+1) -1
              IF ( JSTOP .LT. JSTRT )  GO TO 400
                 DO 300 J = JSTRT, JSTOP
                    NABOR = ADJNCY(J)
                    NBRBLK = BNUM(NABOR)
```

```
                IF ( NBRBLK .GT. K )  GO TO 500
300                 CONTINUE
400             CONTINUE
                FATHER(K) = 0
                GO TO 600
500             FATHER(K) = NBRBLK
600         CONTINUE
            RETURN
        END
```

6.5   The Numerical Subroutines TSFCT (Tree Symmetric FaCTorization) and TSSLV (Tree Symmetric SoLVe)

In this section, we describe the subroutines that implement the numerical factorization and solution for partitioned linear systems associated with quotient trees, stored in the sparse scheme as introduced in Section 6.4.1. The subroutine TSFCT employs the asymmetric version of the factorization, so we begin by first re-examining the asymmetric block factorization procedure of Section 6.1.1 and studying possible improvements.

6.5.1   Computing the Block Modification Matrix

Let the matrix A be partitioned into

$$\begin{pmatrix} B & V \\ V^T & \overline{C} \end{pmatrix}$$

as in Section 6.1.1. Recall that in the factorization scheme, the modification matrix $V^T B^{-1} V$ used to form $C = \overline{C} - V^T B^{-1} V$ is obtained as follows.

$$V^T(L_B^{-T}(L_B^{-1}V)) = V^T(L_B^{-T}W) = V^T\tilde{W} \ .$$

Note also that the modification matrix $V^T\tilde{W}$ is symmetric and that $Nonz(V) \subset Nonz(W)$. We now investigate an efficient way to compute $V^T\tilde{W}$.

Let G be an r by s sparse matrix and H be an s by r matrix. For the i-th row of G, let

$$f_i(G) = \min\{j \mid g_{ij} \neq 0\}, \ 1 \le i \le r \ . \tag{6.5.1}$$

That is, $f_i(G)$ is the column subscript of the first nonzero component in row i of G. Assume that the matrix product GH is *symmetric*. In what follows, we show that only a portion of the matrix H is needed in computing the product. Figure 6.5.1 contains an example with $r = 4$ and $s = 8$. If the product GH is symmetric, the next lemma says that the crosshatched part of H can be ignored in the evaluation of GH.

Figure 6.5.1   Sparse symmetric matrix product.

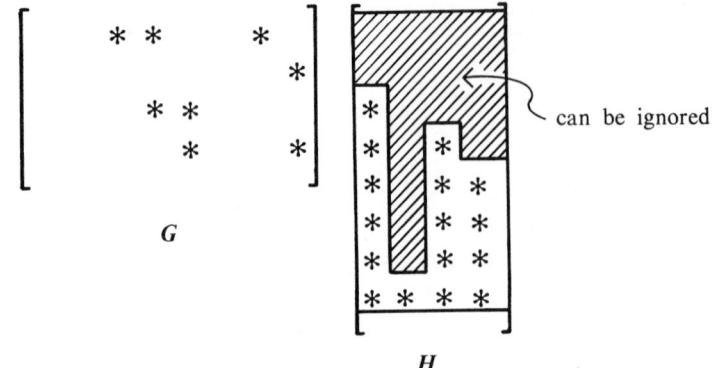

Lemma 6.5.1
If the matrix product GH is symmetric, the product is completely determined by the matrix G and the matrix subset

$$\{h_{jk} \mid f_k(G) \leq j \leq s\}$$

of H.

Proof   It is sufficient to show that every entry in the matrix product can be computed from G and the given subset of H. Since the product is symmetric, its (i,k)-th and (k,i)-th entries are given by

$$\sum_{j=f_i(G)}^{s} g_{ij}h_{jk}$$

or

$$\sum_{j=f_k(G)}^{s} g_{kj}h_{ji} .$$

For definiteness, let $f_k(G) \leq f_i(G)$. The entry can then be obtained using the first expression, which involves components in G and those h_{jk} with $f_k(G) \leq f_i(G) \leq j \leq s$. They belong to the given matrix subset. On the other hand, if $f_k(G) > f_i(G)$, the second expression can be used which involves matrix components in the subset. This proves the lemma.   □

The order in which the components in the product are computed depends on the structure of the matrix (or more specifically, on the column subscripts $f_i(G)$, $1 \leq i \leq r$). For example, in forming GH for the matrices in Figure 6.5.1, the order of computation is given in Figure 6.5.2.

Figure 6.5.2   Illustration of the computation of the product *GH*.

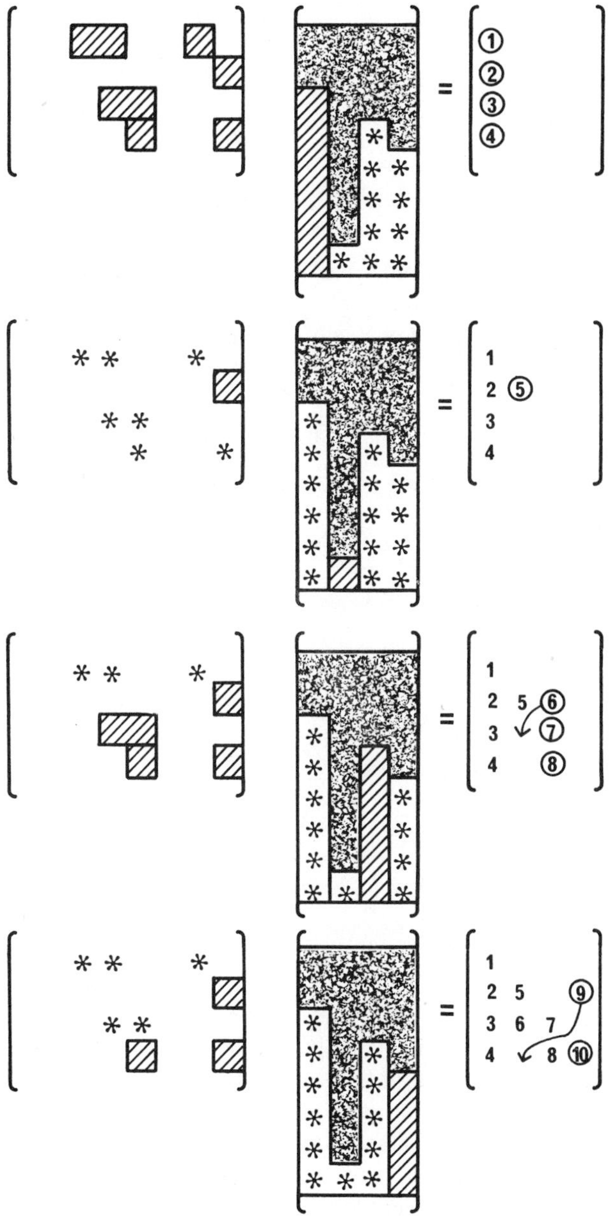

With this framework, we can study changing the submatrix $\overline{C}$ into $C = \overline{C} - V^T B^{-1} V = \overline{C} - V^T (L_B^{-T}(L_B^{-1} V))$. As pointed out in Section 6.1.1, the modification can be carried out one column at a time as follows:

1) Unpack a column $v = V_{*i}$ of V.
2) Solve $B\tilde{w} = v$ by solving the triangular systems $L_B w = v$ and $L_B^T \tilde{w} = w$.
3) Compute the vector $z = V^T \tilde{w}$ and set $C_{*i} = \overline{C}_{*i} - z$.

Now since $V^T \tilde{W}$ is symmetric, Lemma 6.5.1 applies to this modification process, and it is unnecessary to compute the entire vector $\tilde{w}$ from v in Step 2. The components in $\tilde{w}$ above the first nonzero subscript of v do not have to be computed when solving $L_B(L_B^T \tilde{w}) = v$.

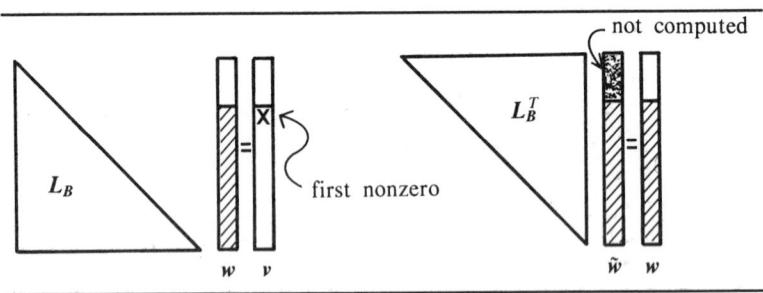

In effect, a smaller system than $L_B(L_B^T \tilde{w}) = v$ needs to be solved. This can have a significant effect on the amount of computation required for the factorization. For example, see Exercise 6.5.1.

6.5.2 *The Subroutine* TSFCT (Tree Symmetric FaCTorization)

The subroutine TSFCT performs the asymmetric block factorization for tree-partitioned systems. The way it computes the block modification matrix is as described in the previous section.

The subroutine accepts as input the tree partitioning information in (NBLKS, XBLK) and FATHER, the data structure in XENV, XNONZ and NZSUBS, and the primary storage vectors DIAG, ENV and NONZ. The vectors DIAG and ENV, on input, contain the nonzeros of the block diagonals of the matrix A. On return, the corresponding nonzeros of the block diagonals of L are overwritten on those of A. Since the implicit scheme is used, the off-block-diagonal nonzeros of A stored in NONZ remain unchanged.

Two temporary vectors of size N are used. The real vector TEMP is used for unpacking off-diagonal block columns so that numerical solution on the unpacked column can be done in the vector TEMP. The second temporary vector FIRST is an integer array used to facilitate indexing into the subscript vector NZSUBS. (See remarks about FIRST below.)

The subroutine TSFCT begins by initializing the temporary vectors TEMP and FIRST (loop DO 100 I = ...). The main loop DO 1600 K = ... is then executed for each block in the partitioning. Within the main loop, the subroutine ESFCT is first called to factor the K-th diagonal block. The next step is to find out where the off-diagonal block is, and it is given by FATHER(K). The loops DO 200 ... and DO 400 ... are then executed to determine the first and last non-null columns respectively in the off-diagonal block so that modification can be performed within these columns. Figure 6.5.3 depicts the role of some of the important local variables in the subroutine.

Figure 6.5.3 Illustration of some of the important local variables used in TSFCT.

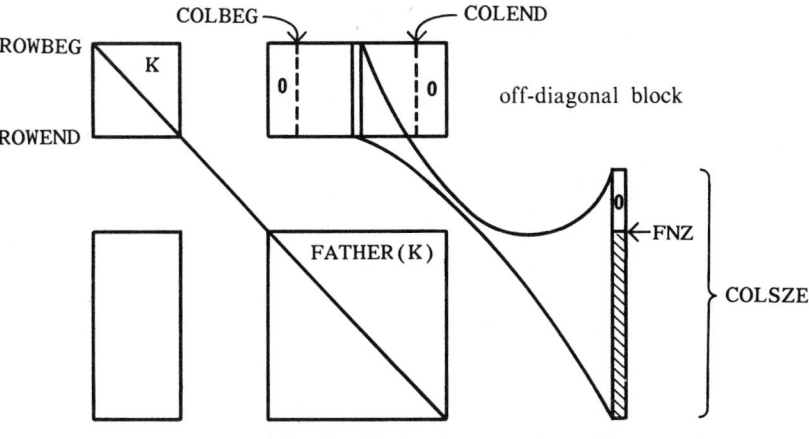

The loop DO 1300 COL = ... applies the modification to the diagonal block given by FATHER(K). Each column in the off-diagonal block is unpacked into the vector TEMP (loop DO 600 J = ...), after which the envelope solvers ELSLV and EUSLV are invoked. The inner loop DO 1100 COL1 = ... then performs the modification in the same manner as discussed in Section 6.5.1.

Before the subroutine proceeds to consider the next block, it updates the temporary vector FIRST for columns in the FATHER(K)-th block, so that the corresponding elements of FIRST point to the next numbers to be used in those columns (loop DO 1500 COL = ...). When all the diagonal blocks have been processed, the subroutine returns.

```
C*****************************************************************
C*****************************************************************
C********        TSFCT ..... TREE SYMMETRIC FACTORIZATION  ******
C*****************************************************************
C*****************************************************************
C
C     PURPOSE - THIS SUBROUTINE PERFORMS THE SYMMETRIC
C               FACTORIZATION OF A TREE-PARTITIONED SYSTEM.
C
C     INPUT PARAMETERS -
C         (NBLKS, XBLK, FATHER) - THE TREE PARTITIONING.
C         XENV - THE ENVELOPE INDEX VECTOR.
C         (XNONZ, NONZ, NZSUBS) - THE OFF-DIAGONAL NONZEROS IN
C               THE ORIGINAL MATRIX.
C
C     UPDATED PARAMETERS -
C         (DIAG, ENV) - STORAGE ARRAYS FOR THE ENVELOPE OF
C               THE DIAGONAL BLOCKS OF THE MATRIX. ON OUTPUT,
C               CONTAINS THE DIAGONAL BLOCKS OF THE FACTOR.
C         IFLAG - THE ERROR FLAG. IT IS SET TO 1 IF A ZERO OR
C               NEGATIVE SQUARE ROOT IS DETECTED DURING THE
C               FACTORIZATION.
C
C     WORKING PARAMETER -
C         TEMP - TEMPORARY ARRAY REQUIRED TO IMPLEMENT THE
C               ASYMMETRIC VERSION OF THE FACTORIZATION.
C         FIRST - TEMPORARY VECTOR USED TO FACILITATE THE
C               INDEXING TO THE VECTOR NONZ (OR NZSUBS)
C               FOR NON-NULL SUBCOLUMNS IN OFF-DIAGONAL
C               BLOCKS.
C
C     PROGRAM SUBROUTINES -
C         ESFCT, ELSLV, EUSLV.
C
C*****************************************************************
C
      SUBROUTINE  TSFCT ( NBLKS, XBLK, FATHER, DIAG, XENV, ENV,
     1                    XNONZ, NONZ, NZSUBS, TEMP, FIRST, IFLAG )
C
C*****************************************************************
C
         DOUBLE PRECISION OPS
         COMMON /SPKOPS/ OPS
         REAL DIAG(1), ENV(1), NONZ(1), TEMP(1), S
         INTEGER FATHER(1), NZSUBS(1), XBLK(1)
         INTEGER FIRST(1), XENV(1), XNONZ(1),
     1           BLKSZE, COL, COL1, COLBEG, COLEND,
     1           COLSZE, FNZ, FNZ1, I, IFLAG, ISTRT, ISTOP,
     1           ISUB, J, JSTOP, JSTRT, K, KENV, KENVO, KFATHR,
     1           NBLKS, NEQNS, ROW, ROWBEG, ROWEND
C
C*****************************************************************
C
C         -------------------
C         INITIALIZATION ...
C         -------------------
          NEQNS = XBLK(NBLKS+1) - 1
          DO 100 I = 1,NEQNS
             TEMP(I) = 0.0E0
             FIRST(I) = XNONZ(I)
  100     CONTINUE
C         ---------------------------
C         LOOP THROUGH THE BLOCKS ...
C         ---------------------------
          DO 1600 K = 1, NBLKS
             ROWBEG = XBLK(K)
             ROWEND = XBLK(K+1) - 1
```

```
            BLKSZE = ROWEND - ROWBEG + 1
            CALL ESFCT ( BLKSZE, XENV(ROWBEG), ENV,
      1                  DIAG(ROWBEG), IFLAG )
            IF ( IFLAG .GT. 0 )  RETURN
C           -----------------------------------------------
C           PERFORM MODIFICATION OF THE FATHER DIAGONAL BLOCK
C           A(FATHER(K),FATHER(K)) FROM THE OFF-DIAGONAL BLOCK
C           A(K,FATHER(K)).
C           -----------------------------------------------
            KFATHR = FATHER(K)
            IF ( KFATHR .LE. 0 )  GO TO 1600
                COLBEG = XBLK(KFATHR)
                COLEND = XBLK(KFATHR+1) - 1
C               -------------------------------------------
C               FIND THE FIRST AND LAST NON-NULL COLUMN IN
C               THE OFF-DIAGONAL BLOCK. RESET COLBEG,COLEND.
C               -------------------------------------------
                DO 200 COL = COLBEG, COLEND
                    JSTRT = FIRST(COL)
                    JSTOP = XNONZ(COL+1) - 1
                    IF ( JSTOP .GE. JSTRT .AND.
      1                 NZSUBS(JSTRT) .LE. ROWEND )  GO TO 300
  200           CONTINUE
  300           COLBEG = COL
                COL = COLEND
                DO 400 COL1 = COLBEG, COLEND
                    JSTRT = FIRST(COL)
                    JSTOP = XNONZ(COL+1) - 1
                    IF ( JSTOP .GE. JSTRT .AND.
      1                 NZSUBS(JSTRT) .LE. ROWEND )  GO TO 500
                    COL = COL - 1
  400           CONTINUE
  500           COLEND = COL
                DO 1300 COL = COLBEG, COLEND
                    JSTRT = FIRST(COL)
                    JSTOP = XNONZ(COL+1) - 1
C                   ---------------------------------------
C                   TEST FOR NULL SUBCOLUMN.  FNZ STORES THE
C                   FIRST NONZERO SUBSCRIPT IN THE BLOCK COLUMN.
C                   ---------------------------------------
                    IF ( JSTOP .LT. JSTRT )  GO TO 1300
                    FNZ = NZSUBS(JSTRT)
                    IF ( FNZ .GT. ROWEND )  GO TO 1300
C                       -----------------------------------
C                       UNPACK A COLUMN IN THE OFF-DIAGONAL BLOCK
C                       AND PERFORM UPPER AND LOWER SOLVES ON THE
C                       UNPACKED COLUMN.
C                       -----------------------------------
                        DO 600 J = JSTRT, JSTOP
                            ROW = NZSUBS(J)
                            IF ( ROW .GT. ROWEND )  GO TO 700
                            TEMP(ROW) = NONZ(J)
  600                   CONTINUE
  700                   COLSZE = ROWEND - FNZ + 1
                        CALL  ELSLV ( COLSZE, XENV(FNZ), ENV,
      1                             DIAG(FNZ), TEMP(FNZ) )
                        CALL  EUSLV ( COLSZE, XENV(FNZ), ENV,
      1                             DIAG(FNZ), TEMP(FNZ) )
C                       -----------------------------------
C                       DO THE MODIFICATION BY LOOPING THROUGH
C                       THE COLUMNS AND FORMING INNER PRODUCTS.
C                       -----------------------------------
                        KENV0 = XENV(COL+1) - COL
                        DO 1100 COL1= COLBEG, COLEND
                            ISTRT = FIRST(COL1)
                            ISTOP = XNONZ(COL1+1) - 1
C                           -------------------------------
C                           CHECK TO SEE IF SUBCOLUMN IS NULL.
```

```
C          ------------------------------------
           FNZ1 = NZSUBS(ISTRT)
           IF ( ISTOP .LT. ISTRT   .OR.
      1         FNZ1 .GT. ROWEND )  GO TO 1100
C          ------------------------------------
C          CHECK IF INNER PRODUCT SHOULD BE DONE.
C          ------------------------------------
           IF ( FNZ1 .LT. FNZ ) GO TO 1100
           IF ( FNZ1 .EQ. FNZ  .AND.
      1         COL1 .LT. COL )  GO TO 1100
           S = 0.0E0
           DO 800 I = ISTRT, ISTOP
              ISUB = NZSUBS(I)
              IF ( ISUB .GT. ROWEND )  GO TO 900
                 S = S + TEMP(ISUB) * NONZ(I)
                 OPS = OPS + 1.0D0
  800      CONTINUE
C          ------------------------------------
C          MODIFY THE ENV OR THE DIAG ENTRY.
C          ------------------------------------
  900      IF ( COL1 .EQ. COL )  GO TO 1000
              KENV = KENV0 + COL1
              IF ( COL1 .GT. COL )
      1          KENV = XENV(COL1+1) - COL1 + COL
              ENV(KENV) = ENV(KENV) - S
              GO TO 1100
 1000         DIAG(COL1) = DIAG(COL1) - S
 1100      CONTINUE
C          ------------------------------------
C          RESET PART OF THE TEMP VECTOR TO ZERO.
C          ------------------------------------
           DO 1200 ROW = FNZ, ROWEND
              TEMP(ROW) = 0.0E0
 1200      CONTINUE
 1300   CONTINUE
C       ------------------------------------
C       UPDATE THE FIRST VECTOR FOR COLUMNS IN
C       FATHER(K) BLOCK, SO THAT IT WILL INDEX TO
C       THE BEGINNING OF THE NEXT OFF-DIAGONAL
C       BLOCK TO BE CONSIDERED.
C       ------------------------------------
        DO 1500 COL = COLBEG, COLEND
           JSTRT = FIRST(COL)
           JSTOP = XNONZ(COL+1) - 1
           IF ( JSTOP .LT. JSTRT )  GO TO 1500
           DO 1400 J = JSTRT, JSTOP
              ROW = NZSUBS(J)
              IF ( ROW .LE. ROWEND )  GO TO 1400
                 FIRST(COL) = J
                 GO TO 1500
 1400         CONTINUE
              FIRST(COL) = JSTOP + 1
 1500      CONTINUE
 1600   CONTINUE
        RETURN
        END
```

6.5.3   The Subroutine TSSLV (Tree Symmetric SoLVe)

The implementation of the solver for tree-partitioned linear systems
does not follow the same execution sequence as specified in
Section 6.2.3. Instead, it uses the alternative decomposition for the
asymmetric factorization as given in Exercise 6.1.2:

$$A = \begin{pmatrix} B & V \\ V^T & \overline{C} \end{pmatrix} = \begin{pmatrix} B & 0 \\ V^T & C \end{pmatrix} \begin{pmatrix} I & \tilde{W} \\ 0 & I \end{pmatrix}, \qquad (6.5.2)$$

where $\tilde{W} = B^{-1}V$ is not explicitly stored, and $C = \overline{C} - V^T B^{-1} V$. Written in this form, the solution to

$$\begin{pmatrix} B & V \\ V^T & \overline{C} \end{pmatrix} \begin{pmatrix} x_1 \\ x_2 \end{pmatrix} = \begin{pmatrix} b_1 \\ b_2 \end{pmatrix}$$

can be computed by solving

$$\begin{pmatrix} B & 0 \\ V^T & C \end{pmatrix} \begin{pmatrix} z_1 \\ z_2 \end{pmatrix} = \begin{pmatrix} b_1 \\ b_2 \end{pmatrix}$$

and

$$\begin{pmatrix} I & \tilde{W} \\ 0 & I \end{pmatrix} \begin{pmatrix} x_1 \\ x_2 \end{pmatrix} = \begin{pmatrix} z_1 \\ z_2 \end{pmatrix}.$$

It is assumed that the submatrices B and C have been factored into $L_B L_B^T$ and $L_C L_C^T$ respectively. The scheme can hence be written as follows.

Forward Solve

Solve $\quad L_B(L_B^T z_1) = b_1$.
Compute $\quad \tilde{b}_2 = b_2 - V^T z_1$.
Solve $\quad L_C(L_C^T z_2) = \tilde{b}_2$.

Backward Solve

Assign $\quad x_2 = z_2$.
Compute $\quad t_1 = V x_2$.
Solve $\quad L_B(L_B^T \tilde{t}_1) = t_1$.
Compute $\quad x_1 = z_1 - \tilde{t}_1$.

This scheme is simply a rearrangement of the operation sequences as given in Section 6.1.2. The only difference is that no temporary vector is required in the forward solve (no real advantage though! Why?). We choose to use this scheme because it simplifies the program organization when the general block p by p tree-partitioned system is being solved.

We now consider the generalization of the above asymmetric scheme. Let A be a p by p tree-partitioned matrix with blocks A_{ij}, $1 \leq i, j \leq p$. Let L_{ij} be the corresponding submatrices of the triangular factor L of A.

Recall from Section 6.2.2 that since A is tree-partitioned, the lower off-diagonal blocks $L_{ij}(i > j)$ are given by

$$L_{ij} = A_{ij}L_{jj}^{-T}.$$ (6.5.3)

We want to define an asymmetric block factorization

$$A = \tilde{L}U$$ (6.5.4)

similar to that of (6.5.2). Obviously the factor $\tilde{L}$ is well defined and its blocks are given by,

$$\tilde{L}_{ij} = \begin{cases} L_{ii}L_{ii}^{T} & \text{if } i = j \\ A_{ij} & \text{if } i > j \\ 0 & \text{otherwise.} \end{cases}$$

The case when $p = 4$ is given below.

$$\tilde{L} = \begin{pmatrix} L_{11}L_{11}^{T} & 0 & 0 & 0 \\ A_{21} & L_{22}L_{22}^{T} & 0 & 0 \\ A_{31} & A_{32} & L_{33}L_{33}^{T} & 0 \\ A_{41} & A_{42} & A_{43} & L_{44}L_{44}^{T} \end{pmatrix}$$

Lemma 6.5.2

$$\tilde{L} = L \begin{pmatrix} L_{11}^{T} & & & 0 \\ & L_{22}^{T} & & \\ & & \cdot & \\ & & & \cdot \\ & & & & \cdot \\ 0 & & & & L_{pp}^{T} \end{pmatrix}$$

Proof   The result follows directly from the relation (6.5.3) between off-diagonal blocks A_{ij} and L_{ij} for tree-partitioned systems.   □

By this lemma, the upper triangular factor U in (6.5.4) can then be obtained simply as

$$U = \begin{pmatrix} L_{11}^T & & & 0 \\ & L_{22}^T & & \\ & & \cdot & \\ & & \cdot & \\ & & & \cdot & \\ 0 & & & L_{pp}^T \end{pmatrix}^{-1} L^T$$

so that we have

$$U_{ik} = \begin{cases} I & \text{if } i = k \\ L_{ii}^{-T} L_{ki}^T & \text{if } i < k \\ 0 & \text{otherwise.} \end{cases}$$

and for $i < k$, the expression can be simplified by (6.5.3) to

$$U_{ik} = L_{ii}^{-T} \left(A_{ki} L_{ii}^{-T} \right)^T$$
$$= L_{ii}^{-T} L_{ii}^{-1} A_{ki}^T$$
$$= \left(L_{ii} L_{ii}^T \right)^{-1} A_{ik} .$$

Therefore, the asymmetric factorization (6.5.4) for the case $p = 4$ can be expressed explicitly as shown by the following:

$$\tilde{L} = \begin{pmatrix} L_{11}L_{11}^T & 0 & 0 & 0 \\ A_{21} & L_{22}L_{22}^T & 0 & 0 \\ A_{31} & A_{32} & L_{33}L_{33}^T & 0 \\ A_{41} & A_{42} & A_{43} & L_{44}L_{44}^T \end{pmatrix}$$

$$U = \begin{pmatrix} I & (L_{11}L_{11}^T)^{-1}A_{12} & (L_{11}L_{11}^T)^{-1}A_{13} & (L_{11}L_{11}^T)^{-1}A_{14} \\ 0 & I & (L_{22}L_{22}^T)^{-1}A_{23} & (L_{22}L_{22}^T)^{-1}A_{24} \\ 0 & 0 & I & (L_{33}L_{33}^T)^{-1}A_{34} \\ 0 & 0 & 0 & I \end{pmatrix} .$$

To consider the actual solution phase on this factorization, we have to relate it to our block storage scheme as described in Section 6.4.1. As before, let

$$V_k = \begin{pmatrix} A_{1k} \\ A_{2k} \\ \cdot \\ \cdot \\ \cdot \\ \cdot \\ A_{k-1,k} \end{pmatrix} , \quad 2 \le k \le p .$$

Since the nonzero components outside the diagonal blocks are stored column by column as in

$$V_2, V_3, \ldots, V_p ,$$

the solution method must be tailored to this storage scheme. The method to be discussed makes use of the observation that

$$\begin{pmatrix} U_{1k} \\ U_{2k} \\ \cdot \\ \cdot \\ \cdot \\ U_{k-1,k} \end{pmatrix} = \begin{pmatrix} (L_{11}L_{11}^T)^{-1} & & & 0 \\ & (L_{22}L_{22}^T)^{-1} & & \\ & & \cdot & \\ & & & \cdot \\ 0 & & & (L_{k-1,k-1}L_{k-1,k-1}^T)^{-1} \end{pmatrix} V_k .$$

Forward Solve $\tilde{L}z = b$

Step 1   Solve $(L_{11}L_{11}^T)z_1 = b_1$.

Step 2   For $k = 2, \ldots, p$ do the following

  2.1)   Compute

$$b_k \leftarrow b_k - V_k^T \begin{pmatrix} z_1 \\ \cdot \\ \cdot \\ \cdot \\ z_{k-1} \end{pmatrix}$$

  2.2)   Solve $(L_{kk}L_{kk}^T)z_k = b_k$.

Backward Solve $Ux = z$

Step 1   Initialize temporary vector $\tilde{z} = \mathbf{0}$,
Step 2 $x_p = z_p$
Step 3   For $k = p - 1, \ p - 2, \ldots, 1$ do the following

   3.1)

$$
\begin{pmatrix} \tilde{z}_1 \\ \cdot \\ \cdot \\ \cdot \\ \tilde{z}_k \end{pmatrix} \leftarrow \begin{pmatrix} \tilde{z}_1 \\ \cdot \\ \cdot \\ \cdot \\ \tilde{z}_k \end{pmatrix} + V_{k+1}x_{k+1}
$$

   3.2)   Solve $(L_{kk}L_{kk}^T)\tilde{x}_k = \tilde{z}_k$.
   3.3)   Compute $x_k = z_k - \tilde{x}_k$.

Note that in the back solve, a temporary vector $\tilde{z}$ is used to accumulate the products and its use is illustrated in Figure 6.5.4.

The subroutine TSSLV implements this solution scheme. Unlike TSFCT, it does not require the FATHER vector of the tree partitioning, although implicitly it depends on it. Inputs to TSSLV are the tree partitioning (NBLKS, XBLK), the diagonal DIAG, the envelope (XENV, ENV) of the diagonal blocks, and the off-diagonal nonzeros in (XNONZ, NONZ, NZSUBS).

There are two main loops in the subroutine TSSLV; one to perform the forward substitution and the other to do the backward solve. In the forward solve, the loop DO 200 ROW = ... is executed to modify the right hand vector before the subroutines ELSLV and EUSLV are called. In the backward solve, the temporary real vector TEMP accumulates the products of off-diagonal blocks and parts of the solution, in preparation for calling ELSLV and EUSLV. At the end, the vector RHS contains the solution vector.

Figure 6.5.4   Backward solve for asymmetric factorization.

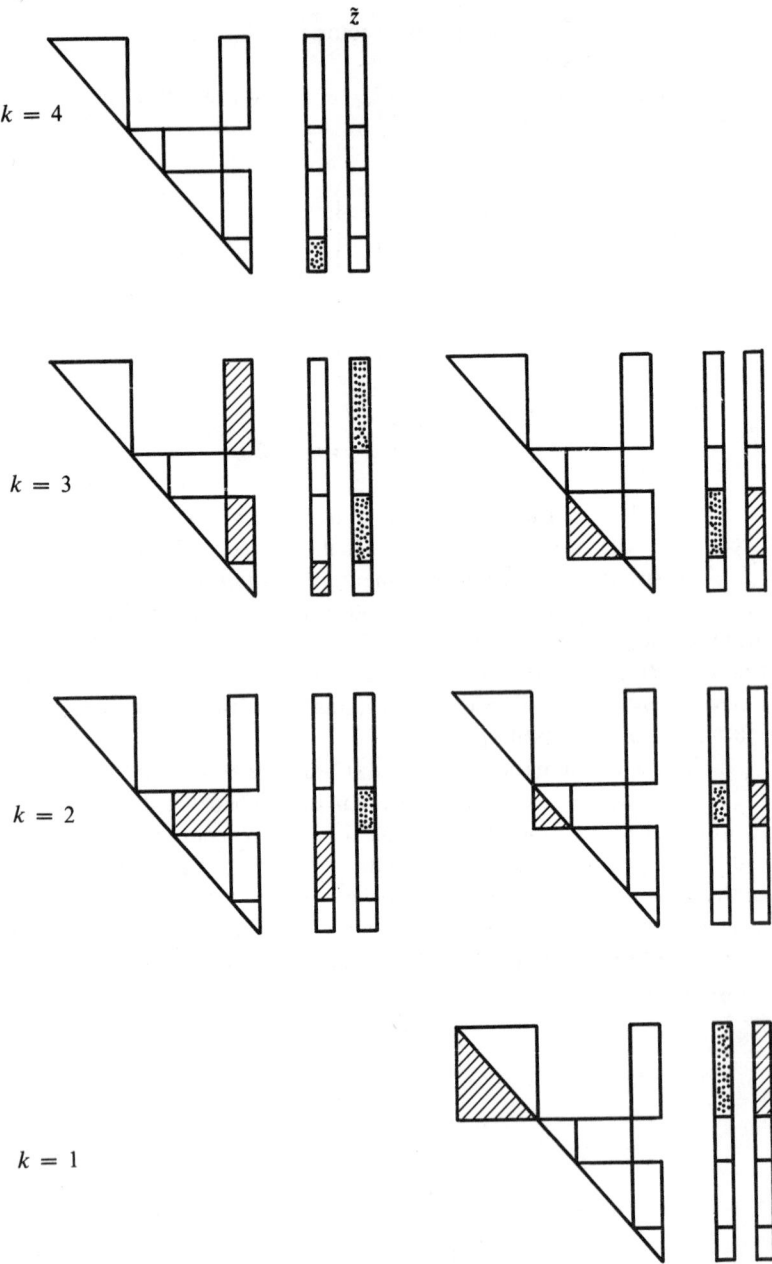

```
C*************************************************************
C*************************************************************
C*********    TSSLV ..... TREE SYMMETRIC SOLVE    *********
C*************************************************************
C*************************************************************
C
C      PURPOSE - TO PERFORM SOLUTION OF A TREE-PARTITIONED
C                FACTORED SYSTEM BY IMPLICIT BACK SUBSTITUTION.
C
C      INPUT PARAMETERS -
C         (NBLKS, XBLK) - THE PARTITIONING.
C         (XENV, ENV) - ENVELOPE OF THE DIAGONAL BLOCKS.
C         (XNONZ, NONZ, NZSUBS) - DATA STRUCTURE FOR THE OFF-
C                BLOCK DIAGONAL NONZEROS.
C
C      UPDATED PARAMETERS -
C         RHS - ON INPUT IT CONTAINS THE RIGHT HAND VECTOR.
C               ON OUTPUT, THE SOLUTION VECTOR.
C
C      WORKING VECTOR -
C         TEMP - TEMPORARY VECTOR USED IN BACK SUBSTITUTION.
C
C      PROGRAM SUBROUTINES -
C         ELSLV, EUSLV.
C
C*************************************************************
C
       SUBROUTINE  TSSLV ( NBLKS, XBLK, DIAG, XENV, ENV,
      1                    XNONZ, NONZ, NZSUBS, RHS, TEMP )
C
C*************************************************************
C
       DOUBLE PRECISION COUNT, OPS
       COMMON /SPKOPS/ OPS
       REAL DIAG(1), ENV(1), NONZ(1), RHS(1), TEMP(1), S
       INTEGER NZSUBS(1), XBLK(1)
       INTEGER XENV(1), XNONZ(1), COL, COL1, COL2, I, J,
      1         JSTOP, JSTRT, LAST, NBLKS, NCOL, NROW, ROW,
      1         ROW1, ROW2
C
C*************************************************************
C
C      -----------------------
C      FORWARD SUBSTITUTION ...
C      -----------------------
       DO 400 I = 1, NBLKS
          ROW1 = XBLK(I)
          ROW2 = XBLK(I+1) - 1
          LAST = XNONZ(ROW2+1)
          IF ( I .EQ. 1 .OR. LAST .EQ. XNONZ(ROW1) ) GO TO 300
C         ----------------------------------------------------
C         MODIFY RHS VECTOR BY THE PRODUCT OF THE OFF-
C         DIAGONAL BLOCK WITH THE CORRESPONDING PART OF RHS.
C         ----------------------------------------------------
          DO 200 ROW = ROW1, ROW2
             JSTRT = XNONZ(ROW)
             IF ( JSTRT .EQ. LAST ) GO TO 300
             JSTOP = XNONZ(ROW+1) - 1
             IF ( JSTOP .LT. JSTRT ) GO TO 200
                S = 0.0E0
                COUNT = JSTOP - JSTRT + 1
                OPS = OPS + COUNT
                DO 100 J = JSTRT, JSTOP
                   COL = NZSUBS(J)
                   S = S + RHS(COL)*NONZ(J)
  100           CONTINUE
                RHS(ROW) = RHS(ROW) - S
```

```
200              CONTINUE
300              NROW = ROW2 - ROW1 + 1
                 CALL ELSLV ( NROW, XENV(ROW1), ENV, DIAG(ROW1),
      1                       RHS(ROW1) )
                 CALL EUSLV ( NROW, XENV(ROW1), ENV, DIAG(ROW1),
      1                       RHS(ROW1) )
400       CONTINUE
C         --------------------
C         BACKWARD SOLUTION ...
C         --------------------
          IF ( NBLKS .EQ. 1 )  RETURN
          LAST = XBLK(NBLKS) - 1
          DO 500 I = 1, LAST
             TEMP(I) = 0.0E0
500       CONTINUE
          I = NBLKS
          COL1 = XBLK(I)
          COL2 = XBLK(I+1) - 1
600       IF ( I .EQ. 1 )  RETURN
          LAST = XNONZ(COL2+1)
          IF ( LAST .EQ. XNONZ(COL1) )   GO TO 900
C         ----------------------------------------------------
C         MULTIPLY OFF-DIAGONAL BLOCK BY THE CORRESPONDING
C         PART OF THE SOLUTION VECTOR AND STORE IN TEMP.
C         ----------------------------------------------------
          DO 800 COL = COL1, COL2
             S = RHS(COL)
             IF ( S .EQ. 0.0E0 )  GO TO 800
                JSTRT = XNONZ(COL)
                IF ( JSTRT .EQ. LAST )  GO TO 900
                JSTOP = XNONZ(COL+1) - 1
                IF ( JSTOP .LT. JSTRT )  GO TO 800
                   COUNT = JSTOP - JSTRT + 1
                   OPS = OPS + COUNT
                   DO 700 J = JSTRT, JSTOP
                      ROW = NZSUBS(J)
                      TEMP(ROW) = TEMP(ROW) + S*NONZ(J)
700                CONTINUE
800          CONTINUE
900       I = I - 1
          COL1 = XBLK(I)
          COL2 = XBLK(I+1) - 1
          NCOL = COL2 - COL1 + 1
          CALL  ELSLV ( NCOL, XENV(COL1), ENV,
      1                  DIAG(COL1), TEMP(COL1) )
          CALL  EUSLV ( NCOL, XENV(COL1), ENV, DIAG(COL1),
      1                  TEMP(COL1) )
          DO 1000 J = COL1, COL2
             RHS(J) = RHS(J) - TEMP(J)
1000      CONTINUE
          GO TO 600
          END
```

Exercises

6.5.1)   Let L and V be as described in Exercise 4.2.5, where V has only 3 nonzeros per column. Compare the operation costs of computing $V^T L^{-T} L^{-1} V$ as $V^T (L^{-T}(L^{-1}V))$ and $(V^T L^{-T})(L^{-1}V)$. Assume n and p are large, so lower order terms can be ignored.

6.6 Additional Notes

The idea of "throwing away" the off-diagonal blocks of the factor L of A, as discussed in this chapter, can be recursively applied (George 1978c). To explain the strategy suppose A is p by p partitioned, with x and b partitioned correspondingly. Let $A_{(k)}$ denote the leading block-k by block-k principal submatrix of A, and let $x_{(k)}$ and $b_{(k)}$ denote the corresponding parts of x and b respectively. Finally, define submatrices of A as in (6.4.1), with L correspondingly partitioned, as shown below for $p = 5$.

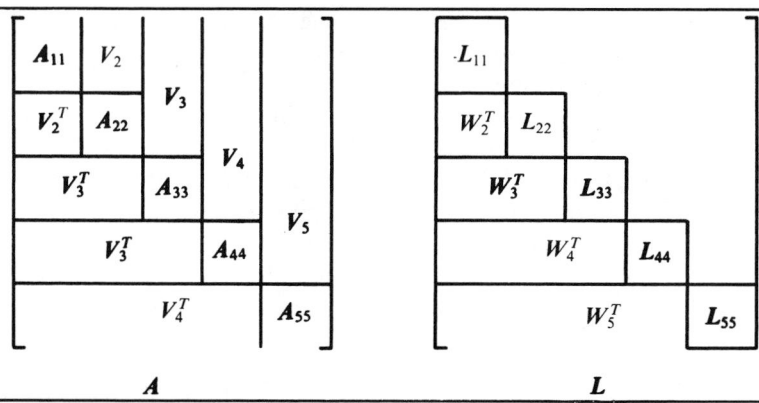

Using this notation, the system $Ax = b$ can be expressed as

$$\begin{pmatrix} A_{(4)} & V_5 \\ V_5^T & A_{55} \end{pmatrix} \begin{pmatrix} x_{(4)} \\ x_5 \end{pmatrix} = \begin{pmatrix} b_{(4)} \\ b_5 \end{pmatrix},$$

and the factorization of A can be expressed as

$$\begin{pmatrix} A_{(4)} & 0 \\ V_5^T & \tilde{A}_{55} \end{pmatrix} \begin{pmatrix} I & A_{(4)}^{-1} V_5 \\ 0 & I \end{pmatrix},$$

where $\tilde{A}_{55} = A_{55} - V_5^T A_{(4)}^{-1} V_5$.

Formally, we can solve $Ax = b$ as follows:

a) *Factorization*: Compute and factor $\tilde{A}_{55}$ into $L_{55} L_{55}^T$. (Note that $V_5^T A_{(4)}^{-1} V_5$ can be computed one column at a time, and the columns discarded after use.)

b) *Solution*:

　　b.1)　Solve $A_{(4)} y_{(4)} = b_{(4)}$.

　　b.2)　Solve $\tilde{A}_{55} x_5 = b_5 - V_5^T y_{(4)}$.

b.3)   Solve $A_{(4)}\tilde{x}_{(4)} = V_5 x_5$.

b.4)   Compute $x_{(4)} = y_{(4)} - \tilde{x}_{(4)}$.

Note that we have only used the ideas presented in Section 6.1 and Exercise 6.1.2 to avoid storing W_5; only V_5 is required. The crucial point is that all that is required for us to solve the five by five partitioned system without storing W_5 is that we be able to solve four by four partitioned systems. Obviously, we can use exactly the same strategy as shown above, to solve the block four by four systems without storing W_4, and so on. Thus, we obtain a method which apparently solves a p by p block system requiring storage only for the diagonal blocks of L and the off-diagonal blocks V_i of the original matrix. However, note that each level of recursion requires a temporary vector $\tilde{x}_{(i)}$ (in Step b.3 above), so there is a point where a finer partitioning no longer achieves a reduction in storage requirement. There are many interesting unexplored questions related to this procedure, and the study of the use of these partitioning and throwaway ideas appears to be a potentially fertile research area.

Partitioning methods have been used successfully in utilizing auxiliary storage (Von Fuchs 1972). The value of p is chosen so that the amount of main storage available is some convenient multiple of $(N/p)^2$. Since A is sparse, some of the blocks will be all zeros. A pointer array is held in main store, with each pointer component either pointing to the current location of the corresponding block, if the block contains nonzeros, or else is zero. If the p by p pointer matrix is itself too large to be held in main store, then it can also be partitioned and the idea recursively applied. This storage management scheme obviously entails a certain amount of overhead, but experience suggests that it is a viable alternative to other out-of-core solution schemes such as band or frontal methods. One advantage is that the actual matrix operations involve simple data structures; only square or rectangular arrays are involved.

Shier (1976) has considered the use of tree partitionings in the context of explicitly inverting a matrix, and provides an algorithm different from ours for finding a tree partitioning of a graph.

7/ One-Way Dissection Methods for Finite Element Problems

7.0 Introduction

In this chapter we consider an ordering strategy designed primarily for problems arising in finite element applications. The strategy is similar to the method of Chapter 6 in that a quotient tree partitioning is obtained, and the computational ideas of implicit solution and asymmetric factorization are exploited. The primary advantage of the one-way dissection algorithm developed in this chapter is that the storage requirements are usually much less than those for either the band or quotient tree schemes described in previous chapters. Indeed, unless the problems are very large, for finite element problems the methods of this chapter are often the best methods in terms of storage requirements of *any* we discuss in this book. They also yield very low solution times, although their factorization times tend to be larger than those of some other methods.

Since the orderings studied in this chapter are quotient tree orderings, the storage and computing methods of Chapter 6 are appropriate, so we do not have to deal with these topics in this chapter. However, the one-way dissection schemes do demand a somewhat more sophisticated storage allocation procedure than that described in Section 6.4.3. This more general allocation procedure is the topic of Section 7.3.

7.1 An Example – The m by ℓ Grid Problem

7.1.1 A One-Way Dissection Ordering

In this section we consider a simple m by ℓ grid problem which motivates the development of the algorithm of Section 7.2. Consider an m by ℓ grid or mesh as shown in Figure 7.1.1, having $N = m\ell$ nodes with $m \leq \ell$. The corresponding finite element matrix problem $Ax = b$ we consider has the property that for some numbering of the equations (nodes) from 1 to N, we have that $a_{ij} \neq 0$ implies node i and node j belong to the same small square.

Figure 7.1.1   An *m* by *ℓ* grid with *m* = 6 and *ℓ* = 11.

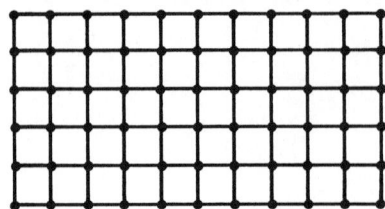

Now let σ be an integer satisfying $1 \leq \sigma \ll \ell$, and choose σ vertical grid lines (which we will refer to as *separators*) which dissect the mesh into $\sigma + 1$ independent blocks of about the same size, as depicted in Figure 7.1.2, where $\sigma = 4$. The $\sigma + 1$ independent blocks are numbered row by row, followed by the separators, as indicated by the arrows in Figure 7.1.2. The matrix structure this ordering induces in the triangular factor L is shown in Figure 7.1.3, where the off-diagonal blocks with fills are hatched. We let

$$\delta = \frac{\ell - \sigma}{\sigma + 1} \ ,$$

that is, the length between dissectors.

Figure 7.1.2   One-way dissection ordering of an *m* by *ℓ* grid, indicated by the arrows. Here $\sigma = 4$, yielding a partitioning with $2\sigma + 1 = 9$ members indicated by the circled numbers.

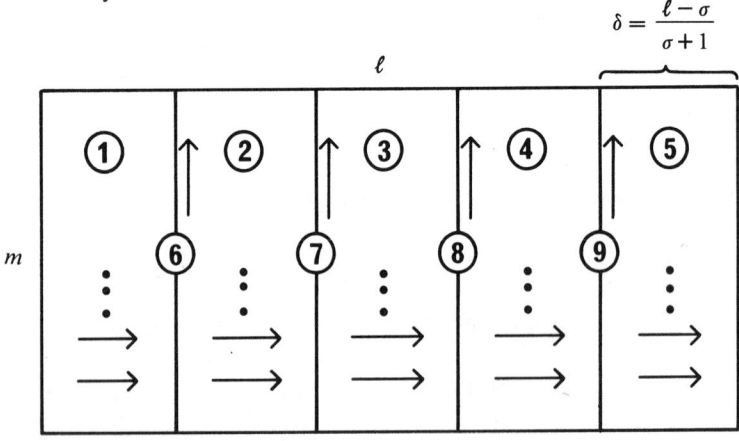

Regarding A and L as partitioned into q^2 submatrices, where $q = 2\sigma + 1$, we first note the dimensions of the various blocks:

Figure 7.1.3   Matrix structure of L induced by the one-way dissection ordering of Figure 7.1.2. The hatched areas indicate where fill occurs in the off-diagonal blocks.

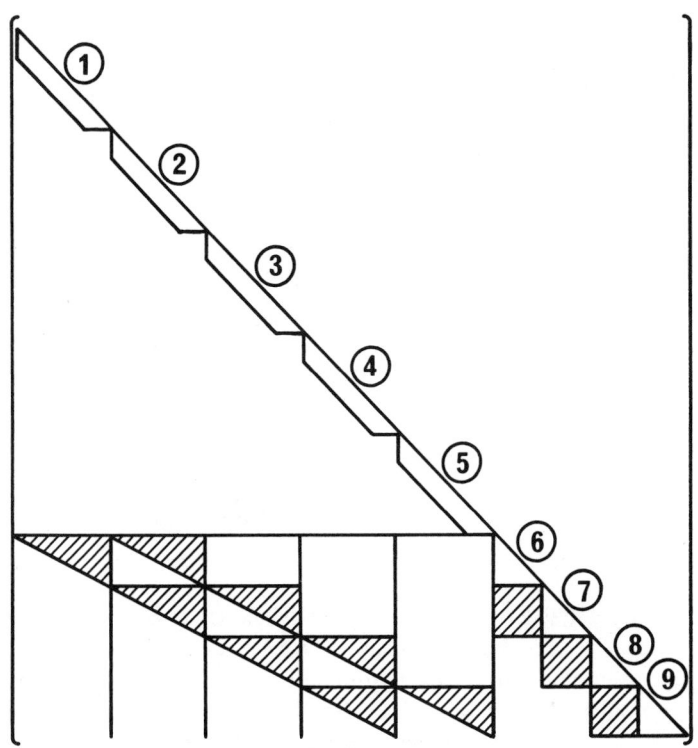

1) A_{kk} is $m\delta$ by $m\delta$ for $1 \le k \le \sigma+1$.

2) A_{kj} is m by $m\delta$ for $k > \sigma+1$ and $j \le \sigma+1$.   (7.1.1)

3) A_{kj} is m by m for $j > \sigma+1$ and $k > \sigma+1$.

Of course in practice σ must be chosen to be an integer, and unless δ is also an integer, the leading $\sigma+1$ diagonal blocks will not all be exactly the same size. However, we will see later that these aberrations are of little practical significance; in any case, our objective in this section is to present some *basic ideas* rather than to study this m by ℓ grid problem in meticulous detail. For our purposes, we assume that σ and δ are integers, and that m and ℓ are large enough that $m << m^2$ and $\ell << \ell^2$.

As we have already stated, the utility of this ordering hinges on using the partitioned matrix techniques developed in Chapter 6.

Indeed, it is not difficult to determine that this ordering is no better or even worse than the standard band ordering if these techniques are not used.   (see Exercises 7.1.2 and 7.1.3.)

The key observation which allows us to use the quotient tree techniques developed in Chapter 6 is that if we view the σ separator blocks as forming a *single* partition member, then the resulting partitioning, now with $p = \sigma + 2$ members, is a *monotonely ordered tree partitioning*.   This is depicted in Figure 7.1.4 for the example of Figure 7.1.2.

Figure 7.1.4   Quotient tree corresponding to one-way dissection ordering, obtained by placing the separators together in one partition.

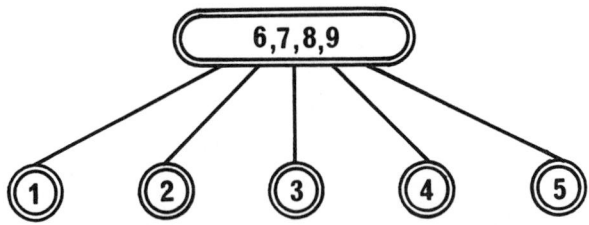

Thus, we will use the storage scheme developed in Section 6.4, and thereby store only the diagonal blocks of L, and the off-diagonal blocks of A. For discussion purposes we will continue to regard A and L as q by q partitioned where $q = 2\sigma + 1$, although the reader should understand that for computational purposes the last σ partition members are combined into one, so that in effect $p = \sigma + 2$ members are involved.

7.1.2   Storage Requirements

Denoting the partitions of L corresponding to A_{ij} by L_{ij}, $1 \le i,j \le 2\sigma + 1$, we now derive an *estimate* for the storage requirements of this one-way dissection ordering, using the implicit storage scheme described in Section 6.4.1. The primary storage requirements are as follows:

i) L_{kk}, $1 \le k \le \sigma + 1$.   The bandwidth of these band matrices is $(\ell + 1)/(\sigma + 1)$, yielding a combined storage requirement of

$$\frac{m(\ell - \sigma)(\ell + 1)}{(\sigma + 1)} \simeq \frac{m\ell^2}{\sigma} .$$

ii) L_{kj}, $\sigma + 1 < j,k \le 2\sigma + 1$, $j < k$. There are $\sigma - 1$ fill blocks, and σ lower triangular blocks, all of which are m by m, yielding a total storage requirement of

$$\frac{(\sigma-1)m^2 + \sigma m(m+1)}{2} \simeq \frac{3\sigma m^2}{2} \; .$$

iii) A_{kj}, $k \le \sigma+1$, $j > \sigma+1$. Except for nodes near the boundary of the grid, all nodes on the separators are connected to 6 nodes in the leading $\sigma+1$ blocks. Thus, primary storage for these matrices totals about $6\sigma m$.

The overhead storage for items i) and ii) is $\ell m + \sigma + 3$ (for the array XENV and XBLK), and about $6\sigma m + \ell m$ for XNONZ and NZSUBS. Thus, if lower order terms are ignored, the storage requirement for this ordering, using the implicit storage scheme of Section 6.4.1, is approximately

$$S(\sigma) = \frac{m\ell^2}{\sigma} + \frac{3\sigma m^2}{2} \; . \tag{7.1.2}$$

If our objective is to minimize storage, then we want to choose σ to minimize $S(\sigma)$. Differentiating with respect to σ, we have

$$\frac{dS}{d\sigma} = -\frac{m\ell^2}{\sigma^2} + \frac{3m^2}{2} \; .$$

Using this, we find that S is approximately minimized by choosing $\sigma = \sigma^*$ where

$$\sigma^* = \ell \left(\frac{2}{3m} \right)^{1/2} \tag{7.1.3}$$

yielding

$$S(\sigma^*) = \sqrt{6} m^{3/2} \ell + O(m\ell) \; . \tag{7.1.4}$$

Note that the corresponding optimal δ^* is given by

$$\delta^* = \left(\frac{3m}{2} \right)^{1/2} \; .$$

It is interesting to compare this result with the storage requirements we would expect if we used a standard band or envelope scheme. Since $m \le \ell$, we would number the grid column by column, yielding a matrix whose bandwidth is $m+1$, and for large m and ℓ the storage required for L would be $m^2\ell + O(m\ell)$. Thus, asymptotically, this one-way dissection scheme reduces storage requirements by a factor of $\sqrt{6/m}$ over the standard schemes of Chapter 4.

7.1.3 Operation Count for the Factorization

Let us now consider the computational requirements for this one-way dissection ordering. Basically, we simply have to count the operations needed to perform the factorization and solution algorithm described in Section 6.2. However, the off-diagonal blocks A_{kj}, for $k \leq \sigma + 1$ and $j > \sigma + 1$, have rather special "pseudo tri-diagonal" structure, which is exploited by the subroutines TSFCT and TSSLV. Thus, determining an approximate operation count is far from trivial. In this Section we consider the factorization; Section 7.1.4 contains a derivation of an approximate operation count for the solution.

It is helpful in the derivation to break the computation into the three categories, where again we ignore low order terms in the calculations.

1. The factorization of the $\sigma + 1$ leading diagonal blocks.

    (In our example of Figures 7.1.1–7.1.3, where $\sigma = 4$, this is the computation of L_{kk}, $1 \leq k \leq 5$.) Observing that the bandwidth of these matrices is $(\ell + 1)/(\sigma + 1)$, and using Theorem 4.1.1, we conclude that the operation count for this category is approximately

$$\frac{m \ell^3}{2\sigma^2} .$$

2. The computation of the L_{kj}, for $k \geq j$ and $j > \sigma + 1$.

    This corresponds to factoring an $m\sigma$ by $m\sigma$ block tri-diagonal matrix having blocks of size m. Using the results of Sections 2.1 and 2.2, we find that operation count is approximately

$$\frac{7\sigma m^3}{6} .$$

3. The modifications to $A_{jj}, A_{j+1,j}$, and $A_{j+1,j+1}$ for $j > \sigma + 1$ involving the off-diagonal blocks A_{kj} and the computed L_{kk}, $k \leq \sigma + 1$, as depicted in Figure 7.1.5.

    The operation count for this computation is discussed below.

In computing the modification matrix in the asymmetric way, we have to compute

$$L_{kk}^{-T}(L_{kk}^{-1}A_{kj}) , \qquad (7.1.5)$$

for $j > \sigma + 1$ and $k \leq \sigma + 1$. In view of the results in Section 6.5, it is not necessary to compute that part of $L_{kk}^{-T}(L_{kk}^{-1}A_{kj})$ which is above the first nonzero in each column of A_{kj}. Thus, when computing $W = L_{kk}^{-1}A_{kj}$, we exploit leading zeros in the columns of A_{kj}, and when computing $\hat{W} = L_{kk}^{-T}W$ we stop the computation as soon as the last

Figure 7.1.5   Matrices which interact with and/or are modified by L_{kk}, A_{kj} and $A_{k,j+1}$, where $k = 3$ and $j = 7$.

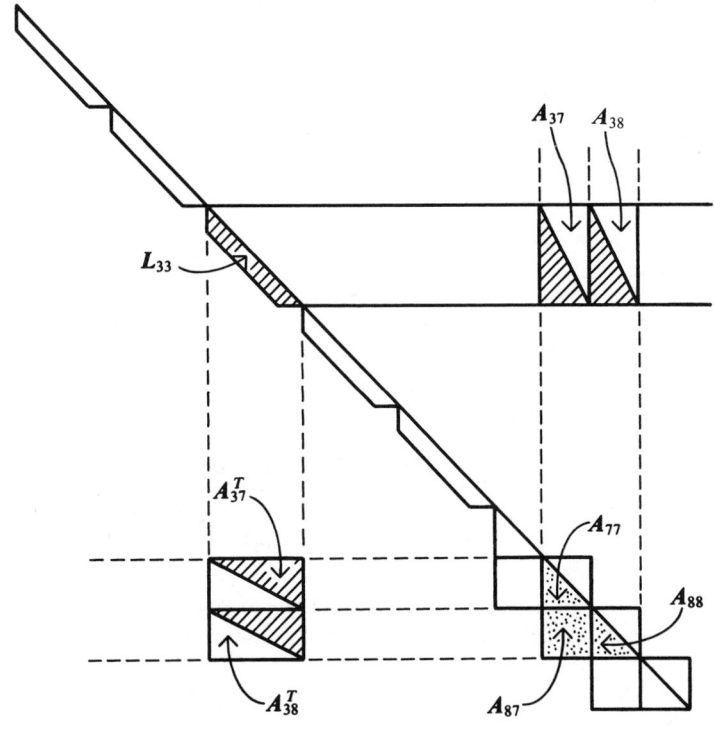

required element of $\tilde{W}$ has been computed, as depicted in Figure 7.1.6.

Figure 7.1.6   Structure of A_{kj} and the part of $\tilde{W}$ that needs to be computed.

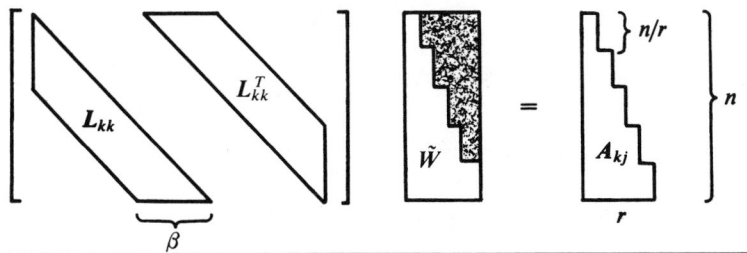

It is straightforward to show that the number of operations required to compute this part of $\tilde{W}$ is approximately given by

$$n(\beta + 1)(r - 1) , \qquad (7.1.6)$$

where A_{kj} is n by r and $L_{kk} + L_{kk}^T$ is an n by n band matrix with bandwidth $\beta \ll n$ (see Exercise 7.1.5). Here, $n \simeq m\ell/\sigma$, $\beta \simeq \ell/\sigma$ and $r = m$; thus the expression (7.1.6) becomes

$$\frac{m^2 \ell^2}{\sigma^2} .$$

Note that there are in total 2σ such off-diagonal blocks, so the computation required to compute all

$$L_{kk}^{-T}(L_{kk}^{-1}A_{kj}) \qquad \text{for } j > \sigma+1 \text{ and } k \leq \sigma+1$$

is approximately

$$\frac{2m^2 \ell^2}{\sigma} . \tag{7.1.7}$$

We now estimate the cost of computing the modifications to A_{kj}, $k > \sigma+1$, $j > \sigma+1$. With (7.1.5) computed, we note that the modification to each entry in the diagonal blocks A_{kk}, $k > \sigma+1$ can be computed in six operations, while that to the off-diagonal blocks $A_{k,k-1}$ requires three operations. That is, the cost for modification is $O(\sigma m^2)$.

Thus, an estimate for the total number of operations required for the factorization, using this one-way dissection ordering, is given by

$$\theta_F(\sigma) = \frac{m\ell^3}{2\sigma^2} + \frac{7\sigma m^3}{6} + \frac{2m^2\ell^2}{\sigma} . \tag{7.1.8}$$

If our objective is to minimize the operation count for the factorization, using this one-way dissection ordering, we want to find the σ_F which minimizes $\theta_F(\sigma)$. For large m and ℓ, it can be shown that choosing

$$\sigma_F = \ell \left(\frac{12}{7m} \right)^{1/2}$$

approximately minimizes (7.1.8), yielding (see Exercise 7.1.6)

$$\theta_F(\sigma_F) = \left(\frac{28}{3} \right)^{1/2} m^{5/2}\ell + O(m^2\ell) . \tag{7.1.9}$$

The corresponding δ_F is given by

$$\delta_F = \left(\frac{7m}{12} \right)^{1/2} .$$

Again it is interesting to compare this result with the operation count if we use a standard band or envelope scheme as described in

Chapter 4. For this m by ℓ grid problem, the factorization operation count would be $\simeq \frac{1}{2}m^3\ell$. Thus, asymptotically this one-way dissection scheme reduces the factorization count by a factor of roughly $4\sqrt{7}/(3m)$.

7.1.4 Operation Count for the Solution

We now derive an estimate of the operation count required to solve $Ax = b$, given the "factorization" as computed in the preceding subsection.

First observe that each of the $\sigma+1$ leading diagonal blocks L_{kk}, $1 \leq k \leq \sigma+1$, is used four times, twice in the lower solve and twice again in the upper solve. This yields an operation count of approximately

$$\frac{4m\,\ell^2}{\sigma} \, .$$

The non-null blocks L_{kj}, for $k > \sigma+1$ and $j > \sigma+1$, are each used twice, for a total operation count of $3\sigma m^2 + O(\sigma m)$. Each matrix A_{kj}, for $k > \sigma+1$ and $j \leq \sigma+1$, is used twice, yielding an operation count of about $12\sigma m$. Thus, an estimate for the operation count associated with the solution, using this one-way dissection ordering, is

$$\theta_S(\sigma) = \frac{4m\,\ell^2}{\sigma} + 3\sigma m^2 \, . \tag{7.1.10}$$

If we wish to minimize θ_S with respect to σ, we find σ should be approximately

$$\sigma_S = \frac{2\ell}{\sqrt{3m}} \, ,$$

whence

$$\theta_S(\sigma_S) = 4\sqrt{3}m^{3/2}\ell + O(m\,\ell) \, . \tag{7.1.11}$$

Again it is interesting to compare (7.1.11) with the corresponding operation count if we were to use standard band or envelope schemes, which would be about $2m^2\ell$. Thus, asymptotically, the one-way dissection ordering reduces the solution operation count by a factor of about $2\sqrt{3/m}$.

Of course in practice we cannot choose σ to simultaneously minimize storage, factorization operation count, and solution operation count; σ must be fixed. Since the main attraction for these methods is their low storage requirements, in the algorithm of the next section σ is chosen to attempt to minimize storage.

Exercises

7.1.1)   What are the coefficients of the high order terms in (7.1.8) and (7.1.11) if σ is chosen to be σ^*, given by (7.1.3)?

7.1.2)   Suppose we use the one-way dissection ordering of this section with σ chosen to be $O(\ell/\sqrt{m}\,)$, but we do not use the implicit storage technique; that is, we actually store the off-diagonal blocks L_{ij}, $i > \sigma+1$, $j \leq \sigma+1$. What would the storage requirements be then? If we used these blocks in the solution scheme, what would the operation count corresponding to θ_S now be?

7.1.3)   Suppose we use the one-way dissection ordering of this section with σ chosen to be $\ell/\sqrt{m}$, but we use the symmetric version of the factorization scheme rather than the asymmetric version. (See Section 6.1.1) Show that now the factorization operation count is $O(m^3\ell)$ rather than $O(m^{5/2}\ell)$. How much temporary storage is required to carry out the computation?

7.1.4)   Throughout Section 7.1 we assume that $m \leq \ell$, although we did not explicitly use that fact anywhere. Do our results still apply for $m > \ell$? Why did we assume $m \leq \ell$?

7.1.5)   Let M be an n by n symmetric positive definite band matrix with bandwidth $\beta << n$ and Cholesky factorization LL^T. Let V be an n by r ($r << n$) "pseudo-tridiagonal" matrix, for which the leading nonzero in column i is in position

$$\mu_i = \left\lceil \frac{(i-1)(n-1)}{r-1} \right\rceil$$

and let $\tilde{W} = L^{-T}(L^{-1}V)$. Show that the number of operations required to compute $\tilde{w}_{ij}$, $1 \leq i \leq r$, $\mu_i \leq j \leq n$, is approximately $n(\beta+1)(r-1)$. (Note that this is approximately the pseudo-lower triangle of $\tilde{W}$, described in Exercise 2.2.8.)

7.1.6)   Let σ_F minimize $\theta_F(\sigma)$ in (7.1.8). Show that a lower bound for σ_F is given by

$$\bar{\sigma}_F = \ell \left(\frac{12}{7m} \right)^{1/2},$$

whence

$$\theta_F(\bar{\sigma}_F) = \frac{7}{24} m^2\ell + 2 \left(\frac{7}{3} \right)^{1/2} m^{5/2}\ell \ .$$

7.1.7) In the description of the one-way dissection ordering of the m by ℓ grid given in this section, the separator blocks were numbered "end to end." It turns out that the order in which these separator blocks are numbered is important. For example, the blocks in the example of Figure 7.1.2 might have been numbered as indicated in the diagram below.

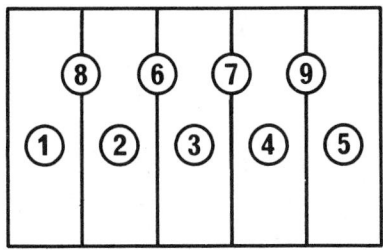

a) Draw a figure similar to Figure 7.1.5 showing the structure of L corresponding to this ordering. Is there more or fewer fill blocks than the ordering shown in Figure 7.1.2?

b) For the one-way dissection ordering of the m by ℓ grid shown in Figure 7.1.2, the number of fill blocks is $\sigma - 1$. Show that for some orderings of the separator blocks, as many as $2\sigma - 3$ fill blocks may result.

7.2 An Algorithm for Finding One-Way Dissection Orderings of Irregular Mesh Problems

7.2.1 The Algorithm

The description and analysis of the previous section suggests that in general terms, we would like to find a set of "parallel" separators having relatively few nodes. These separators should disconnect the graph or mesh into components which can be ordered so as to have small envelopes. This is essentially what the following heuristic algorithm attempts to do.

The algorithm operates on a given graph $G = (X, E)$, which we assume to be connected. The extension to disconnected graphs is obvious. Recall from Chapter 3 that the set $Y \subset X$ is a *separator* of the connected graph G if the section graph $G(X - Y)$ is disconnected.

We now give a step-by-step description of the algorithm, followed by some explanatory remarks for the important steps. In the algorithm $N = |X|$, and m and ℓ correspond roughly to m and $\ell - 1$ in Section 7.1. The algorithm attempts to choose σ to minimize storage, but it can easily be modified so as to attempt to minimize the factorization or solution operation count.

Step 1   (Generate level structure) Find a pseudo-peripheral node x by the algorithm of Section 4.3.2, and generate the level structure rooted at x.

$$\mathcal{L}(x) = \{L_0, \ L_1, \ \ldots, \ L_\ell\} \ .$$

Step 2   (Estimate δ) Calculate $m = N/(\ell + 1)$, and set

$$\delta = \left(\frac{3m + 13}{2} \right)^{1/2} .$$

Step 3   (Limiting case) If $\delta < \ell/2$, and $|X| > 50$ go to Step 4. Otherwise, set $p = 1$, $Y_p = X$ and go to Step 6.

Step 4   (Find separator) Set $i = 1$, $j = \lfloor \delta + 0.5 \rfloor$, and $T = \varnothing$. While $j < \ell$ do the following

    4.1)   Choose $T_i = \{x \in L_j \mid Adj(x) \cap L_{j+1} \neq \varnothing\}$
    4.2)   Set $T \leftarrow T \cup T_i$
    4.3)   Set $i \leftarrow i + 1$ and $j = \lfloor i\delta + 0.5 \rfloor$.

Step 5   (Define blocks) Let Y_k, $k = 1, \ldots, p - 1$ be the connected components of the section graph $G(X - T)$, and set $Y_p = T$.

Step 6   (Internal numbering) Number each Y_k, $k = 1, \ldots, p$ consecutively using the method described in Section 6.4.2.

Step 1 of the algorithm produces a (hopefully) long, narrow level structure. This is desirable because the separators are selected as subsets of some of the levels L_i.

The calculation of the numbers m and ℓ computed in Step 2 is motivated directly by the analysis of the m by ℓ grid in Section 7.1. Since m is the average number of nodes per level, it serves as a measure of the width of the level structure. The derivation of σ^* given in (7.1.3) was obtained in a fairly crude way, since our objective was simply to convey the basic ideas. A more careful analysis along with some experimentation suggests that a better value for σ^* is

$$\ell \left(\frac{2}{3m + 13} \right)^{1/2} .$$

The corresponding δ^* is given by the formula used in Step 2.

Step 3 is designed to handle anomalous situations where $m \ll \ell$, or when N is simply too small to make the use of the one-way dissection method worthwhile. Experiments indicate that for small finite element problems, and/or "long slender" problems, the methods of Chapter 4 are more efficient, regardless of the basis for comparison. In these cases, the entire graph is processed as one block ($p = 1$).

That is, an ordinary envelope ordering as discussed in Chapter 4 is produced for the graph.

Step 4 performs the actual selection of the separators, and is done essentially as though the graph corresponds to an m by ℓ grid as studied in Section 7.1. As noted earlier, each L_i of $\mathcal{L}$ is a separator of G. In Step 4, approximately equally spaced levels are chosen from $\mathcal{L}$, and subsets of these levels (the T_i) which are possibly smaller separators are then found.

Finally, in Step 6 the $p \geq \sigma + 1$ independent blocks created by removing the separators from the graph are numbered, using the internal renumbering scheme described in Section 6.4.2.

Although the choice of σ and the method of selection of the separators seems rather crude, we have found that attempts at more sophistication do not often yield significant benefits (except for some unrealistic, contrived examples). Just as in the regular rectangular grid case, the storage requirement, as a function of σ, is very flat near its minimum. Even relatively large perturbations in the value of σ, and in the selection of the separators, usually produce rather small changes in storage requirements.

In Figure 7.2.1 we have an example of an irregular mesh problem, along with some indications of the steps carried out by the algorithm. (For purposes of this illustration, we assume that the test for $|X| > 50$ in Step 3 of the algorithm has been removed.) Figure 7.2.1 (a) contains the level numbers of the original level structure, while Figure 7.2.1 (b) displays the nodes chosen as the separators. Here $m = N/(\ell + 1) = 25/11 \simeq 2.27$, $\delta = \sqrt{9.91} \simeq 3.12$. The levels chosen from which to pick the separators are levels 4 and 8.

7.2.2 Subroutines for Finding a One-Way Dissection Partitioning

The set of subroutines which implements the one-way dissection algorithm is given in the control diagram in Figure 7.2.2. The subroutines FNROOT and ROOTLS are used together to determine a pseudo-peripheral node of a connected component in a given graph. They have been discussed in detail in Section 4.3.3. The subroutine REVRSE is a utility subroutine that is used to reverse an integer array. The execution of the calling statement

CALL REVRSE(NV, V)

will interchange the entries in the integer vector V of size NV in the following way:

$$V(i) \leftrightarrow V(NV - i + 1) \text{ for } 1 \leq i \leq \lfloor NV/2 \rfloor .$$

The remaining two subroutines GEN1WD and FN1WD are described in detail below.

Figure 7.2.1 Diagram of an irregular mesh showing the separators chosen by the algorithm.

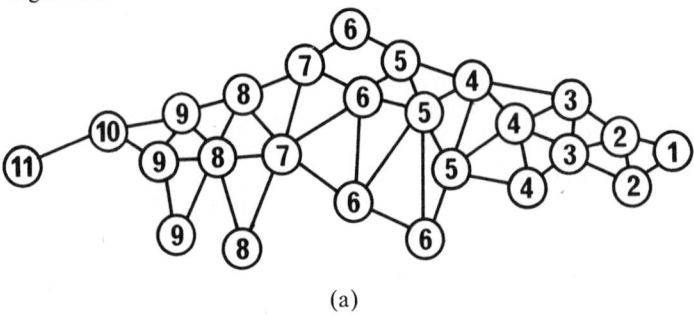

(a)

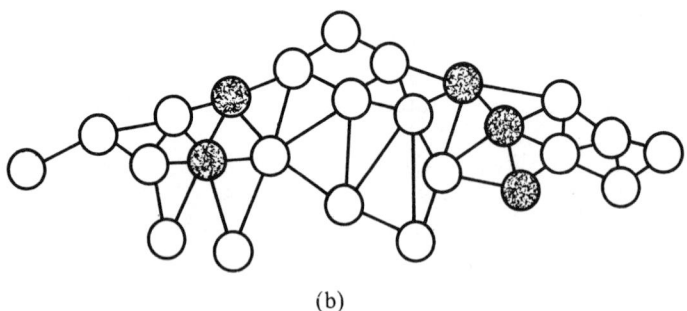

(b)

GEN1WD (GENeral 1-Way Dissection)

This is the driver subroutine for finding a one-way dissection partitioning of a general disconnected graph. The input and output parameters of GEN1WD follow the same notations as the implementations of other ordering algorithms. The parameters NEQNS, XADJ and ADJNCY are for the adjacency structure of the given graph. Returned from the subroutine are the one-way dissection ordering in the vector PERM, and the partitioning information in (NBLKS, XBLK). Three working vectors MASK, XLS and LS are used by GEN1WD. The array pair (XLS, LS) is used by FN1WD to store a level structure rooted at a pseudo-peripheral node, and the vector MASK is used by the subroutine to mask off nodes that have been numbered.

The subroutine begins by initializing the vector MASK so that all nodes are considered unnumbered. It then goes through the graph and obtains a node i not yet numbered. The node defines an unnumbered connected component in the graph and the subroutine FN1WD is called to find a one-way dissector for the component. The set of dissecting nodes forms a block in the partitioning. Each component in the

Figure 7.2.2   Control relation of subroutines for the one-way dissection algorithm.

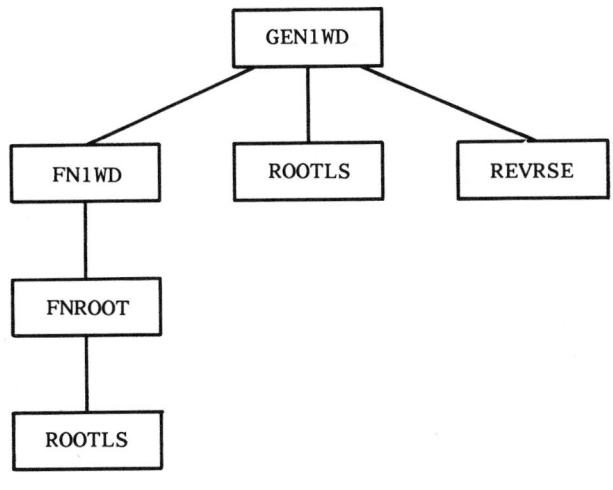

remainder of the dissected subgraph also constitutes a block, and they are found by calling the subroutine ROOTLS.

After going through all the connected components in the graph, the subroutine reverses the permutation vector PERM and block index vector XBLK, since the one-way dissectors which are found first should be ordered after the remaining nodes.

```
C****************************************************************
C****************************************************************
C********     GEN1WD ..... GENERAL ONE-WAY DISSECTION  ******
C****************************************************************
C****************************************************************
C
C     PURPOSE - GEN1WD FINDS A ONE-WAY DISSECTION PARTITIONING
C        FOR A GENERAL GRAPH.  FN1WD IS USED FOR EACH CONNECTED
C        COMPONENT.
C
C     INPUT PARAMETERS -
C        NEQNS - NUMBER OF EQUATIONS.
C        (XADJ, ADJNCY) - THE ADJACENCY STRUCTURE PAIR.
C
C     OUTPUT PARAMETERS -
C        (NBLKS, XBLK) - THE PARTITIONING FOUND.
C        PERM - THE ONE-WAY DISSECTION ORDERING.
C
C     WORKING VECTORS -
C        MASK - IS USED TO MARK VARIABLES THAT HAVE
C               BEEN NUMBERED DURING THE ORDERING PROCESS.
C        (XLS, LS) - LEVEL STRUCTURE USED BY ROOTLS.
C
C     PROGRAM SUBROUTINES -
C        FN1WD, REVRSE, ROOTLS.
C
```

```
C*****************************************************************
C
      SUBROUTINE  GEN1WD ( NEQNS, XADJ, ADJNCY, MASK,
     1                     NBLKS, XBLK, PERM, XLS, LS )
C
C*****************************************************************
C
      INTEGER ADJNCY(1), LS(1), MASK(1), PERM(1),
     1        XBLK(1), XLS(1)
      INTEGER XADJ(1), CCSIZE, I, J, K, LNUM,
     1        NBLKS, NEQNS, NLVL, NODE, NSEP,
     1        NUM, ROOT
C
C*****************************************************************
C
      DO 100 I = 1, NEQNS
         MASK(I) = 1
 100  CONTINUE
      NBLKS = 0
      NUM   = 0
      DO 400 I = 1, NEQNS
         IF ( MASK(I) .EQ. 0 )  GO TO 400
C            ---------------------------------------------
C            FIND A ONE-WAY DISSECTOR FOR EACH COMPONENT.
C            ---------------------------------------------
             ROOT = I
             CALL  FN1WD ( ROOT, XADJ, ADJNCY, MASK,
     1                     NSEP, PERM(NUM+1), NLVL, XLS, LS )
             NUM = NUM + NSEP
             NBLKS = NBLKS + 1
             XBLK(NBLKS) = NEQNS - NUM + 1
             CCSIZE = XLS(NLVL+1) - 1
C            ---------------------------------------------
C            NUMBER THE REMAINING NODES IN THE COMPONENT.
C            EACH COMPONENT IN THE REMAINING SUBGRAPH FORMS
C            A NEW BLOCK IN THE PARTITIONING.
C            ---------------------------------------------
             DO 300 J = 1, CCSIZE
                NODE = LS(J)
                IF ( MASK(NODE) .EQ. 0 )  GO TO 300
                    CALL  ROOTLS ( NODE, XADJ, ADJNCY, MASK,
     1                             NLVL, XLS, PERM(NUM+1) )
                    LNUM = NUM + 1
                    NUM  = NUM + XLS(NLVL+1) - 1
                    NBLKS = NBLKS + 1
                    XBLK(NBLKS) = NEQNS - NUM + 1
                    DO 200 K = LNUM, NUM
                       NODE = PERM(K)
                       MASK(NODE) = 0
 200                CONTINUE
                    IF ( NUM .GT. NEQNS )  GO TO 500
 300         CONTINUE
 400  CONTINUE
C     -------------------------------------------------------
C     SINCE DISSECTORS FOUND FIRST SHOULD BE ORDERED LAST,
C     ROUTINE REVRSE IS CALLED TO ADJUST THE ORDERING
C     VECTOR, AND THE BLOCK INDEX VECTOR.
C     -------------------------------------------------------
 500  CALL  REVRSE ( NEQNS, PERM )
      CALL  REVRSE ( NBLKS, XBLK )
      XBLK(NBLKS+1) = NEQNS + 1
      RETURN
      END
```

FN1WD (FiNd 1-Way Dissection ordering)

This subroutine applies the one-way dissection algorithm described in Section 7.2.1 to a connected component of a subgraph. It operates on a component specified by the input parameters ROOT, MASK, XADJ and ADJNCY. Output from this subroutine is the set of dissecting nodes given by (NSEP, SEP).

The first step in the subroutine is to find a level structure rooted at a pseudo-peripheral node which it does by calling FNROOT. Based on the characteristics of the level structure (NLVL, the number of levels and WIDTH, the average width), the subroutine determines the level increment DELTA to be used. If the number of levels NLVL or the size of the component is too small, the whole component is returned as the "dissector".

With DELTA determined, the subroutine then marches along the level structure picking up levels, subsets of which form the set of parallel dissectors.

```
C***************************************************************
C***************************************************************
C*******       FN1WD .....  FIND ONE-WAY DISSECTORS       ******
C***************************************************************
C***************************************************************
C
C     PURPOSE - THIS SUBROUTINE FINDS ONE-WAY DISSECTORS OF
C        A CONNECTED COMPONENT SPECIFIED BY MASK AND ROOT.
C
C     INPUT PARAMETERS -
C        ROOT - A NODE THAT DEFINES (ALONG WITH MASK) THE
C               COMPONENT TO BE PROCESSED.
C        (XADJ, ADJNCY) - THE ADJACENCY STRUCTURE.
C
C     OUTPUT PARAMETERS -
C        NSEP - NUMBER OF NODES IN THE ONE-WAY DISSECTORS.
C        SEP - VECTOR CONTAINING THE DISSECTOR NODES.
C
C     UPDATED PARAMETER -
C        MASK - NODES IN THE DISSECTOR HAVE THEIR MASK VALUES
C               SET TO ZERO.
C
C     WORKING PARAMETERS-
C        (XLS, LS) - LEVEL STRUCTURE USED BY THE ROUTINE FNROOT.
C
C     PROGRAM SUBROUTINE -
C        FNROOT.
C
C***************************************************************
C
        SUBROUTINE  FN1WD ( ROOT, XADJ, ADJNCY, MASK,
     1                      NSEP, SEP, NLVL, XLS, LS )
C
C***************************************************************
C
        INTEGER ADJNCY(1), LS(1), MASK(1), SEP(1), XLS(1)
        INTEGER XADJ(1), I, J, K, KSTOP, KSTRT, LP1BEG, LP1END,
     1          LVL, LVLBEG, LVLEND, NBR, NLVL, NODE,
     1          NSEP, ROOT
        REAL DELTP1, FNLVL, WIDTH
C
```

```
C********************************************************************
C
          CALL   FNROOT ( ROOT, XADJ, ADJNCY, MASK,
     1                     NLVL, XLS, LS )
          FNLVL = FLOAT(NLVL)
          NSEP  = XLS(NLVL + 1) - 1
          WIDTH = FLOAT(NSEP) / FNLVL
          DELTP1 = 1.0 + SQRT((3.0*WIDTH+13.0)/2.0)
          IF  (NSEP .GE. 50 .AND. DELTP1 .LE. 0.5*FNLVL) GO TO 300
C         ----------------------------------------------------------
C         THE COMPONENT IS TOO SMALL, OR THE LEVEL STRUCTURE
C         IS VERY LONG AND NARROW. RETURN THE WHOLE COMPONENT.
C         ----------------------------------------------------------
             DO 200 I = 1, NSEP
                NODE = LS(I)
                SEP(I) = NODE
                MASK(NODE) = 0
  200        CONTINUE
             RETURN
C         ----------------------------
C         FIND THE PARALLEL DISSECTORS.
C         ----------------------------
  300     NSEP = 0
          I = 0
  400     I = I + 1
             LVL = IFIX (FLOAT(I)*DELTP1 + 0.5)
             IF ( LVL .GE. NLVL )   RETURN
             LVLBEG = XLS(LVL)
             LP1BEG = XLS(LVL + 1)
             LVLEND = LP1BEG - 1
             LP1END = XLS(LVL + 2) - 1
             DO 500 J = LP1BEG, LP1END
                NODE = LS(J)
                XADJ(NODE) =  - XADJ(NODE)
  500        CONTINUE
C         ---------------------------------------------------------
C         NODES IN LEVEL LVL ARE CHOSEN TO FORM DISSECTOR.
C         INCLUDE ONLY THOSE WITH NEIGHBORS IN LVL+1 LEVEL.
C         XADJ IS USED TEMPORARILY TO MARK NODES IN LVL+1.
C         ---------------------------------------------------------
             DO 700 J = LVLBEG, LVLEND
                NODE = LS(J)
                KSTRT = XADJ(NODE)
                KSTOP = IABS(XADJ(NODE+1)) - 1
                DO 600 K = KSTRT, KSTOP
                   NBR = ADJNCY(K)
                   IF ( XADJ(NBR) .GT. 0 )   GO TO 600
                      NSEP = NSEP + 1
                      SEP(NSEP) = NODE
                      MASK(NODE) = 0
                      GO TO 700
  600           CONTINUE
  700        CONTINUE
             DO 800 J = LP1BEG, LP1END
                NODE = LS(J)
                XADJ(NODE) = - XADJ(NODE)
  800        CONTINUE
           GO TO 400
       END
```

7.3 On Finding the Envelope Structure of Diagonal Blocks

In Chapter 4, the envelope structure of a symmetric matrix A has been studied. It has been shown that the envelope structure is preserved under symmetric factorization; in other words, if F is the filled matrix of A, then

$$Env(A) = Env(F) \ .$$

In this section, we consider the envelope structure of the diagonal block submatrices of the filled matrix with respect to a given partitioning. This is important in setting up the data structure for the storage scheme described in Section 6.4.1.

7.3.1 Statement of the Problem

Let A be a sparse symmetric matrix partitioned as

$$A = \begin{pmatrix} A_{11} & A_{12} & \cdots & A_{1p} \\ A_{12}^T & A_{22} & \cdots & A_{2p} \\ \cdot & & & \cdot \\ & & & \\ \cdot & & & \cdot \\ A_{1p}^T & A_{2p}^T & \cdots & A_{pp} \end{pmatrix} , \qquad (7.3.1)$$

where each A_{kk} is a square submatrix. The *block diagonal matrix* of A with respect to the given partitioning is defined to be

$$Bdiag(A) = \begin{pmatrix} A_{11} & & & 0 \\ & A_{22} & & \\ & & \cdot & \\ & & & \cdot \\ & & & & \cdot \\ 0 & & & A_{pp} \end{pmatrix} . \qquad (7.3.2)$$

Let the triangular factor L of A be correspondingly partitioned as

$$L = \begin{pmatrix} L_{11} & & & & & 0 \\ L_{21} & L_{22} & & & & \\ & & \cdot & & \cdot & \\ & & \cdot & & & \\ & & \cdot & & & \\ & & \cdot & & & \\ L_{p1} & L_{p2} & \cdot & \cdot & \cdot & L_{pp} \end{pmatrix} .$$

Then the associated block diagonal matrix of the filled matrix F will be

$$Bdiag(F) = \begin{pmatrix} F_{11} & & & & 0 \\ & F_{22} & & & \\ & & \cdot & & \\ & & & \cdot & \\ & & & & \cdot \\ 0 & & & & F_{pp} \end{pmatrix}$$

where $F_{kk} = L_{kk} + L_{kk}^T$ for $1 \le k \le p$.

Our objective is to determine the envelope structure of $Bdiag(F)$. Although $Env(A) = Env(F)$, the result does not hold in general for $Bdiag(A)$ and $Bdiag(F)$ due to the possible creation of nonzeros outside $Env(Bdiag(A))$ during the factorization.

7.3.2 Characterization of the Block Diagonal Envelope via Reachable Sets

Recall from Chapter 4 that the envelope structure of a matrix A is characterized by the column subscripts

$$f_i(A) = \min\{j \mid a_{ij} \ne 0\} , \quad 1 \le i \le N .$$

In terms of the associated graph $G^A = (X^A, E^A)$, where $X^A = \{x_1, \ldots, x_N\}$, these numbers are given by

$$f_i(A) = \min\{s \mid x_s \in Adj(x_i) \cup \{x_i\}\} . \tag{7.3.3}$$

In this subsection, we shall study the envelope structure of $Bdiag(F)$ by relating the first nonzero column subscript with the corresponding graph structure.

Let $G^A = (X^A, E^A)$ and $G^F = (X^F, E^F)$ be the undirected graphs associated with the symmetric matrices A and F respectively. Let $P = \{Y_1, Y_2, \ldots, Y_p\}$ be the set partitioning of X^A that corresponds to the matrix partitioning of A. It is useful to note that

$$G^{A_{kk}} = G^A(Y_k) \, ,$$

$$G^{F_{kk}} = G^F(Y_k) \, ,$$

and

$$G^{Bdiag(F)} = (X^A, E^{Bdiag(F)})$$

where

$$E^{Bdiag(F)} = \cup \{E^F(Y_k) \mid 1 \leq k \leq p\} \, .$$

In what follows, we shall use f_i to stand for $f_i(Bdiag(F))$. Let row i belong to the k-th block in the partitioning. In other words, we let $x_i \in Y_k$. In terms of the filled graph, the quantity f_i is given by

$$f_i = \min\{s \mid s = i \text{ or } \{x_s, x_i\} \in E^F(Y_k)\} \, .$$

We now relate it to the original graph G^A through the use of reachable sets introduced in Section 5.1.2. By Theorem 5.1.2 which characterizes the fill via reachable sets, we have

$$f_i = \min\{s \mid x_s \in Y_k, x_i \in Reach(x_s, \{x_1, \ldots, x_{s-1}\}) \cup \{x_s\}\} \, .$$
$$(7.3.4)$$

In Theorem 7.3.2, we prove a stronger result.

Lemma 7.3.1
Let $x_i \in Y_k$, and let

$$S = Y_1 \cup \cdots \cup Y_{k-1} \, .$$

That is, S contains all the nodes in the first $k-1$ blocks. Then

$$x_i \in Reach(x_{f_i}, S) \cup \{x_{f_i}\} \, .$$

Proof   By definition of f_i, $\{x_i, x_{f_i}\} \in E^F$ so that by Theorem 5.1.2, $x_i \in Reach(x_{f_i}, \{x_1, \ldots, x_{f_i-1}\})$. We can then find a path $x_i, x_{r_1}, \ldots, x_{r_t}, x_{f_i}$ where $\{x_{r_1}, \ldots, x_{r_t}\} \subset \{x_1, \ldots, x_{f_i-1}\}$.

We now prove that x_i can also be reached from x_{f_i} through S, which is a subset of $\{x_1, \ldots, x_{f_i-1}\}$. If $t = 0$, clearly $x_i \in Reach(x_{f_i}, S)$. On the other hand, if $t \neq 0$, let x_{r_s} be the node with the largest index number in $\{x_{r_1}, \ldots, x_{r_t}\}$. Then $x_i, x_{r_1}, \ldots, x_{r_{s-1}}, x_{r_s}$ is a path from x_i to x_{r_s} through $\{x_1, x_2, \ldots, x_{r_s-1}\}$ so that

$$\{x_i, x_{r_s}\} \in E^F \, .$$

But $r_s < f_i$, so by the definition of f_i we have $x_{r_s} \notin Y_k$, or in other words $x_{r_s} \in S$. The choice of r_s implies

$$\{x_{r_1}, \ldots, x_{r_t}\} \subset S$$

and thus $x_i \in Reach(x_{f_i}, S)$.   □

Theorem 7.3.2
Let $x_i \in Y_k$ and $S = Y_1 \cup \cdots \cup Y_{k-1}$. Then

$$f_i = \min\{s \mid x_s \in Y_k, \; x_i \in Reach(x_s, S) \cup \{x_s\}\} .$$

Proof   By Lemma 7.3.1, it remains to show that $x_i \notin Reach(x_r, S)$
for $x_r \in Y_k$ and $r < f_i$. Assume for contradiction that we can find
$x_r \in Y_k$ with $r < f_i$ and $x_i \in Reach(x_r, S)$. Since

$$S \subset \{x_1, \ldots, x_{r-1}\} ,$$

we have $x_i \in Reach(x_r, \{x_1, \ldots, x_{r-1}\})$ so that $\{x_i, x_r\} \in E^F(Y_k)$.
This contradicts the definition of f_i.   □

Corollary 7.3.3
Let x_i and S be as in Theorem 7.3.2. Then

$$f_i = \min\{s \mid x_s \in Reach(x_i, S) \cup \{x_i\}\} .$$

Proof   It follows from Theorem 7.3.2 and the symmetry of the
"*Reach*" operator.   □

It is interesting to compare this result with that given by (7.3.4). To
illustrate the result, we consider the partitioned matrix example in
Figure 7.3.1.
Consider $Y_2 = \{x_5, x_6, x_7, x_8\}$. Then the associated set S is $\{x_1, x_2, x_3, x_4\}$.
We have

$$Reach(x_5, S) = \{x_{10}, x_{11}\}$$
$$Reach(x_6, S) = \{x_7, x_8, x_9, x_{10}\}$$
$$Reach(x_7, S) = \{x_6, x_8\}$$
$$Reach(x_8, S) = \{x_6, x_7, x_{10}, x_{11}\} .$$

By Corollary 7.3.3,

$$f_5(Bdiag(F)) = 5$$
$$f_6(Bdiag(F)) = f_7(Bdiag(F)) = f_8(Bdiag(F)) = 6 .$$

Figure 7.3.1   An 11 by 11 partitioned matrix A.

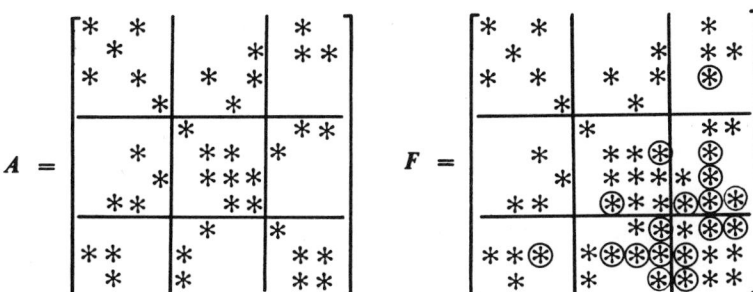

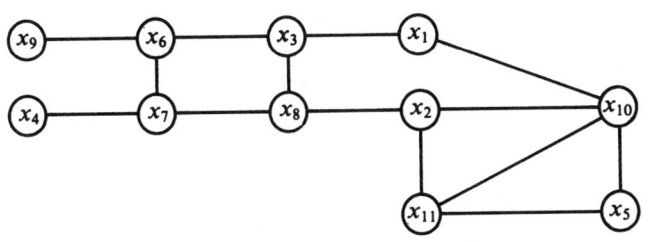

7.3.3   An Algorithm and Subroutines for Finding Diagonal Block Envelopes

Corollary 7.3.3 readily provides a method for finding $f_i(Bdiag(F))$ and hence the envelope structure of $Bdiag(F)$. However, in the actual implementation, Lemma 7.3.1 is more easily applied. The algorithm can be described as follows.

Let $P = \{Y_1, \ldots, Y_p\}$ be the partitioning. For each block k in the partitioning, do the following:

Step 1   (Initialization) $S = Y_1 \cup \cdots \cup Y_{k-1}$, $T = S \cup Y_k$.
Step 2   (Main loop) For each node x_r in Y_k do:

    2.1)   Determine $Reach(x_r, S)$ in the subgraph $G(T)$.
    2.2)   For each $x_i \in Reach(x_r, S)$, set $f_i = r$.
    2.3)   Reset $T \leftarrow T - (Reach(x_r, S) \cup \{x_r\})$.

The implementation of this algorithm consists of two subroutines, which are discussed below.

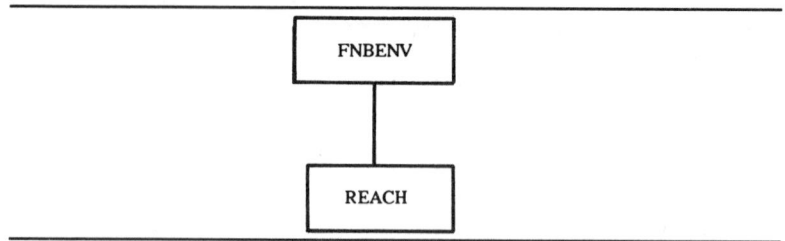

REACH (find REACHable sets)

Given a set S and a node $x \notin S$ in a graph. To study the reachable set $Reach(x,S)$, it is helpful to introduce the related notion of neighborhood set. Formally, the *neighborhood set of x in S* is defined to be

$Nbrhd(x,S) =$

$\{s \in S \mid s$ is reachable from x through a subset of $S\}$.

Reachable and neighborhood sets are related by the following lemma.

Lemma 7.3.4

$$Reach(x,S) = Adj(Nbrhd(x,S) \cup \{x\}) .$$

The subroutine REACH applies this simple relation to determine the reachable set of a node via a subset in a given *subgraph*. The subgraph is specified by the input parameters XADJ, ADJNCY and MARKER, where a node belongs to the subgraph if its MARKER value is zero. The subset S is specified by the mask vector SMASK, where a node belongs to S if its SMASK value is nonzero. The variable ROOT is the input node, whose reachable set is to be determined.

It returns the reachable set in (RCHSZE, RCHSET). As a by-product, the neighborhood set in (NHDSZE, NBRHD) is also returned. On exit, nodes in the reachable or neighborhood sets will have their MARKER value set to ROOT.

After initialization, the subroutine loops through the neighbors of the given ROOT. Neighbors not in the subset S are included in the reach set, while neighbors in the subset S are put into the neighborhood set. Furthermore, each neighbor in the subset S is examined to obtain new reachable nodes. This process is repeated until no neighbors in S can be found.

```
C*************************************************************
C*************************************************************
C*************    REACH ..... REACHABLE SET    ***********
C*************************************************************
C*************************************************************
C
C     PURPOSE - THIS SUBROUTINE IS USED TO DETERMINE THE
C     REACHABLE SET OF A NODE Y THROUGH A SUBSET S
C     (I.E. REACH(Y,S) ) IN A GIVEN SUBGRAPH.  MOREOVER,
C     IT RETURNS THE NEIGHBORHOOD SET OF Y IN S, I.E.
C     NBRHD(Y,S), THE SET OF NODES IN S THAT CAN BE
C     REACHED FROM Y THROUGH A SUBSET OF S.
C
C     INPUT PARAMETERS -
C     ROOT - THE GIVEN NODE NOT IN THE SUBSET S.
C     (XADJ, ADJNCY) - THE ADJACENCY STRUCTURE PAIR.
C     SMASK - THE MASK VECTOR FOR THE SET S.
C          = 0,  IF THE NODE IS NOT IN S,
C          > 0,  IF THE NODE IS IN S.
C
C     OUTPUT PARAMETERS -
C     (NHDSZE, NBRHD) - THE NEIGHBORHOOD SET.
C     (RCHSZE, RCHSET) - THE REACHABLE SET.
C
C     UPDATED PARAMETERS -
C     MARKER - THE MARKER VECTOR USED TO DEFINE THE SUBGRAPH,
C              NODES IN THE SUBGRAPH HAVE MARKER VALUE 0.
C              ON RETURN, THE REACHABLE AND NEIGHBORHOOD NODE
C              SETS HAVE THEIR MARKER VALUES RESET TO ROOT.
C
C*************************************************************
C
      SUBROUTINE  REACH ( ROOT, XADJ, ADJNCY, SMASK, MARKER,
     1                    RCHSZE, RCHSET, NHDSZE, NBRHD )
C
C*************************************************************
C
      INTEGER ADJNCY(1), MARKER(1), NBRHD(1), RCHSET(1),
     1        SMASK(1)
      INTEGER XADJ(1), I, ISTOP, ISTRT, J, JSTOP, JSTRT,
     1        NABOR, NBR, NHDBEG, NHDPTR, NHDSZE, NODE,
     1        RCHSZE, ROOT
C
C*************************************************************
C
C         ------------------
C         INITIALIZATION ...
C         ------------------
          NHDSZE = 0
          RCHSZE = 0
          IF ( MARKER(ROOT) .GT. 0 )  GO TO 100
             RCHSZE = 1
             RCHSET(1) = ROOT
             MARKER(ROOT) = ROOT
100       ISTRT = XADJ(ROOT)
          ISTOP = XADJ(ROOT+1) - 1
          IF ( ISTOP .LT. ISTRT )  RETURN
C         --------------------------------------
C         LOOP THROUGH THE NEIGHBORS OF ROOT ...
C         --------------------------------------
          DO 600 I = ISTRT, ISTOP
             NABOR =  ADJNCY(I)
             IF ( MARKER(NABOR) .NE. 0 )  GO TO 600
                IF ( SMASK(NABOR) .GT. 0 )  GO TO 200
C               --------------------------------------------------
C               NABOR IS NOT IN S, INCLUDE IT IN THE REACH SET.
C               --------------------------------------------------
```

```
                      RCHSZE = RCHSZE + 1
                      RCHSET(RCHSZE) = NABOR
                      MARKER(NABOR) =   ROOT
                      GO TO 600
C                     --------------------------------------------------
C                     NABOR IS IN SUBSET S, AND HAS NOT BEEN CONSIDERED.
C                     INCLUDE IT INTO THE NBRHD SET AND FIND THE NODES
C                     REACHABLE FROM ROOT THROUGH THIS NABOR.
C                     --------------------------------------------------
      200             NHDSZE = NHDSZE + 1
                      NBRHD(NHDSZE) = NABOR
                      MARKER(NABOR) = ROOT
                      NHDBEG = NHDSZE
                      NHDPTR = NHDSZE
      300             NODE    = NBRHD(NHDPTR)
                      JSTRT   = XADJ(NODE)
                      JSTOP   = XADJ(NODE+1) - 1
                      DO 500 J = JSTRT, JSTOP
                         NBR = ADJNCY(J)
                         IF ( MARKER(NBR) .NE. 0 )  GO TO 500
                            IF ( SMASK(NBR)  .EQ. 0 )  GO TO 400
                            NHDSZE = NHDSZE + 1
                            NBRHD(NHDSZE) = NBR
                            MARKER(NBR) = ROOT
                            GO TO 500
      400                   RCHSZE = RCHSZE + 1
                            RCHSET(RCHSZE) = NBR
                            MARKER(NBR) = ROOT
      500             CONTINUE
                      NHDPTR = NHDPTR + 1
                      IF ( NHDPTR .LE. NHDSZE )  GO TO 300
      600       CONTINUE
                RETURN
            END
```

FNBENV (FiNd diagonal Block ENVelope)

This subroutine serves the same purpose as FNTENV in Section 6.4.3. They are both used to determine the envelope structure of the factored diagonal blocks in a partitioned matrix. Unlike FNTENV, this subroutine FNBENV finds the *exact* envelope structure. Although it works for general partitioned matrices, it is more expensive to use than FNTENV, and for the orderings provided by the RQT algorithm, the output from FNTENV is satisfactory. However, for one-way dissection orderings the more sophisticated FNBENV is essential.

Inputs to FNBENV are the adjacency structure (XADJ, ADJNCY), the ordering (PERM, INVP) and the partitioning (NBLKS, XBLK). The subroutine will produce the envelope structure in the index vector XENV and the variable MAXENV will contain the size of the envelope.

Three temporary vectors are required. The vector SMASK is used to specify those nodes in the subset S (see the above algorithm). On the other hand, the nodes in the set T are given by those with MARKER value 0. The vector MARKER is also used temporarily to store the first neighbor in each row of a block. The third temporary vector RCHSET is used to contain both the reachable and neighborhood sets. Since the two sets do not overlap, we can organize the vector RCHSET as follows.

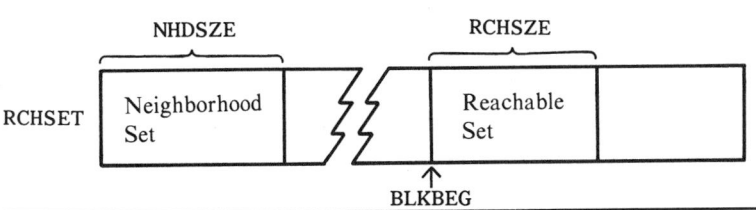

The subroutine begins by initializing the temporary vectors SMASK and MARKER. The main loop goes through and processes each block. For each block, its nodes are added to the subgraph by turning their MARKER values to zeros. For each node i in the block, the subroutine REACH is called so that nodes in the thus-determined reachable sets will have node i as their first neighbor. Before the next block is processed, the MARKER values are reset and nodes in the current block are added to the subset S.

```
C***********************************************************
C***********************************************************
C*******      FNBENV  .....  FIND DIAGONAL BLOCK ENVELOPE  *******
C***********************************************************
C***********************************************************
C
C     PURPOSE - THIS SUBROUTINE FINDS THE EXACT ENVELOPE
C         STRUCTURE OF THE DIAGONAL BLOCKS OF THE CHOLESKY
C         FACTOR OF A PERMUTED PARTITIONED MATRIX.
C
C     INPUT PARAMETERS -
C         (XADJ, ADJNCY) - ADJACENCY STRUCTURE OF THE GRAPH.
C         (PERM, INVP) - THE PERMUTATION VECTOR AND ITS INVERSE.
C         (NBLKS, XBLK) - THE PARTITIONING.
C
C     OUTPUT PARAMETERS _
C         XENV - THE ENVELOPE INDEX VECTOR.
C         ENVSZE - THE SIZE OF THE ENVELOPE.
C
C     WORKING PARAMETERS -
C         SMASK - MARKS NODES THAT HAVE BEEN CONSIDERED.
C         MARKER - IS USED BY ROUTINE REACH.
C         RCHSET - IS USED BY THE SUBROUTINE REACH.
C             STORES BOTH REACHABLE AND NEIGHBORHOOD SETS.
C
C     PROGRAM SUBROUTINES -
C         REACH.
C
C***********************************************************
C
        SUBROUTINE  FNBENV ( XADJ, ADJNCY, PERM, INVP, NBLKS, XBLK,
     1                       XENV, ENVSZE, SMASK, MARKER, RCHSET )
C
C***********************************************************
C
        INTEGER ADJNCY(1), INVP(1), MARKER(1), PERM(1),
     1          RCHSET(1), SMASK(1), XBLK(1)
        INTEGER XADJ(1), XENV(1), BLKBEG, BLKEND, I,
     1          IFIRST, INHD, K, ENVSZE, NBLKS, NEQNS,
     1          NEWNHD, NHDSZE, NODE, RCHSZE
C
C***********************************************************
```

```
C
C           ------------------
C           INITIALIZATION ...
C           ------------------
            NEQNS = XBLK(NBLKS+1) - 1
            ENVSZE = 1
            DO 100 I = 1, NEQNS
               SMASK(I) = 0
               MARKER(I) = 1
   100      CONTINUE
C           -----------------------
C           LOOP OVER THE BLOCKS ...
C           -----------------------
            DO 700 K = 1, NBLKS
               NHDSZE = 0
               BLKBEG = XBLK(K)
               BLKEND = XBLK(K+1) - 1
               DO 200 I = BLKBEG, BLKEND
                  NODE = PERM(I)
                  MARKER(NODE) = 0
   200         CONTINUE
C              --------------------------------------------
C              LOOP THROUGH THE NODES IN CURRENT BLOCK ...
C              --------------------------------------------
               DO 300 I = BLKBEG, BLKEND
                  NODE = PERM(I)
                  CALL   REACH ( NODE, XADJ, ADJNCY, SMASK,
     1                           MARKER, RCHSZE, RCHSET(BLKBEG),
     1                           NEWNHD, RCHSET(NHDSZE+1) )
                  NHDSZE = NHDSZE + NEWNHD
                  IFIRST = MARKER(NODE)
                  IFIRST = INVP(IFIRST)
                  XENV(I) = ENVSZE
                  ENVSZE = ENVSZE + I - IFIRST
   300         CONTINUE
C              ----------------------------------------
C              RESET MARKER VALUES OF NODES IN NBRHD SET.
C              ----------------------------------------
               IF ( NHDSZE .LE. 0 ) GO TO 500
                  DO 400 INHD = 1, NHDSZE
                     NODE = RCHSET(INHD)
                     MARKER(NODE) = 0
   400            CONTINUE
C              ----------------------------------------
C              RESET MARKER AND SMASK VALUES OF NODES IN
C              THE CURRENT BLOCK.
C              ----------------------------------------
   500            DO 600 I = BLKBEG, BLKEND
                     NODE = PERM(I)
                     MARKER(NODE) = 0
                     SMASK(NODE) = 1
   600            CONTINUE
   700      CONTINUE
            XENV(NEQNS+1) = ENVSZE
            ENVSZE = ENVSZE - 1
            RETURN
         END
```

7.3.4   Execution Time Analysis of the Algorithm

For general partitioned matrices, the complexity of the diagonal block envelope algorithm depends on the partitioning factor p, the sparsity of the matrix and the way blocks are connected. However, for one-way dissection partitionings, we have the following result.

Theorem 7.3.5

Let $G = (X,E)$ and $P = \{Y_1, \ldots, Y_p\}$ be a one-way dissection partitioning. The complexity of the algorithm FNBENV is $O(|E|)$.

Proof   For a node x_i in the first $p - 1$ blocks, the subroutine REACH, when called, merely looks through the adjacency list for the node x_i. On the other hand, when nodes in the last block Y_p are processed, the adjacency lists for all the nodes in the graph are inspected at most once. Hence, in the entire algorithm, the adjacency structure is gone through at most twice.   □

Exercises

7.3.1)   Construct an example of a tree-partitioned matrix structure A to show that FNTENV is not adequate to determine the *exact* envelope structure of the block diagonal matrix $Bdiag(F)$, where F is the filled matrix of A.

7.3.2)   Give an example to show that Theorem 7.3.5 does not hold for all tree-partitionings P.

7.3.3)   This question involves solving a sequence of $m\ell$ by $m\ell$ finite element matrix problems $Ax = b$ of the type studied in Section 7.2, with $m = 5, 10, 15$, and 20, and $\ell = 2m$. Set the diagonal elements of A to 8, the off-diagonal elements to -1, and arrange the right hand side b so that the solution to the system is a vector of all ones. Use the programs provided in Chapter 4 to solve these problems, taking care to record the storage required and the execution times for each phase of the solution of each problem. Repeat the procedure using the one-way dissection ordering subroutines provided in this chapter, along with the appropriate subroutines from Chapter 6. Compare the two methods for solving these problems with respect to the criteria discussed in Section 3 of Chapter 2.

7.4 Additional Notes

It is interesting to speculate about more sophisticated ways of choosing the one-way dissectors. For example, instead of using a fixed δ, one might instead use a sequence δ_i, $i = 1, 2, \ldots$, where δ_i is obtained from local information about the part of the level structure that remains to be processed after the first $i - 1$ dissectors have been chosen. Investigations such as these, aimed at the development of robust heuristics, are good candidates for senior projects and masters theses.

The fundamental idea that makes the one-way dissection method effective is the use of the "throw-away" technique introduced in Section 6.2. This technique can be recursively applied, as described in the *additional notes* at the end of Chapter 6, which implies that the one-way dissection scheme of this chapter may also be similarly generalized. In its simplest form the idea is to also apply the one-way dissection technique to the $\sigma + 1$ independent blocks, rather than ordering them using the RCM algorithm. The basic approach for this two level scheme is depicted in Figure 7.4.1.

Of course the idea can be generalized to more than two levels, but apparently in practice using more than two levels does not yield significant benefit. It can be shown that for an n by n grid problem ($m = \ell = n$), if the optimal σ_1 and σ_2 are chosen, the storage and operation counts for this two-level ordering are $O(n^{7/3})$ and $O(n^{10/3})$ respectively, compared to $O(n^{5/2})$ and $O(n^{7/2})$ for the ordinary (one-level) one-way dissection scheme as described in this chapter (Ng 1979).

Figure 7.4.1   A two-level one-way dissection ordering, having $\sigma_1 = 2$ level-1 dissectors, $\sigma_2 = 3$ level-2 dissectors, and $(\sigma_1 + 1)(\sigma_2 + 1) = 12$ independent blocks, which are numbered grid column by column.

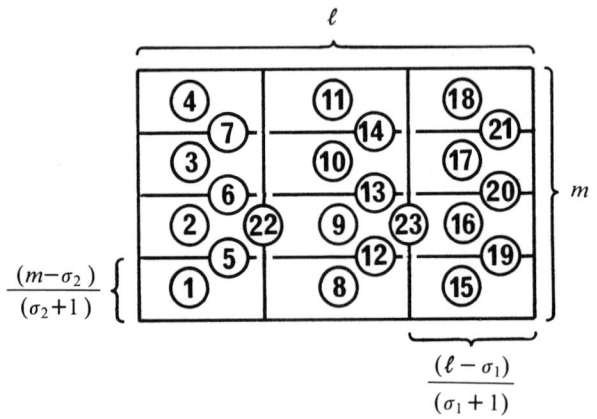

8/ Nested Dissection Methods

8.0 Introduction

In Chapter 7, we have studied the so-called one-way dissection method, and we have seen that it lends itself readily to the implicit tree-partitioning scheme of Chapter 6. In this chapter, we consider a different dissection method, which attempts to minimize fill, just as the minimum degree algorithm described in Chapter 5 attempts to do.

The nested dissection method of this section is appropriate primarily for matrix problems arising in finite difference and finite element applications. The main advantage of the algorithm of Section 8.2, compared to the minimum degree algorithm, is its speed, and its modest and predictable storage requirements. The orderings produced are similar in nature to those provided by the minimum degree algorithm, and for this reason we do not deal with a storage scheme, allocation procedure, or numerical subroutines in this chapter. Those of Chapter 5 are appropriate for nested dissection orderings.

Separators, which we defined in Section 3.1, play a central role in the study of sparse matrix factorization. Let A be a symmetric matrix and G^A be its associated undirected graph. Consider a separator S in G^A, whose removal disconnects the graph into two parts whose node sets are C_1 and C_2.

If the nodes in S are numbered after those of C_1 and C_2, this induces a partitioning on the correspondingly ordered matrix and it has the form shown in Figure 8.0.1. The crucial observation is that the zero block in the matrix remains zero after the factorization. Since one of the primary purposes in the study of sparse matrix computation is to preserve as many zero entries as possible, the use of separators in this way is central. When appropriately chosen, a (hopefully large) submatrix is guaranteed to stay zero. Indeed, the idea can be recursively applied, so that zeros can be preserved in the same manner in the submatrices.

The recursive application of this basic observation has come to be known as the *nested dissection method*. George in 1973 applied this technique to sparse systems associated with an n by n regular grid or

Figure 8.0.1   Use of a separator to partition a matrix.

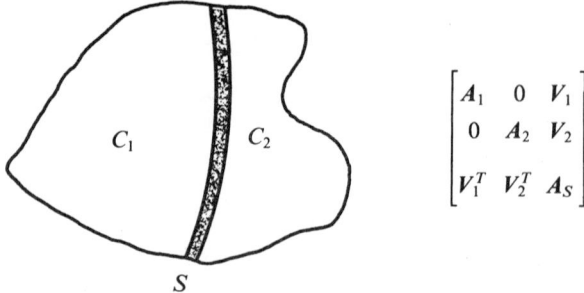

$$\begin{bmatrix} A_1 & 0 & V_1 \\ 0 & A_2 & V_2 \\ V_1^T & V_2^T & A_S \end{bmatrix}$$

mesh consisting of $(n-1)^2$ small elements. In the next section, we shall give a careful analysis of the method for this special problem.

8.1   Nested Dissection of a Regular Grid

8.1.1   The Ordering

Let X be the set of vertices of the n by n regular grid. Let S^0 consist of the vertices on a mesh line which as nearly as possible divides X into two equal parts R^1 and R^2. Figure 8.1.1 shows the case when $n = 10$. If we number the nodes of the two components R^1 and R^2 row by row, followed by those in S^0, a matrix structure as shown in Figure 8.1.2 is obtained. Let us call this the *one-level dissection ordering*.

Figure 8.1.1   A one-level dissection ordering of a 10 by 10 grid.

86	87	88	89	90	100	40	39	38	37
81	82	83	84	85	99	36	35	34	33
76	77	78	79	80	98	32	31	30	29
71	72	73	74	75	97	28	27	26	25
66	67	68	69	70	96	24	23	22	21
61	62	63	64	65	95	20	19	18	17
56	57	58	59	60	94	16	15	14	13
51	52	53	54	55	93	12	11	10	9
46	47	48	49	50	92	8	7	6	5
41	42	43	44	45	91	4	3	2	1

To get a nested dissection ordering, we continue dissecting the remaining two components. Choose vertex sets

Figure 8.1.2 Matrix structure associated with a one-level dissection ordering.

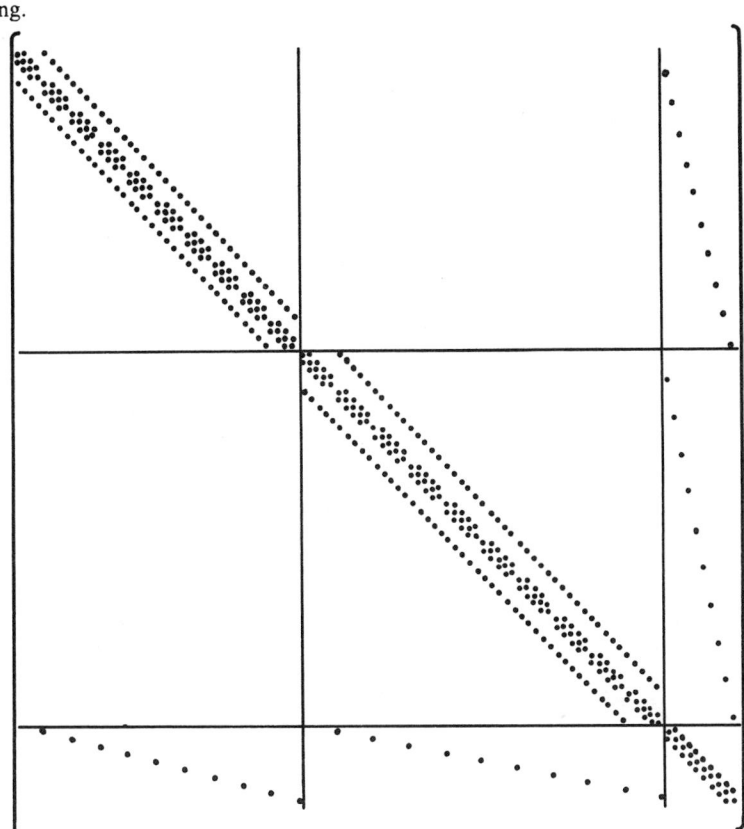

$$S^j \subset R^j, \ j = 1, 2$$

consisting of nodes lying on mesh lines which as nearly as possible divide R^j into equal parts. If the variables associated with vertices in $R^j - S^j$ are numbered before those associated with S^j, we induce in the two leading principal submatrices exactly the same structure as that of the overall matrix.

The process can be repeated until the components left are not dissectable. This yields a *nested dissection ordering*. Figure 8.1.3 shows such an ordering on the 10 by 10 grid problem and Figure 8.1.4 shows the correspondingly ordered matrix structure. Note the recursive pattern in the matrix structure.

Figure 8.1.3 A nested dissection ordering of a 10 by 10 grid.

78	77	85	68	67	100	29	28	36	20
76	75	84	66	65	99	27	26	35	19
80	79	83	70	69	98	31	30	34	21
74	73	82	64	63	97	25	24	33	18
72	71	81	62	61	96	23	22	32	17
90	89	88	87	86	95	40	39	38	37
54	53	60	46	45	94	10	9	16	3
52	51	59	44	43	93	8	7	15	2
56	55	58	48	47	92	12	11	14	4
50	49	57	42	41	91	6	5	13	1

8.1.2 Storage Requirements

Nested dissection employs a strategy commonly known as *divide and conquer*. The strategy splits a problem into smaller subproblems whose individual solutions can be combined to yield the solution to the original problem. Moreover, the subproblems have structures similar to the original one so that the process can be repeated recursively until the solutions to the subproblems are trivial.

In the study of such strategies, some forms of recursive equations need to be solved. We now provide some results in preparation for the analysis of the storage requirement for nested dissection orderings. The proofs of these are left as exercises.

Lemma 8.1.1
Let $f(n) = 4f(n/2) + kn^2 + O(n)$. Then

$$f(n) = kn^2\log_2 n + O(n^2) .$$

Lemma 8.1.2
Let $g(n) = g(n/2) + kn^2\log_2 n + O(n^2)$. Then

Figure 8.1.4   Matrix structure associated with a nested dissection ordering.

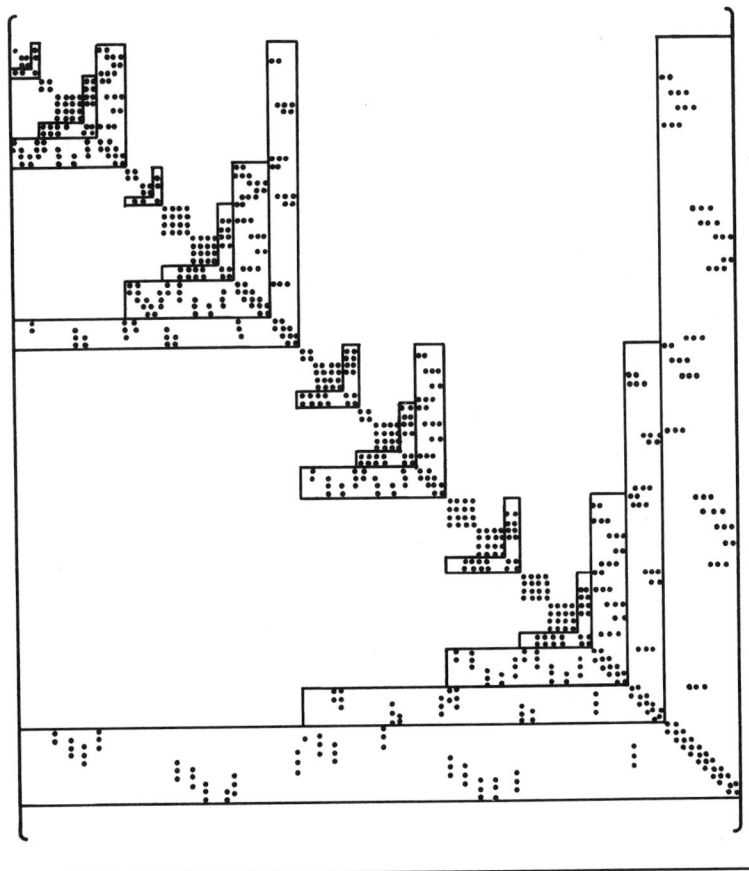

$$g(n) = \frac{4}{3}kn^2\log_2 n + O(n^2) .$$

Lemma 8.1.3

Let $h(n) = 2h(n/2) + kn^2\log_2 n + O(n^2)$. Then

$$h(n) = 2kn^2\log_2 n + O(n^2) .$$

In order to give an analysis of the nested dissection orderings recursively, we introduce bordered n by n grids. A *bordered n by n grid* contains an n by n subgrid, where one or more sides of this subgrid is bordered by an additional grid line. Figure 8.1.5 contains some examples of bordered 3 by 3 grids.

We are now ready to analyze the storage requirement for the nested dissection ordering. Let $S(n,i)$ be the number of nonzeros in the

Figure 8.1.5   Some bordered 3 by 3 grids.

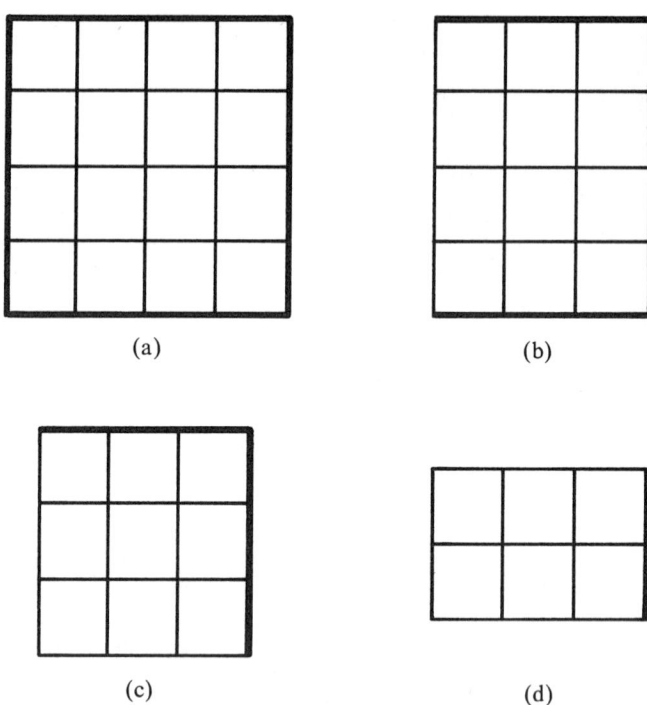

(a)   (b)

(c)   (d)

factor of a matrix associated with an n by n grid ordered by nested dissection, where the grid is bordered along i sides. Clearly, what we are after is the quantity $S(n, 0)$. For our purposes, when $i = 2$, we always refer to the one as shown in Figure 8.1.5(c), rather than the following grid:

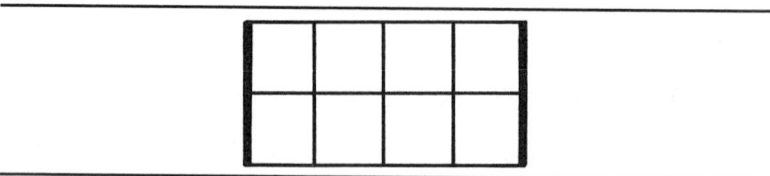

In what follows, we relate the quantities $S(n, i)$, $0 \le i \le 4$. Consider first $S(n, 0)$. In Figure 8.1.6, a "+" shaped separator is used to divide the n by n grid into 4 smaller subgrids. The variables in regions ①, ②, ③ and ④ are to be numbered before those in ⑤ so that a matrix structure of the form in Figure 8.1.7 is induced.

Figure 8.1.6   Dissection of an n by n grid.

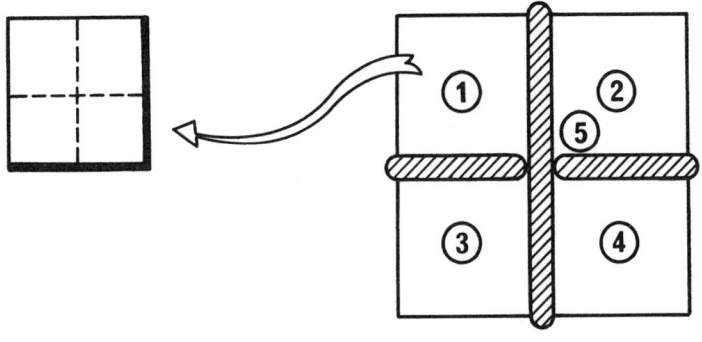

The number of nonzeros in the factor comes from the L_{ii}'s ($1 \leq i \leq 4$) and the L_{5i} for $1 \leq i \leq 5$. Now since the strategy is applied recursively on the smaller subgrids, we have

$$\eta(L_{ii}) + \eta(L_{5i}) \simeq S(n/2, 2)$$

for $1 \leq i \leq 4$. As for L_{55} which corresponds to the nodes in the "+" separator, we can determine the number of nonzeros using Theorem 5.1.2. It is given by

$$\eta(L_{55}) = 2 \sum_{i=n}^{3n/2} i + n^2/2 + O(n) = 7n^2/4 + O(n) \, .$$

Thus, we obtain the first recursion equation:

$$S(n, 0) = 4S(\frac{n}{2}, 2) + \frac{7n^2}{4} + O(n) \, . \qquad (8.1.1)$$

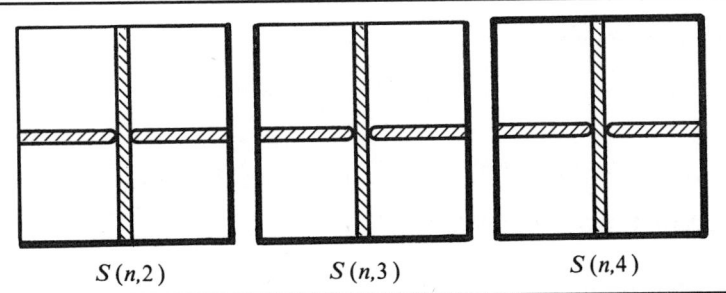

$S(n,2)$ $S(n,3)$ $S(n,4)$

The other recursion equations can be established in the same way. In general, it can be expressed as $S(n,i) = $ cost to store the 4 bordered $n/2$ by $n/2$ subgrids + cost to store the "+" separator. We leave it to

Figure 8.1.7 Matrix structure for dissection in Figure 8.1.6.

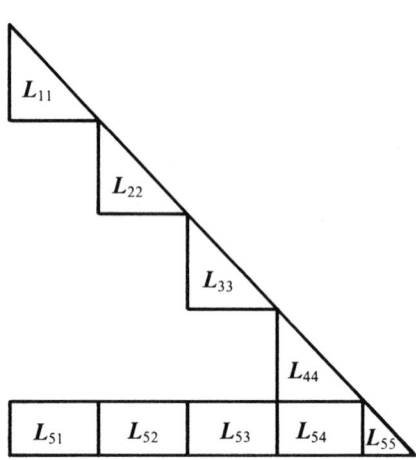

the reader to verify the following results.

$$S(n, 2) = S(n/2, 2) + 2S(n/2, 3) + S(n/2, 4) + 19n^2/4 + O(n)$$
$$(8.1.2)$$

$$S(n, 3) = 2S(n/2, 3) + 2S(n/2, 4) + 25n^2/4 + O(n) \qquad (8.1.3)$$

$$S(n, 4) = 4S(n/2, 4) + 31n^2/4 + O(n) . \qquad (8.1.4)$$

Theorem 8.1.4

The number of nonzeros in the triangular factor L of a matrix associated with a regular n by n grid ordered by nested dissection is given by

$$\eta(L) = 31(n^2\log_2 n)/4 + O(n^2) .$$

Proof   The result follows from the recurrence relations (8.1.1)–(8.1.4). Applying Lemma 8.1.1 to equation (8.1.4), we get

$$S(n,4) = 31(n^2\log_2 n)/4 + O(n^2) ,$$

so that (8.1.3) becomes

$$S(n,3) = 2S(n/2,3) + 31(n^2\log_2 n)/8 + O(n^2) .$$

The solution to it gives, by Lemma 8.1.3,

$$S(n,3) = 31(n^2\log_2 n)/4 + O(n^2) .$$

Substituting $S(n,3)$ and $S(n,4)$ into equation (8.1.2), we have

$$S(n,2) = S(n/2,2) + 93(n^2\log_2 n)/16 + O(n^2) .$$

Again, the solution is

$$S(n,2) = 31(n^2\log_2 n)/4 + O(n^2)$$

so that

$$\eta(L) = S(n,0) = 31(n^2\log_2 n)/4 + O(n^2) .$$

$\square$

It is interesting to note from the proof of Theorem 8.1.4 that the asymptotic bounds for $S(n,i)$, $i = 0, 2, 3, 4$ are all $31(n^2\log_2 n)/4$. (What about $i = 1$?)

8.1.3 Operation Counts

Let A be a matrix associated with an n by n grid ordered by nested dissection. To estimate the number of operations required to factor A, we can follow the same approach as used in the previous section. We first state some further results on recursive equations.

Lemma 8.1.5
Let $f(n) = f(n/2) + kn^3 + O(n^2\log_2 n)$. Then

$$f(n) = 8kn^3/7 + O(n^2\log_2 n) .$$

Lemma 8.1.6
Let $g(n) = 2g(n/2) + kn^3 + O(n^2\log_2 n)$. Then

$$g(n) = 4kn^3/3 + O(n^2\log_2 n) .$$

Lemma 8.1.7
Let $h(n) = 4h(n/2) + kn^3 + O(n^2)$. Then

$$h(n) = 2kn^3 + O(n^2\log_2 n) .$$

In parallel to $S(n,i)$, we introduce $\theta(n,i)$ to be the number of operations required to factor a matrix associated with an n by n grid ordered by nested dissection, where the grid is bordered on i sides. To determine $\theta(n, 0)$, we again consider Figure 8.1.6; clearly $\theta(n, 0)$ is the cost of eliminating the four $n/2$ by $n/2$ bordered subgrids, together with the cost of eliminating the nodes in the "+" dissector. Applying Theorem 2.1.2, we have

$$\theta(n, 0) \simeq 4\theta(n/2, 2) + \sum_{i=n}^{3n/2} i^2 + \frac{1}{2} \sum_{i=1}^{n} i^2 \tag{8.1.5}$$

$$= 4\theta(n/2, 2) + 19n^3/24 + n^3/6 + O(n^2)$$

$$= 4\theta(n/2, 2) + 23n^3/24 + O(n^2) .$$

We leave it to the reader to verify the following equations:

$$\theta(n, 2) = \theta(n/2, 2) + 2\theta(n/2, 3) + \theta(n/2, 4) + 35n^3/6 + O(n^2) \tag{8.1.6}$$

$$\theta(n, 3) = 2\theta(n/2, 3) + 2\theta(n/2, 4) + 239n^3/24 + O(n^2) \tag{8.1.7}$$

$$\theta(n, 4) = 4\theta(n/2, 4) + 371n^3/24 + O(n^2) . \tag{8.1.8}$$

Theorem 8.1.8

The number of operations required to factor a matrix associated with an n by n grid ordered by nested dissection is given by

$$829n^3/84 + O(n^2 \log_2 n) .$$

Proof   All that is required is to determine $\theta(n, 0)$. Applying Lemma 8.1.7 to equation (8.1.8), we obtain

$$\theta(n, 4) = 371n^3/12 + O(n^2 \log_2 n) .$$

This means equation (8.1.7) can be rewritten as

$$\theta(n, 3) = 2\theta(n/2, 3) + 849n^3/48 + O(n^2 \log_2 n) .$$

By Lemma 8.1.6, we have

$$\theta(n, 3) = 283n^3/12 + O(n^2 \log_2 n) .$$

Substituting $\theta(n, 3)$ and $\theta(n, 4)$ into (8.1.6), we get

$$\theta(n, 2) = \theta(n/2, 2) + 1497n^3/96 + O(n^2 \log_2 n) ,$$

which is, by Lemma 8.1.5,

$$\theta(n, 2) = 499n^3/28 + O(n^2 \log_2 n) .$$

Finally, from equation (8.1.5),

$$\theta(n, 0) = 829n^3/84 + O(n^2\log_2 n) \ .$$

□

8.1.4 Optimality of the Ordering

In this section, we establish lower bounds on the number of nonzero entries in the factor (primary storage) and the number of operations required to effect the symmetric factorization for *any* ordering of the matrix system associated with an n by n regular grid. We show that at least $O(n^3)$ operations are required for its factorization and the corresponding lower triangular factor must have at least $O(n^2\log_2 n)$ nonzero components. The nested dissection ordering described in Section 8.1.1 attains these lower bounds, so that the ordering can be regarded as optimal in the order of magnitude sense.

We first consider the lower bound on operations.

Lemma 8.1.9
Let $G = (X,E)$ be the graph associated with the n by n grid. Let $x_1, x_2, \ldots, x_N$ be any ordering on G. Then there exists an x_i such that

$$|\,Reach\,(x_i, \{x_1, \ldots, x_{i-1}\})| \ge n - 1 \ .$$

Proof   Let x_i be the *first* node to be removed which completely vacates a row or column of the grid. For definiteness, let it be a column {row}. At this stage, there are at least $(n-1)$ mesh *rows* {columns} with uneliminated nodes. At least one in each of these rows {columns} can be reached from x_i through the subset $\{x_1, \ldots, x_{i-1}\}$. This proves the lemma.   □

Theorem 8.1.10
The factorization of a matrix associated with an n by n grid requires at least $O(n^3)$ operations.

Proof   By Lemma 8.1.9, there exists an x_i such that

$$Reach\,(x_i, \{x_1, \ldots, x_{i-1}\}) \cup \{x_i\}$$

is a clique of size at least n in the filled graph $G^{F(A)}$ (see Exercise 5.1.4). This corresponds to a full n by n submatrix in the filled matrix F so that symmetric factorization requires at least $n^3/6 + O(n^2)$ operations.   □

The proof for the lower bound on primary storage follows a different argument. For each k by k subgrid, the following lemma identifies a special edge in the resulting filled graph.

Lemma 8.1.11
Consider any k by k subgrid in the given n by n grid. There exists an edge in G^F joining a pair of parallel boundary lines in the subgrid.

Proof   There are four boundary mesh lines in the k by k subgrid. Let x_i be the first boundary node in the subgrid to be removed that completely vacates a boundary line (not including the corner vertices).

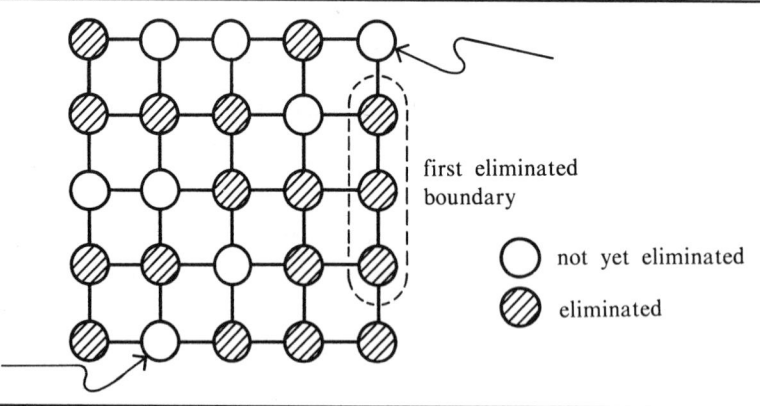

first eliminated
boundary

○ not yet eliminated

◍ eliminated

Then there always exist two nodes in the remaining parallel boundary lines that are linked through

$$\{x_1,\ x_2,\ \ldots,\ x_{i-1},\ x_i\}\ .$$

(See the nodes pointed to in the figure above.) In other words, there is an edge joining them in G^F.   □

Theorem 8.1.12
The triangular factor of a matrix associated with an n by n grid has at least $O(n^2\log_2 n)$ nonzeros.

Proof   Consider each subgrid of size k. It follows from Lemma 8.1.11 that there is an edge in G^F joining a pair of parallel boundary lines in the subgrid. Each such edge can be chosen for at most k subgrids of size k. Since the number of subgrids of size k is $(n-k+1)^2$, the number of such distinct edges is bounded below by

$$\frac{(n-k+1)^2}{k}\ .$$

Futhermore, for subgrids of different sizes, the corresponding edges must be different. So, we have

$$|E^F| \geq \sum_{k=1}^{n} \frac{(n-k+1)^2}{k} \simeq n^2 \log_2 n .$$

□

Exercises

8.1.1)   Let A be the matrix associated with an n by n grid, ordered by the one-level dissection scheme. Show that

     a)   the number of operations required to perform the symmetric factorization is $\frac{13}{24}n^4 + O(n^3)$

     b)   the number of nonzeros in the factor L is $n^3 + O(n^2)$.

8.1.2)   Prove the recursive equations in Lemmas 8.1.1–8.1.3 and Lemmas 8.1.5–8.1.7.

8.1.3)   In establishing equation (8.1.6) for $\theta(n, 2)$, we assumed that the "+" separator is ordered as in (a)

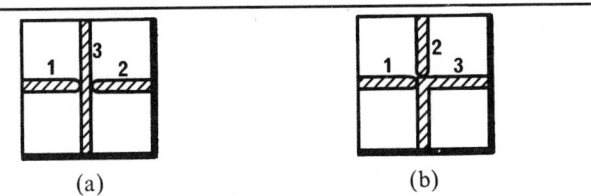

(a)   (b)

     Assume $\theta'(n, 2)$ is the corresponding cost if (b) is used. Show that

$$\theta'(n, 2) = \theta'(n/2, 2) + 2\theta(n/2, 3) + \theta(n/2, 4) + 125n^3/24 + O(n^2) .$$

     How does it compare to $\theta(n, 2)$?

8.1.4)   Prove results similar to Theorems 8.1.4 and 8.1.8 for an m by ℓ grid where m is large and $m < \ell$.

8.1.5)   Prove that any ordering of an n by n grid must yield a matrix whose bandwidth is at least $n - 1$.

8.1.6)   Consider the n by n grid. It is known that the associated graph $G = (X,E)$ satisfies the *isoparametric inequality*: for any subset S, if $|S| \leq n^2/2$ then $|Adj(S)| \geq |S|^{1/2}$. Prove that any ordering on G yields a profile of at least $O(n^3)$.

8.1.7)   Suppose one carries out "incomplete nested dissection" on the n by n grid problem (George et al. 1978c). That is, one only carries out the dissection ℓ levels, where $\ell < \log_2 n$, and numbers the remaining independent grid subarrays row by row. Show that if $\ell \geq \log_2(\sqrt{n})$ then the operation count for this ordering remains $O(n^3)$. Show that the number of nonzeros in the corresponding factor L is $O(n^2\sqrt{n})$.

8.1.8)   Using a method due to Strassen (1969), and extended by Bunch and Hopcroft (1974), it is possible to solve a dense m by m system of linear equations, and to multiply two dense m by m matrices together, in $O(m^{\log_2 7})$ operations. Using this result, along with modifications to Lemmas 8.1.5–8.1.7, show that the n by n grid problem can be solved in $O(n^{\log_2 7})$ operations, using the nested dissection ordering (Rose 1976b).

8.2   Nested Dissection of General Problems

8.2.1   A Heuristic Algorithm

The optimality of the nested dissection ordering for the n by n grid problem has been established in the previous section. The underlying idea of splitting the grid into two pieces of roughly equal size with a small separator is clearly important. In this section, we describe a heuristic algorithm that applies this strategy for orderings of general graphs.

How do we find a small separator to disconnect a given graph into components of approximately equal size? The method is to generate a long level structure of the graph and then choose a small separator from a "middle" level. The overall dissection ordering algorithm is described below. Let $G = (X,E)$ be the given graph.

Step 1   (Initialization) Set $R = X$, and $N = |X|$.

Step 2   (Generate a level structure) Find a connected component $G(C)$ in $G(R)$ and construct a level structure of the component $G(C)$ rooted at a pseudo-peripheral node r:

$$\mathcal{L}(r) = \{L_0, L_1, \ldots, L_\ell\} .$$

Step 3   (Find separator) If $\ell \leq 2$, set $S = C$ and go to Step 4. Otherwise let $j = \lfloor (\ell + 1)/2 \rfloor$, and determine the set $S \subset L_j$, where

$$S = \{y \in L_j \mid Adj(y) \cap L_{j+1} \neq \emptyset\} .$$

Step 4   (Number separator and loop) Number the nodes in the separator S from $N - |S| + 1$ to N. Reset $R \leftarrow R - S$ and

$N \leftarrow N - |S|$. If $R \neq \emptyset$, go to Step 2.

In Step 3 of the algorithm, the separator set S can be obtained by simply discarding nodes in L_j which are not adjacent to any node in L_{j+1}. In many cases, this reduces the size of the separators.

8.2.2   Computer Implementation

The set of subroutines which implements the nested dissection ordering algorithm consists of the following:

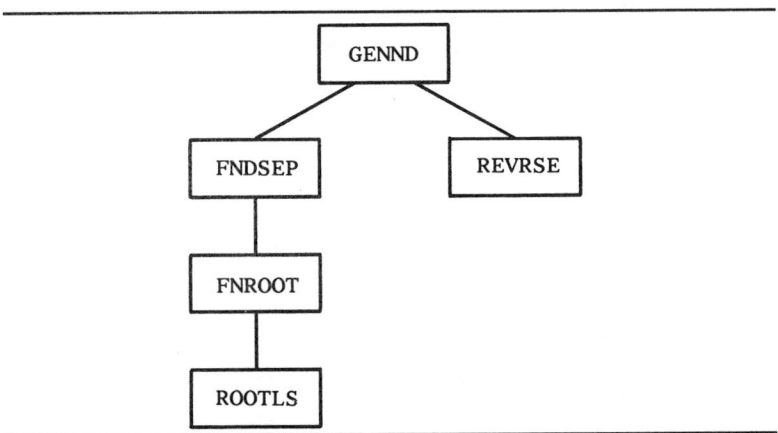

The subroutines FNROOT and ROOTLS have been described in Section 4.3.3 and the utility subroutine REVRSE was described in Section 7.2.2. The other two are described below.

GENND (GENeral Nested Dissection ordering)

This is the driver subroutine for this set of subroutines. It is used to determine a nested dissection ordering for a general disconnected graph. The input graph is given by NEQNS and (XADJ, ADJNCY), and the output ordering is returned in the vector PERM. The working vector MASK is used to mask off nodes that have been numbered during the ordering process. Two more working vectors (XLS, LS) are required and they are used by the called subroutine FNDSEP.

The subroutine begins by initializing the vector MASK. It then goes through the graph until it finds a node i not yet numbered. This node i defines a component in the unnumbered portion of the graph. The subroutine FNDSEP is then called to find a separator in the component. Note that the separator is collected in the vector PERM starting at position NUM+1. So, after all nodes have been numbered, the vector PERM has to be reversed to get the final ordering.

```
C**************************************************************
C**************************************************************
C********    GENND ..... GENERAL NESTED DISSECTION    ********
C**************************************************************
C**************************************************************
C
C     PURPOSE - SUBROUTINE GENND FINDS A NESTED DISSECTION
C        ORDERING FOR A GENERAL GRAPH.
C
C
C     INPUT PARAMETERS -
C        NEQNS - NUMBER OF EQUATIONS.
C        (XADJ, ADJNCY) - ADJACENCY STRUCTURE PAIR.
C
C     OUTPUT PARAMETERS -
C        PERM - THE NESTED DISSECTION ORDERING.
C
C     WORKING PARAMETERS -
C        MASK - IS USED TO MASK OFF VARIABLES THAT HAVE
C               BEEN NUMBERED DURING THE ORDERNG PROCESS.
C        (XLS, LS) - THIS LEVEL STRUCTURE PAIR IS USED AS
C               TEMPORARY STORAGE BY FNROOT.
C
C     PROGRAM SUBROUTINES -
C        FNDSEP, REVRSE.
C
C**************************************************************
C
      SUBROUTINE  GENND ( NEQNS, XADJ, ADJNCY, MASK,
     1                     PERM, XLS, LS )
C
C**************************************************************
C
      INTEGER ADJNCY(1), MASK(1), LS(1), PERM(1),
     1        XLS(1)
      INTEGER XADJ(1), I, NEQNS, NSEP, NUM, ROOT
C
C**************************************************************
C
         DO 100 I = 1, NEQNS
            MASK(I) = 1
  100    CONTINUE
         NUM    = 0
         DO 300 I = 1, NEQNS
C           ----------------------------
C           FOR EACH MASKED COMPONENT ...
C           ----------------------------
  200       IF ( MASK(I) .EQ. 0 )  GO TO 300
            ROOT = I
C           ---------------------------------------------
C           FIND A SEPARATOR AND NUMBER THE NODES NEXT.
C           ---------------------------------------------
            CALL  FNDSEP ( ROOT, XADJ, ADJNCY, MASK,
     1                     NSEP, PERM(NUM+1), XLS, LS )
            NUM  = NUM + NSEP
            IF ( NUM .GE. NEQNS )  GO TO 400
            GO TO 200
  300    CONTINUE
C        ------------------------------------------------
C        SINCE SEPARATORS FOUND FIRST SHOULD BE ORDERED
C        LAST, ROUTINE REVRSE IS CALLED TO ADJUST THE
C        ORDERING VECTOR.
C        ------------------------------------------------
  400    CALL  REVRSE ( NEQNS, PERM )
         RETURN
      END
```

FNDSEP (FiND SEParator)

This subroutine is used by GENND to find a separator for a connected subgraph. The connected component is specified by the input parameters ROOT, XADJ, ADJNCY and MASK. Returned from FNDSEP is the separator in (NSEP, SEP). The array pair (XLS, LS) is used to store a level structure of the component.

The subroutine first generates a level structure rooted at a pseudo-peripheral node by calling FNROOT. If the number of levels is less than 3, the whole component is returned as the "separator." Otherwise, a middle level, given by MIDLVL is determined. The loop DO 500 I = ... goes through the nodes in this middle level. A node is included in the separator if it has some neighbor in the next level. The separator is then returned in (NSEP, SEP).

```
C**************************************************************
C**************************************************************
C***********      FNDSEP .....  FIND SEPARATOR     **********
C**************************************************************
C**************************************************************
C
C       PURPOSE - THIS ROUTINE IS USED TO FIND A SMALL
C                 SEPARATOR FOR A CONNECTED COMPONENT SPECIFIED
C                 BY MASK IN THE GIVEN GRAPH.
C
C       INPUT PARAMETERS -
C          ROOT - IS THE NODE THAT DETERMINES THE MASKED
C                 COMPONENT.
C          (XADJ, ADJNCY) - THE ADJACENCY STRUCTURE PAIR.
C
C       OUTPUT PARAMETERS -
C          NSEP - NUMBER OF VARIABLES IN THE SEPARATOR.
C          SEP - VECTOR CONTAINING THE SEPARATOR NODES.
C
C       UPDATED PARAMETER -
C          MASK - NODES IN THE SEPARATOR HAVE THEIR MASK
C                 VALUES SET TO ZERO.
C
C       WORKING PARAMETERS -
C          (XLS, LS) - LEVEL STRUCTURE PAIR FOR LEVEL STRUCTURE
C                 FOUND BY FNROOT.
C
C       PROGRAM SUBROUTINES -
C          FNROOT.
C
C**************************************************************
C
        SUBROUTINE  FNDSEP ( ROOT, XADJ, ADJNCY, MASK,
     1                       NSEP, SEP, XLS , LS )
C
C**************************************************************
C
        INTEGER ADJNCY(1), LS(1), MASK(1), SEP(1), XLS(1)
        INTEGER XADJ(1), I, J, JSTOP, JSTRT, MIDBEG,
     1          MIDEND, MIDLVL, MP1BEG, MP1END,
     1          NBR, NLVL, NODE, NSEP, ROOT
C
C**************************************************************
```

```
C
              CALL   FNROOT ( ROOT, XADJ, ADJNCY, MASK,
         1                     NLVL, XLS, LS )
C             -------------------------------------------------
C             IF THE NUMBER OF LEVELS IS LESS THAN 3, RETURN
C             THE WHOLE COMPONENT AS THE SEPARATOR.
C             -------------------------------------------------
              IF ( NLVL .GE. 3 )  GO TO 200
                 NSEP = XLS(NLVL+1) - 1
                 DO 100 I = 1, NSEP
                    NODE = LS(I)
                    SEP(I) = NODE
                    MASK(NODE) = 0
     100         CONTINUE
                 RETURN
C             --------------------------------------------------------
C             FIND THE MIDDLE LEVEL OF THE ROOTED LEVEL STRUCTURE.
C             --------------------------------------------------------
     200      MIDLVL = (NLVL + 2)/2
             MIDBEG = XLS(MIDLVL)
             MP1BEG = XLS(MIDLVL + 1)
             MIDEND = MP1BEG - 1
             MP1END = XLS(MIDLVL+2) - 1
C             --------------------------------------------------------
C             THE SEPARATOR IS OBTAINED BY INCLUDING ONLY THOSE
C             MIDDLE-LEVEL NODES WITH NEIGHBORS IN THE MIDDLE+1
C             LEVEL. XADJ IS USED TEMPORARILY TO MARK THOSE
C             NODES IN THE MIDDLE+1 LEVEL.
C             --------------------------------------------------------
              DO 300 I = MP1BEG, MP1END
                 NODE = LS(I)
                 XADJ(NODE) = - XADJ(NODE)
     300      CONTINUE
              NSEP  = 0
              DO 500 I = MIDBEG, MIDEND
                 NODE = LS(I)
                 JSTRT = XADJ(NODE)
                 JSTOP = IABS(XADJ(NODE+1)) - 1
                 DO 400 J = JSTRT, JSTOP
                    NBR = ADJNCY(J)
                    IF ( XADJ(NBR) .GT. 0 )  GO TO 400
                       NSEP = NSEP + 1
                       SEP(NSEP) = NODE
                       MASK(NODE) = 0
                       GO TO 500
     400         CONTINUE
     500      CONTINUE
C             -------------------------------
C             RESET XADJ TO ITS CORRECT SIGN.
C             -------------------------------
              DO 600 I = MP1BEG, MP1END
                 NODE = LS(I)
                 XADJ(NODE) = - XADJ(NODE)
     600      CONTINUE
              RETURN
          END
```

Exercises

8.2.1)   This problem involves modifying GENND and FNDSEP to implement a form of "incomplete nested dissection." Add a parameter MINSZE to both subroutines, and modify FNDSEP so that
it only dissects the component given to it if the number of
nodes in the component is greater than MINSZE. Otherwise,
the component should be numbered using the RCM subroutine
from Chapter 4.   Conduct an experiment to investigate
whether the result you are asked to prove in Exercise 8.1.7
appears to hold for the heuristic orderings produced by the
algorithm of this section.   One way to do this would be to
solve a sequence of problems of increasing size, such as the
test set #2 from Chapter 9, with MINSZE set to $\sqrt{N}$. (For
the n by n grid problem, note that $\ell \geq \log_2(\sqrt{N})$ implies that
the final level independent blocks have $O(n)$ nodes. That is,
$O(\sqrt{N})$ nodes, where $N = n^2$.) Monitor the operation
counts for these problems, and compare them to the
corresponding values for the original (complete) dissection
algorithm. Similarly, you could compare storage requirements to see if they appear to grow as $N\sqrt{N}$ for your
incomplete dissection algorithm.

8.2.2)   Show that in the algorithm of Section 8.2.1, the number of
fills and factorization operation count is independent of the
order the nodes in the separator are numbered.

8.3   Additional Notes

Lipton, Tarjan and Rose (1979) have provided a major advance in the
development of automatic nested dissection algorithms. The key to
their algorithm is a fundamental result by Lipton and Tarjan (1977)
showing that the nodes of any N–node planar graph can be partitioned
into three sets A, B, and C where $Adj(A) \cap B = \emptyset$, $|C|$ is $O(\sqrt{N})$,
and $|A|$ and $|B|$ are bounded by $2N/3$. They also provided an algo-
rithm which finds A, B, and C in $O(N)$ time. Using this result Lipton
et al. have developed an ordering algorithm for two dimensional finite
element problems for which the $O(N^{3/2})$ operation and $O(N\log_2 N)$
storage bounds are guaranteed. Moreover, the ordering algorithm itself runs in $O(N\log_2 N)$ time. On the negative side, their algorithm
appears to be substantially more complicated than the simple heuristic
one given in this chapter. A practical approach might be to combine
the two methods, and use their more sophisticated scheme only if the
simple approach in this chapter yields a "bad" separator.

The use of nested dissection ideas has been shown to be effective for problems associated with three dimensional structures. (George 1973, Duff and Reid 1976, Rose and Whitten 1976, Eisenstat 1976.) Thus, research into automatic nested dissection algorithms for these non-planar problems appears to be a potentially fertile area.

The use of dissection methods on parallel and vector computers has been investigated by numerous researchers (Calahan 1975, 1976, George et al. 1978g, Lambiotte 1975). Vector computers tend to be most efficient if they can operate on "long" vectors, but the use of dissection techniques tend to produce short vectors, unless some unconventional methods of arranging the data are employed. Thus, the main issue in these studies involves balancing several conflicting criteria to produce the best solution time. Often this does not correspond at all closely to minimizing the arithmetic performed.

9/ Numerical Experiments

9.0 Introduction

In Chapter 1 we asserted that the success of algorithms for sparse matrix computations depends crucially on the quality of their computer implementations. This is why we have included computer implementations of the algorithms discussed in the previous chapters, and have provided a detailed discussion of how those programs work. In this chapter we provide results from numerical experiments where these subroutines have been used to solve some test problems.

Our primary objective here is to provide some concrete examples which illustrate the points made in Section 2.3, where "practical considerations" were discussed, and where it was pointed out how complicated it is to compare different methods. Data structures vary in their complexity, and the execution time for solving a problem consists of several components whose importance varies with the ordering strategy and the problem. The numerical results provided in this chapter give the user information to gauge the significance of some of these points.

As an attractive byproduct, the reader is supplied with data about the absolute time and storage requirements for some representative sparse matrix computations on a typical computer.

The test problems are of one specific type, typical of those arising in finite element applications. Our justification for this is that we are simply trying to provide some evidence illustrating the practical points made earlier; we regard it as far too ambitious to attempt to gather evidence about the relative merits of different methods over numerous classes of problems. It is more or less self-evident that for some classes of problems, one method may be uniformly better than all others, or that the relative merits of the methods in our book may be entirely different for other classes of problems. Restricting our attention to problems of one class simply removes one of the variables in an already complicated study.

Nevertheless, the test problems do represent a large and important application area for sparse matrix techniques, and have the additional advantage that they are associated with physical objects (meshes) which provide us with a picture (graph) of the matrix problem.

An outline of the remaining parts of this chapter is as follows. In Section 9.1 we describe the test problems, and in Section 9.2 we describe the information supplied in some of the tables, along with the reasons for providing it. These tables, containing the "raw" experimental data, appear at the end of Section 9.2. In Section 9.3 we review the main criteria used in comparing methods, and then proceed to compare five methods, according to these criteria, when applied to the test problems. Finally, in Section 9.4 we consider the influence of the different storage schemes on the storage and computational efficiency of the numerical subroutines.

9.1   Description of the Test Problems

The two sets of test problems are positive definite matrix equations typical of those which might arise in structural analysis or the study of heat conduction (Zienkiewicz 1977). (For an excellent tutorial see Chapter 6 of Strang 1973.) The problems are derived from the triangular meshes shown in Figure 9.1.1 as follows. The basic meshes shown are subdivided by a factor s in the obvious way, yielding a mesh having s^2 as many triangles as the original, as shown in Figure 9.1.2 for the pinched hole domain with $s = 3$. Providing a basic mesh along with a subdivision factor determines a new mesh having N nodes. Then, for some labelling of these N nodes, we generate an N by N symmetric positive definite matrix problem $Ax = b$, where $a_{ij} \neq 0$ if and only if nodes of the mesh are joined by an edge. Thus, the generated meshes can be viewed as the graphs of the corresponding matrix problem.

The two sets of test problems are derived from these meshes. Test set #1 is simply the nine mesh problems, subdivided by an appropriate factor so that the resulting matrix problems have about 1000–1500 equations, as shown in Table 9.1.1. The second set of problems is a sequence of nine graded-L problems obtained by subdividing the initial graded-L mesh of Figure 9.1.1 by subdivision factors $s = 4, 5, \ldots, 12$, as indicated in Table 9.1.2.

9.2   The Numbers Reported and What They Mean

In Chapters 4 through 8 we have described five methods, which in this chapter we refer to by the mnemonics RCM (reverse Cuthill-McKee), RQT (refined quotient tree), 1WD (one-way dissection), QMD (quotient minimum degree), and ND (nested dissection). Recall that we described only three basic data structures and corresponding

Figure 9.1.1   Mesh problems with $s = 1$.

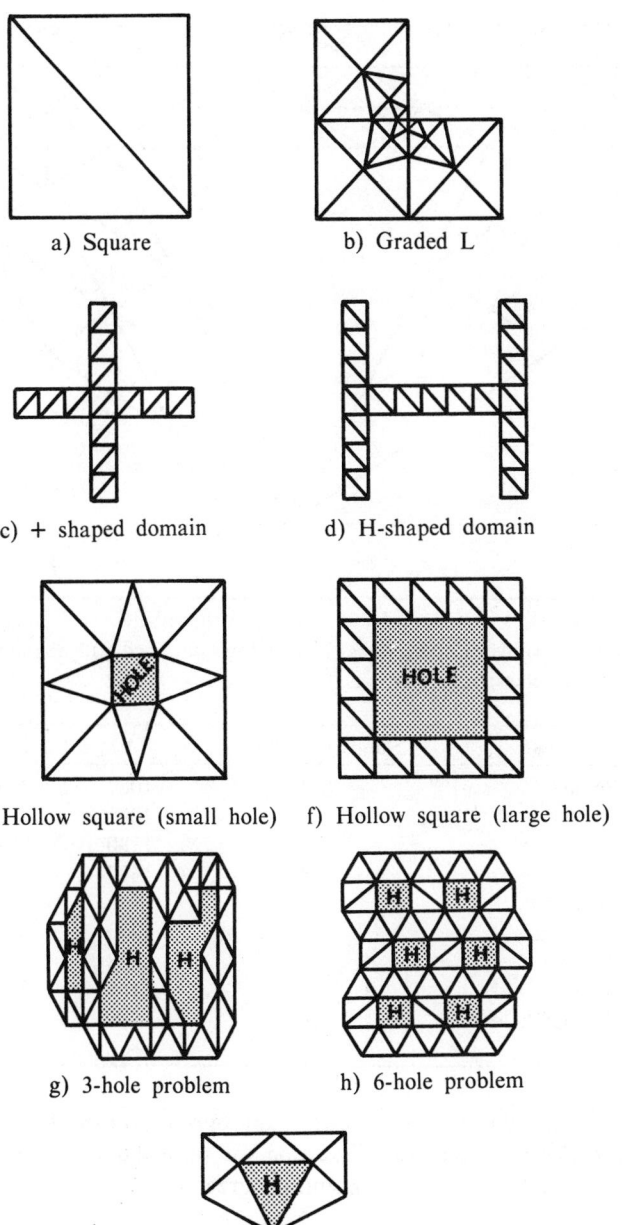

a) Square

b) Graded L

c) + shaped domain

d) H-shaped domain

e) Hollow square (small hole)

f) Hollow square (large hole)

g) 3-hole problem

h) 6-hole problem

i) Pinched hole problem

Figure 9.1.2   Pinched hole domain with subdivision factor $s = 3$.

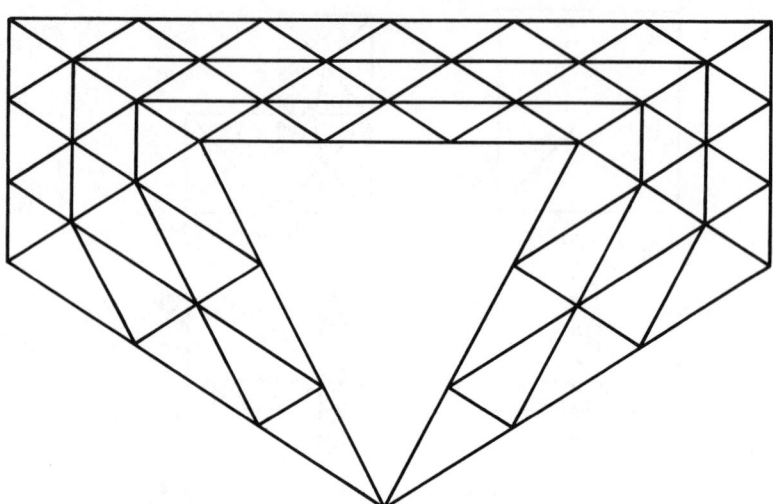

Table 9.1.1   Data on test problem set #1 with the subdivision factors used to generate the problems, the number of equations obtained, and the number of edges in the corresponding graphs.

Problem	Subdivision factor	N	$|E|$
Square	32	1089	3136
Graded L	8	1009	2928
+ domain	9	1180	3285
H domain	8	1377	3808
Small hole	12	936	2665
Large hole	9	1440	4032
3 holes	6	1138	3156
6 holes	6	1141	3162
Pinched hole	19	1349	3876

numerical subroutines, because it is appropriate to use the same data structures with the one-way dissection and refined quotient tree orderings, and similarly for the minimum degree and nested dissection orderings.

In the tables at the end of this section, *operations* mean *multiplicative* operations (multiplications and divisions). For reasons already discussed in Chapter 2, we regard this as a reasonable measure of the amount of arithmetic performed, since arithmetic operations in matrix computations typically occur in multiply-add pairs. *Execution*

Table 9.1.2   Data on test problem set #2, which is derived from the Graded-L mesh with subdivision factors $s = 4, 5, \ldots, 12$.

Subdivision factor	N	$|E|$
4	265	744
5	406	1155
6	577	1656
7	778	2247
8	1009	2928
9	1270	3699
10	1561	4560
11	1882	5511
12	2233	6552

time is reported in seconds on an **IBM 3031** computer, a fairly recent architecture using high speed cache memory, and on which typical operations take from .4 microseconds for a simple fixed-point register-to-register operation, to about 7 microseconds for a floating-point division. As is usual in multiprogrammed operating system environments, accurate timing results are difficult to obtain and may be in error by up to 10 percent. We have attempted to reduce these errors somewhat by making multiple runs, and running when the computer was lightly loaded. The programs were all compiled using the optimizing version of the compiler, which usually generates very efficient machine code.

Recall that we concluded in Section 2.3 that in some comparisons of ordering strategies, it might be reasonable to ignore one or more of the four basic steps in the overall solution procedure. For this reason, in the numerical experiments we report execution times for each of the four individual steps: order, allocate, factor, and solve.

There are four storage statistics reported in the tables: *order storage, allocation storage, total (solution) storage*, and *overhead storage*. All our experiments were performed within the framework of a sparse matrix package called **SPARSPAK** (George 1978a, 1979a) which allocates all array storage from a single one dimensional array. The order storage, allocation storage, and solution storage reported is the amount of storage used from that array. Thus, we feel that these numbers represent the amount of storage required when the various subroutines are used in a *practical setting*, rather than the irreducible minimum necessary to execute the subroutines. To illustrate this point, note that one does not *need* to preserve the original graph when one uses the QMD ordering subroutine, (which destroys its input graph during execution,) but in most practical applications one *would* preserve the graph since it is required for the subsequent symbolic factorization step. Thus, the ordering storage entries under QMD in the tables *include* the space necessary to preserve the original graph.

As another example, it is obviously not necessary to preserve PERM and INVP after the allocation has been performed, since the numerical factorization and solution subroutines do not use these arrays. However, in most situations the arrays would be saved in order to place the numerical values of *A* and *b* in the appropriate places in the data structure, and to replace the values of *x* in the original order after the (permuted) solution has been computed. In Table 9.2.1 we list the arrays included in our storage reporting, for the different phases of the computation (order, allocate, factorization, solution), and for the five methods.   The notation $A(B)$ in Table 9.2.1 means arrays A and B use the same storage space, in sequence.

Table 9.2.1   Arrays included in reported storage requirements for each phase of the five methods. Storage required for the underlined arrays in the *Solution* column is reported as "overhead storage."

	Order	Allocate	Solution
RCM	XADJ, ADJNCY, PERM, XLS, MASK	XADJ, ADJNCY, PERM, INVP, XENV	PERM, INVP, RHS, XENV, ENV, DIAG
1WD	XADJ, ADJNCY, PERM, BNUM, LS(SUBG), XBLK, MASK, XLS	XADJ, ADJNCY, PERM, INVP, XBLK, MASK, MARKER, FATHER, XENV, NZSUBS, RCHSET(XNONZ)	PERM, INVP, RHS, XENV, ENV, DIAG, XNONZ, NZSUBS, NONZ, TEMPV, FIRST
RQT	XADJ, ADJNCY, PERM, XBLK, MASK, NODLVL(BNUM), XLS, LS(SUBG)	XADJ, ADJNCY, PERM, INVP, XBLK, MASK, FATHER, XENV, XNONZ, NZSUBS	same as above
ND	XADJ, ADJNCY, PERM, LS, XLS, MASK	XADJ, ADJNCY, PERM, INVP, XLNZ, XNZSUB, NZSUB, MRGLNK, RCHLNK, MARKER	PERM, INVP, RHS, XNZSUB, NZSUB, XLNZ, LNZ, DIAG, LINK, FIRST, TEMPV
QMD	XADJ, 2 copies of ADJNCY, PERM, MARKER, DEG, RCHSET, NBRHD, QSIZE, QLINK	same as above	same as above

Strictly speaking, we should distinguish between factorization storage and triangular solution storage, since several arrays required by TSFCT and GSFCT are not required by their respective solvers. However, the storage for these arrays will usually be relatively small, compared to the total storage required for the triangular solution. Thus, we report only "solution storage" in our tables.

So far our discussion about storage reporting has dealt only with the first three categories: ordering, allocation, and numerical solution. The fourth category is "overhead storage," which is included in order to illustrate how much of the total storage used during the factorization/solution phase is occupied by data *other than the nonzeros in L and the right hand side b* (which is overwritten by *x*). If a storage location is not being used to store a component of *L* or *b*, then we count it as overhead storage. The arrays making up the overhead storage entries are underlined in Table 9.2.1. Note that solution storage *includes* overhead storage.

There is another reason for reporting overhead storage as a separate item. On computers having a large word size, it may be sensible to pack several integers per word. Indeed, some computer manufacturers provide short integer features directly in their Fortran languages. For example, IBM Fortran allows one to declare integers as INTEGER∗2 or INTEGER∗4, which will be represented using 16 or 32 bits respectively. Since much of the overhead storage involves integer data, the reader can gauge the potential storage savings to be realized if the Fortran processor one is using provides these short integer features. However, note that all the experiments were performed on an IBM 3031 in single precision, and both integers and floating point numbers are represented using 32 bits.

Table 9.2.2   Results of the RCM method applied to test set #1. (Operations and storage scaled by 10^{-4})

Problem	Order		Allocation		Solution					
	Time	Store	Time	Store	Time		Operations		Store	
					Fact	Solve	Fact	Solve	Total	Ovrhd
936	0.21	0.91	0.04	0.91	2.85	0.40	30.18	4.55	2.65	0.28
1009	0.27	0.99	0.04	0.99	3.43	0.46	37.49	5.25	3.03	0.30
1089	0.24	1.06	0.05	1.06	3.25	0.45	34.46	5.11	2.99	0.33
1440	0.32	1.38	0.06	1.38	4.74	0.62	53.77	7.23	4.19	0.43
1180	0.32	1.13	0.06	1.13	2.86	0.44	31.87	5.17	3.06	0.35
1377	0.30	1.31	0.06	1.31	1.99	0.40	18.64	4.34	2.72	0.41
1138	0.30	1.09	0.06	1.09	2.81	0.45	28.88	4.92	2.92	0.34
1141	0.25	1.09	0.05	1.09	4.40	0.52	54.08	6.75	3.83	0.34
1349	0.36	1.31	0.06	1.31	5.74	0.69	64.95	7.95	4.52	0.40

Table 9.2.3   Results of the 1WD method applied to test set #1.   (Operations and storage scaled by 10^{-4})

Problem	Order		Allocation		Solution						
	Time	Store	Time	Store	Time		Operations		Store		
					Fact	Solve	Fact	Solve	Total	Ovrhd	
936	0.38	1.19	0.25	1.24	3.39	0.36	26.60	3.06	1.72	0.61	
1009	0.47	1.29	0.28	1.35	5.47	0.40	44.90	3.66	1.94	0.66	
1089	0.45	1.39	0.30	1.46	5.23	0.42	41.48	3.62	2.03	0.72	
1440	0.60	1.81	0.39	1.89	4.43	0.52	33.04	4.32	2.51	0.94	
1180	0.55	1.48	0.33	1.56	3.35	0.42	24.50	3.48	2.03	0.78	
1377	0.57	1.73	0.37	1.82	3.25	0.49	21.41	3.57	2.21	0.92	
1138	0.52	1.43	0.30	1.49	3.17	0.41	23.56	3.48	1.98	0.75	
1141	0.46	1.43	0.31	1.49	5.10	0.44	41.08	3.77	2.07	0.74	
1349	0.61	1.72	0.36	1.79	7.03	0.56	58.38	4.79	2.57	0.88	

Table 9.2.4   Results of the RQT method applied to test set #1.   (Operations and storage scaled by 10^{-4})

Problem	Order		Allocation		Solution						
	Time	Store	Time	Store	Time		Operations		Store		
					Fact	Solve	Fact	Solve	Total	Ovrhd	
936	0.18	1.19	0.15	1.19	4.17	0.53	32.84	4.83	2.14	0.75	
1009	0.26	1.29	0.16	1.30	4.88	0.59	39.32	5.49	2.38	0.81	
1089	0.22	1.39	0.17	1.40	4.68	0.61	37.04	5.43	2.45	0.88	
1440	0.29	1.81	0.22	1.81	6.75	0.82	55.46	7.44	3.29	1.15	
1180	0.31	1.48	0.19	1.49	2.39	0.54	13.52	3.41	2.04	0.95	
1377	0.28	1.73	0.21	1.74	2.30	0.58	11.69	3.53	2.27	1.12	
1138	0.30	1.43	0.17	1.43	4.32	0.60	31.98	5.13	2.41	0.91	
1141	0.23	1.43	0.17	1.42	7.18	0.72	63.98	6.92	2.86	0.91	
1349	0.36	1.72	0.21	1.72	7.79	0.86	67.68	8.30	3.42	1.08	

Table 9.2.5   Results of the ND method applied to test set #1.   (Operations and storage scaled by 10^{-4})

Problem	Order		Allocation		Solution						
	Time	Store	Time	Store	Time		Operations		Store		
					Fact	Solve	Fact	Solve	Total	Ovrhd	
936	0.77	1.00	0.24	1.78	2.19	0.29	16.25	2.96	2.73	1.15	
1009	0.95	1.09	0.25	1.97	3.77	0.37	31.11	4.00	3.38	1.29	
1089	0.92	1.17	0.27	2.10	3.40	0.37	26.82	3.91	3.43	1.36	
1440	1.35	1.53	0.34	2.68	2.69	0.41	19.05	4.06	3.90	1.72	
1180	1.10	1.25	0.28	2.17	2.05	0.31	14.22	3.15	3.09	1.40	
1377	1.20	1.45	0.32	2.51	2.34	0.36	15.71	3.59	3.54	1.61	
1138	1.15	1.20	0.27	2.11	2.32	0.32	16.89	3.32	3.14	1.37	
1141	1.14	1.20	0.27	2.13	2.35	0.32	17.23	3.34	3.16	1.38	
1349	1.39	1.45	0.33	2.60	4.41	0.48	35.48	4.98	4.31	1.69	

Table 9.2.6   Results of the QMD method applied to test set #1.   (Operations and storage scaled by 10^{-4})

Problem	Order		Allocation		Solution					
	Time	Store	Time	Store	Time		Operations		Store	
					Fact	Solve	Fact	Solve	Total	Ovrhd
936	1.47	1.91	0.21	1.80	2.27	0.30	19.34	3.11	2.83	1.18
1009	1.57	2.08	0.24	1.97	3.37	0.37	30.91	3.95	3.36	1.29
1089	1.78	2.23	0.26	2.12	3.00	0.36	26.31	3.85	3.42	1.38
1440	2.33	2.91	0.34	2.74	2.70	0.42	21.62	4.21	4.04	1.79
1180	1.89	2.38	0.27	2.19	1.44	0.27	9.86	2.72	2.90	1.42
1377	2.09	2.76	0.30	2.52	1.50	0.30	10.00	2.99	3.25	1.62
1138	1.97	2.29	0.27	2.15	1.82	0.29	13.80	3.04	3.04	1.41
1141	2.07	2.29	0.27	2.17	2.03	0.31	16.24	3.20	3.14	1.43
1349	2.07	2.76	0.32	2.62	3.70	0.46	32.41	4.82	4.26	1.71

Table 9.2.7   Results of the RCM method applied to test set #2.   (Operations and storage scaled by 10^{-4})

Problem	Order		Allocation		Solution					
	Time	Store	Time	Store	Time		Operations		Store	
					Fact	Solve	Fact	Solve	Total	Ovrhd
265	0.07	0.25	0.01	0.25	0.33	0.07	2.97	0.75	0.48	0.08
406	0.12	0.39	0.02	0.39	0.71	0.13	6.62	1.39	0.86	0.12
577	0.16	0.56	0.03	0.56	1.30	0.21	12.88	2.32	1.39	0.17
778	0.22	0.76	0.03	0.76	2.15	0.32	22.78	3.59	2.10	0.23
1009	0.27	0.99	0.05	0.99	3.43	0.46	37.49	5.25	3.03	0.30
1270	0.34	1.25	0.06	1.25	5.19	0.62	58.37	7.36	4.19	0.38
1561	0.42	1.54	0.07	1.54	7.50	0.83	86.95	9.97	5.61	0.47
1882	0.51	1.86	0.09	1.86	10.62	1.11	124.90	13.14	7.32	0.56
2233	0.62	2.20	0.10	2.20	14.44	1.40	174.10	16.91	9.35	0.67

Table 9.2.8   Results of the 1WD method applied to test set #2.   (Operations and storage scaled by 10^{-4})

Problem	Order		Allocation		Solution					
	Time	Store	Time	Store	Time		Operations		Store	
					Fact	Solve	Fact	Solve	Total	Ovrhd
265	0.12	0.33	0.07	0.35	0.65	0.09	4.38	0.69	0.42	0.18
406	0.18	0.52	0.11	0.54	1.29	0.15	9.25	1.18	0.69	0.27
577	0.27	0.74	0.17	0.77	2.39	0.22	18.11	1.87	1.04	0.38
778	0.35	0.99	0.20	1.04	3.69	0.32	29.22	2.68	1.45	0.51
1009	0.48	1.29	0.28	1.35	5.46	0.41	44.90	3.66	1.94	0.66
1270	0.60	1.63	0.35	1.70	7.76	0.54	66.00	4.87	2.53	0.83
1561	0.70	2.00	0.43	2.09	11.09	0.68	97.14	6.36	3.25	1.02
1882	0.91	2.42	0.51	2.51	14.73	0.85	131.13	7.90	4.00	1.22
2233	1.05	2.87	0.62	2.98	19.64	1.04	177.17	9.91	4.89	1.45

Table 9.2.9   Results of RQT method applied to test set #2.   (Operations and storage scaled by 10^{-4})

Problem	Order		Allocation		Solution						
	Time	Store	Time	Store	Time		Operations		Store		
					Fact	Solve	Fact	Solve	Total	Ovrhd	
265	0.07	0.33	0.04	0.34	0.57	0.12	3.24	0.81	0.47	0.21	
406	0.10	0.52	0.07	0.52	1.12	0.20	7.11	1.49	0.78	0.33	
577	0.16	0.74	0.09	0.74	1.94	0.29	13.70	2.46	1.19	0.46	
778	0.20	0.99	0.12	1.00	3.15	0.43	24.03	3.78	1.72	0.63	
1009	0.26	1.29	0.16	1.30	4.86	0.60	39.32	5.49	2.38	0.81	
1270	0.33	1.63	0.20	1.63	7.15	0.78	60.94	7.67	3.19	1.02	
1561	0.40	2.00	0.25	2.01	10.14	1.04	90.43	10.35	4.15	1.25	
1882	0.48	2.42	0.30	2.43	14.06	1.31	129.48	13.59	5.28	1.51	
2233	0.56	2.87	0.35	2.88	19.20	1.67	179.99	17.44	6.60	1.79	

Table 9.2.10   Results of the ND method applied to test set #2.   (Operations and storage scaled by 10^{-4})

Problem	Order		Allocation		Solution						
	Time	Store	Time	Store	Time		Operations		Store		
					Fact	Solve	Fact	Solve	Total	Ovrhd	
265	0.20	0.28	0.06	0.49	0.46	0.07	3.25	0.72	0.70	0.32	
406	0.33	0.43	0.10	0.77	0.93	0.12	6.93	1.27	1.17	0.50	
577	0.49	0.62	0.14	1.10	1.62	0.19	12.56	1.99	1.77	0.72	
778	0.68	0.84	0.20	1.51	2.58	0.29	20.10	2.88	2.50	0.99	
1009	0.95	1.09	0.26	1.97	3.80	0.37	31.11	4.00	3.38	1.29	
1270	1.22	1.38	0.32	2.49	5.30	0.49	44.45	5.27	4.39	1.63	
1561	1.56	1.69	0.39	3.08	7.34	0.63	62.57	6.82	5.58	2.01	
1882	1.93	2.04	0.50	3.74	9.73	0.78	83.85	8.54	6.90	2.44	
2233	2.35	2.43	0.58	4.45	12.79	0.94	111.18	10.57	8.42	2.92	

Table 9.2.11   Results of the QMD method applied to test set #2.   (Operations and storage scaled by 10^{-4})

Problem	Order		Allocation		Solution						
	Time	Store	Time	Store	Time		Operations		Store		
					Fact	Solve	Fact	Solve	Total	Ovrhd	
265	0.39	0.54	0.06	0.49	0.36	0.07	2.65	0.65	0.67	0.32	
406	0.72	0.83	0.09	0.78	0.78	0.11	6.35	1.19	1.14	0.50	
577	1.01	1.18	0.13	1.12	1.26	0.18	10.35	1.83	1.70	0.73	
778	1.39	1.60	0.19	1.52	2.22	0.26	19.65	2.80	2.48	1.00	
1009	1.55	2.08	0.24	1.97	3.43	0.38	30.91	3.95	3.36	1.29	
1270	2.48	2.62	0.34	2.53	4.59	0.48	42.56	5.15	4.37	1.66	
1561	2.56	3.23	0.39	3.09	5.90	0.60	55.43	6.53	5.45	2.03	
1882	3.32	3.90	0.48	3.76	8.36	0.76	80.10	8.49	6.91	2.47	
2233	3.53	4.63	0.55	4.43	12.06	0.98	119.13	10.71	8.47	2.89	

9.3   Comparison of the Methods

9.3.1   Criteria for Comparing Methods

In this section we shall not attempt to answer the question "which method should we use?". Sparse matrices vary a great deal, and the collection of test problems is of only one special class. Our objective here is to illustrate, using the data reported in Section 9.2, the issues involved in answering the question, given a particular problem or class of problems. These issues have already been discussed, or at least mentioned, in Section 2.3.

The main criteria were a) storage requirements, b) execution time, and c) cost. In some contexts, keeping storage requirements low is of overwhelming importance, while in other situations, low execution time is of primary concern. Perhaps most frequently, however, we are interested in choosing the method which results in the lowest computer charges. This charging function is typically a fairly complicated multi-parameter function of storage used (S), execution time (T), amount of input and output performed, . . . etc. For our class of problems and the methods we treat, this charging function can usually be quite well approximated by a function of the form

$$COST(S,T) = T \times p(S),$$

where $p(S)$ is a polynomial of degree d, usually equal to 1. (However, sometimes $d = 0$, and in other cases where large storage demands are discouraged, $d = 2$.) For purposes of illustration, in this book we assume $p(S) = S$.

Recall from Section 2.3 that the relative importance of the ordering and allocation, factorization, and solution depends on the context in which the sparse matrix problem arises. In some situations only one problem of a particular structure is to be solved, so any comparison of methods should certainly include ordering and allocation costs. In other situations where many problems having identical structure must be solved, it may be sensible to ignore the ordering and allocation costs. Finally, in still other contexts where numerous systems differing only in their right hand side must be solved, it may be appropriate to consider only the time and/or storage associated with the triangular solution, *given the factorization.*

In some of the tables appearing later we report a "minimum" and "maximum" total cost. The distinction is that the maximum cost is computed assuming that the storage used by any of the four phases (order, allocate, factor, solve) is equal to the maximum storage required by any of them (usually the factorization step). The minimum cost is obtained by assuming that the storage used by each phase is the minimum required by that phase (as specified in

Table 9.2.1). We report both costs to show that for some methods and problems, the costs are quite different and it is therefore worthwhile to segment the computation into its constituent parts, and use only the requisite storage for each phase.

9.3.2   Comparison of the Methods Applied to Test Set #1

Now consider Table 9.3.1, which we obtained by averaging the results in the tables of Section 9.2 for test set #1, and then computing the various costs. One of the most important things that it shows is that for the nine problems of this test set, the method of choice depends very much on the criterion we wish to optimize. For example, if total execution time is the basis for choice, then RCM should be chosen. If solution time, or factorization plus solution time or factorization plus solution cost, is of primary importance, then QMD should be chosen. If storage requirements, solve cost, or total cost are the most important criteria, then 1WD is the method of choice.

Several other aspects of Table 9.3.1 are noteworthy. Apparently, QMD yields a somewhat better ordering than ND, which is reflected in lower execution times and costs for the factorization and solution, and lower storage requirements. However, the fact that the ordering time for ND is substantially lower than that for QMD results in lower *total* costs and execution time for ND, compared to QMD.

Another interesting aspect of the ND and QMD total cost entries is the substantial difference between the maximum and minimum costs. Recall from Section 9.3.1 that the maximum cost is computed assuming that the storage used during any of the phases (order, allocate, factor, solve) is equal to the maximum used by any of them, while the minimum cost is computed assuming that each phase uses only what is normally required, as prescribed by Table 9.2.1. These numbers suggest that even for "one-shot" problems, segmenting the computation into its natural components, and using only the storage required for each phase, is well worthwhile.

Table 9.3.1   Average values of the various criteria for problem set #1. (Costs and storage scaled by 10^{-4})

Method	Cost				Storage	Execution Time		
	Total (Max)	Total (Min)	Fact+ Solve	Solve		Total	Fact+ Solve	Solve
RCM	14.60	13.86	13.47	1.64	3.32	4.39	4.06	0.49
1WD	12.22	11.72	10.45	0.95	2.12	5.77	4.94	0.45
RQT	15.60	15.11	14.44	1.68	2.58	6.04	5.59	0.65
ND	15.63	12.92	10.89	1.22	3.41	4.59	3.19	0.36
QMD	16.66	14.52	9.30	1.15	3.36	4.96	2.77	0.34

After examining Table 9.3.1, the reader might wonder whether methods such as RQT and ND have any merit, compared to the three other methods, since they fail to show up as winners according to any of the criteria we are considering. However, averages tend to hide differences among the problems, and to illustrate that each method does have a place, Table 9.3.2 contains a frequency count of which method was best, based on the various criteria, for the problems of set #1. Note that no row in the table is all zeros. This suggests that even within a particular class of problems, and for a fixed criterion (e.g., execution time, storage), the method of choice varies considerably across problems. One should also keep in mind that special combinations of criteria may make any of the methods look best, for almost any of the problems.

Table 9.3.2   Frequency counts of which method was best on the basis of various criteria for test problem set #1.

Method	Cost				Storage	Execution Time		
	Total (Max)	Total (Min)	Fact+ Solve	Solve		Total	Fact+ Solve	Solve
RCM	3	3	1	0	0	4	0	0
1WD	4	4	2	7	9	0	0	0
RQT	1	1	0	0	0	1	0	0
ND	1	1	1	0	0	3	2	3
QMD	0	0	5	2	0	1	7	6

One rather striking aspect of Table 9.3.2 is the very strong showing of 1WD in terms of cost and storage.

9.3.3 Comparison of the Methods Applied to Test Set #2

In order to illustrate some additional points, we include Tables 9.3.3 and 9.3.4, generated from Tables 9.2.7–9.2.11 of Section 9.2, which contain the experimental results for test problem set #2. Table 9.3.3 contains the same information as Table 9.3.1, for the Graded-L problem with $s = 4$, (yielding $N = 265$). Table 9.3.4 is also the same, except the subdivision factor is $s = 12$, yielding $N = 2233$.

First note that for $N = 265$, the RCM method displays a considerable advantage in most categories, and is very competitive in the remaining ones. However, for $N = 2233$, it has lost its advantage in all except total execution time. (Some other experiments show that it loses to ND in this category also for somewhat larger graded-L problems.) One of the main points we wish to make here is that even for essentially similar problems such as these, the *size* of the problem can influence the method of choice. Roughly speaking, for "small problems," the more sophisticated methods simply do not pay.

It is interesting to again note how very effective the 1WD method is in terms of storage and cost.

Notice also that the relative cost of the ordering and allocation steps, compared to the total cost, is going down as N increases, for all the methods.   For the RCM, 1WD and RQT methods, these first two steps have become relatively unimportant in the overall cost and execution time when N reaches about 2000. However, for the ND and QMD methods, even for N as large as 2233, the ordering and allocation steps still account for a significant fraction of the total execution time. Since these steps in general require less storage than the numerical computation steps, the difference between MAX cost and MIN cost remains important even for $N = 2233$.

Table 9.3.3   Values of the various criteria for the Graded-L problem with $s = 4$, yielding $N = 265$.   (Costs and storage scaled by 10^{-4})

Method	Cost				Storage	Execution Time		
	Total (Max)	Total (Min)	Fact+ Solve	Solve		Total	Fact+ Solve	Solve
RCM	0.23	0.21	0.19	0.04	0.48	0.48	0.40	0.07
1WD	0.39	0.37	0.31	0.04	0.42	0.93	0.73	0.09
RQT	0.38	0.36	0.32	0.06	0.47	0.81	0.69	0.12
ND	0.56	0.46	0.37	0.05	0.70	0.79	0.53	0.07
QMD	0.59	0.53	0.29	0.04	0.67	0.88	0.43	0.07

Table 9.3.4   Values of the various criteria for the Graded-L problem with $s = 12$, yielding $N = 2233$.   (Costs and storage scaled by 10^{-4})

Method	Cost				Storage	Execution Time		
	Total (Max)	Total (Min)	Fact+ Solve	Solve		Total	Fact+ Solve	Solve
RCM	154.83	149.69	148.10	13.09	9.35	16.56	15.84	1.40
1WD	109.41	106.10	101.25	5.11	4.89	22.35	20.69	1.04
RQT	143.69	140.30	137.67	10.99	6.60	21.78	20.87	1.67
ND	140.45	124.03	115.74	7.95	8.42	16.67	13.74	0.94
QMD	145.07	129.28	110.49	8.33	8.47	17.13	13.04	0.98

9.4 The Influence of Data Structures

In several places in this book we have emphasized the importance of data structures (storage schemes) for sparse matrices. In Section 2.3 we distinguished between primary storage and overhead storage, and through a simple example showed that primary storage requirements may not be a reliable indicator of the storage actually required by different computer programs, because of differences in overhead storage. We also pointed out in Section 2.3 that differences in data structures could lead to substantial differences in the arithmetic-operations-per-second output of the numerical subroutines. The main objective of this section is to provide some experimental evidence which supports these contentions, and to illustrate the potential magnitude of the differences involved.

9.4.1 Storage Requirements

In Table 9.4.1 we have compiled the primary and total storage requirements for the five methods, applied to test problem set #2. Recall that primary storage is that used for the numerical values of L and b, and overhead storage is "everything else," consisting mainly of integer pointer data associated with maintaining a compact representation of L.

Table 9.4.1 Primary and total storage for each method, applied to test problem set #2. (Numbers are scaled by 10^{-4})

	Primary Storage								
	Number of Equations								
Method	265	406	577	778	1009	1270	1561	1882	2233
RCM	0.40	0.74	1.22	1.87	2.73	3.81	5.14	6.76	8.68
1WD	0.24	0.42	0.66	0.94	1.28	1.70	2.23	2.78	3.45
RQT	0.26	0.45	0.73	1.10	1.57	2.17	2.90	3.77	4.80
ND	0.39	0.67	1.05	1.52	2.10	2.76	3.57	4.46	5.51
QMD	0.35	0.64	0.97	1.48	2.08	2.70	3.42	4.43	5.58

	Total Storage								
	Number of Equations								
Method	265	406	577	778	1009	1270	1561	1882	2233
RCM	0.48	0.86	1.39	2.10	3.03	4.19	5.61	7.32	9.35
1WD	0.42	0.69	1.04	1.45	1.94	2.53	3.25	4.00	4.89
RQT	0.47	0.78	1.19	1.72	2.38	3.19	4.15	5.28	6.60
ND	0.70	1.17	1.77	2.50	3.38	4.39	5.58	6.90	8.42
QMD	0.67	1.14	1.70	2.48	3.36	4.37	5.45	6.91	8.47

The numbers in Table 9.4.1 illustrate some important practical points:

1. For some methods, the overhead component in the storage requirements is very substantial, even for the larger problems where the relative importance of overhead is diminished somewhat. For example, for the QMD method, the ratio (*overhead storage*)/(*total storage*) ranges from about .48 to .34 as N goes from 265 to 2233. Thus, while the ratio *is* decreasing, it is still very significant for even fairly large problems.

   By way of comparison, for the RCM method, which utilizes a very simple data structure, the (*overhead storage*)/(*total storage*) ratio ranges from about .17 to .07 over the same problems.

2. Another point, (essentially a consequence of 1 above,) is that primary storage is a very unreliable indicator of a program's array storage requirements. For example, if we were comparing the RCM and QMD methods on the basis of primary storage requirements for the problems of test set #2, then QMD would be the method of choice for all N. However, in terms of *actual* storage requirements, the RCM method is superior until N is about 1500!

   This comparison also illustrates the potential importance of being able to use less storage for integers than that used for floating point numbers. In many circumstances, the number of binary digits used to represent floating point numbers is at least twice that necessary to represent integers of a sufficient magnitude. If it is convenient to exploit this fact, the significance of the overhead storage component will obviously be diminished. For example, if integers required only half as much storage as floating point numbers, the cross-over point between RCM and QMD would be at $N \simeq 600$, rather than $\simeq 1500$ as stated above.

3. Generally speaking, the information in Table 9.4.1 shows that while the more sophisticated ordering algorithms do succeed in reducing primary storage over their simpler counterparts, since they also necessitate the use of correspondingly more sophisticated storage schemes, the *net* reduction in storage requirements over the simpler schemes is not as pronounced as the relative differences in primary storage indicate. For example, primary storage requirements indicate that 1WD enjoys a storage saving of more than 50 percent over RCM, for $N \geq 778$, and that the advantage increases with N. However, the total storage requirements, while they still indicate that the storage advantage of 1WD over RCM increases with N, also show that the point at which a 50 percent savings occurs has still not been reached at $N = 2233$.

9.4.2   Execution Time

In Table 9.4.2 we have computed and tabulated the operations-per-second performance of the factorization and solution subroutines for the five methods, applied to test problem set #2. The information in the table suggests the following:

1. Generally speaking, the efficiency (i.e., operations-per-second) of the subroutines tends to improve with increasing N. This is to be expected since loops, once initiated, will tend to be executed more often as N increases. Thus, there will be less loop initialization overhead per arithmetic operation performed.

    In this connection, note that the relative improvement from $N = 265$ to $N = 2233$ varies considerably over the six different subroutines and five different orderings involved. (Recall that the 1WD and RQT methods use the same numerical subroutines, as do the ND and QMD methods.) For example, the operations-per-second output for the ND solver (GSSLV) only improved from 9.44×10^4 to 10.15×10^4 over the full range of N, while the RQT solver (TSSLV) improved from 6.07×10^4 to 9.50×10^4. These differences in improvement appear to be due to the variation in the number of auxiliary subroutines used. For example, TSFCT uses subroutines ELSLV, EUSLV, and ESFCT (which in turn uses ELSLV), while GSFCT uses none at all. These subroutine calls contribute a large low order component to the execution time.

    These differences in the performance of the numerical subroutines illustrate how unrealistic it is to conclude much of a practical nature from a study of operation counts alone. For example, if we were to compare the RCM method to the QMD method on the basis of factorization operation counts, for the problems of test set #2, we would choose QMD for all the problems. However, in terms of execution time, QMD does not win until N reaches about 1600.

2. We have already observed that efficiency varies across the different subroutines, and varies with N. It is also interesting that for a fixed problem and subroutine, efficiency varies with the ordering used. As an example, consider the factorization entries for the 1WD and RQT methods, for $N = 2233$. (Remember that both methods employ the subroutine TSFCT.) This difference in efficiency can be understood by observing that unlike the subroutines ESFCT and GSFCT, where the majority of the numerical computation is isolated in a single loop, the numerical computation performed by TSFCT is distributed among three auxiliary subroutines (which vary in efficiency), in addition to a major computational loop of its own. Thus, one ordering may yield a more efficient TSFCT than another simply because a larger proportion of the

Table 9.4.2   Operations-per-second for each method applied to test problem #2.   (Operations are scaled by 10^{-4})

	Factorization								
	Number of Equations								
Method	265	406	577	778	1009	1270	1561	1882	2233
RCM	9.01	9.37	9.91	10.58	10.92	11.24	11.59	11.76	12.06
1WD	6.78	7.19	7.57	7.93	8.22	8.51	8.76	8.90	9.02
RQT	5.68	6.33	7.07	7.63	8.10	8.53	8.92	9.21	9.37
ND	7.07	7.42	7.75	7.78	8.20	8.38	8.52	8.62	8.69
QMD	7.30	8.10	8.19	8.87	9.01	9.27	9.39	9.59	9.88

	Solution								
	Number of Equations								
Method	265	406	577	778	1009	1270	1561	1882	2233
RCM	10.21	10.69	10.87	11.21	11.50	11.81	11.97	11.87	12.08
1WD	7.96	8.05	8.50	8.27	9.00	9.01	9.31	9.30	9.50
RQT	6.79	7.45	8.38	8.78	9.16	9.83	9.95	10.35	10.47
ND	10.25	10.28	10.49	9.81	10.71	10.83	10.89	11.00	11.20
QMD	9.77	10.52	9.98	10.65	10.50	10.65	10.89	11.12	10.89

associated computation is performed by the most efficient auxiliary subroutines or loop of TSFCT.

A final complicating factor in this study of the 1WD and RQT factorization entries is that apparently, the proportions of the computation performed by the different computational loops of TSFCT varies with N, and the variation with N is different for the one-way dissection ordering than it is for the refined quotient tree ordering. For small problems, TSFCT operates more efficiently on the one-way dissection ordering than on the refined quotient tree ordering, but as N increases, the situation is reversed, with the cross-over point occurring at about $N = 1700$.

We should caution the reader not to infer too much from this particular example.   On a different computer with a different compiler and instruction set, the relative efficiencies of the computational loop in TSFCT and its auxiliary subroutines may be quite different. However, this example *does* illustrate that efficiency is not only a function of the data structure used, but may depend in a rather subtle way on the ordering used with that data structure, and on the problem size.

A/ Some Hints on Using the Subroutines

A.1 Sample Skeleton Drivers

Different sparse methods have been described in Chapters 4–8 for solving linear systems. They differ in storage schemes, ordering strategies, data structures, and/or numerical subroutines. However, the overall procedure in using these methods is the same. Four distinct phases can be identified:

Step 1 Ordering
Step 2 Data structure set-up
Step 3 Factorization
Step 4 Triangular solution

Subroutines required to perform these steps for each method are included in Chapters 4–8. In Figures A.1.1–A.1.3, three skeleton drivers are provided; they are for the envelope method (Chapter 4), the tree partitioning method (Chapter 6) and the nested dissection method (Chapter 8) respectively. They represent the sequence in which the subroutines should be called in the solution of a given sparse system by the selected scheme. Note that these are just *skeleton* programs; the various arrays are assumed to have been appropriately declared, and no checking for possible errors is performed.

When an ordering subroutine is called, the zero-nonzero pattern of the sparse matrix A is assumed to be in the adjacency structure pair (XADJ, ADJNCY). It is rare that the user has this representation provided for him. Thus, the user must create this structure prior to the execution of the ordering step.

The creation of the adjacency structure is not a trivial task, especially in situations where the (i,j) pairs for which $a_{ij} \neq 0$ become available in random order. We shall not concern ourselves with this problem here. Exercises 3.3.1 and 3.3.6 in Chapter 3 indicate how part of this problem can be solved. The package SPARSPAK to be discussed in Appendix B provides ways to generate the adjacency structure pair in the (XADJ, ADJNCY) format.

In the skeleton drivers, there are two subroutines that have not been discussed before. The subroutine INVRSE, called after the

---

Figure A.1.1   Skeleton driver for the envelope method.

```
C       ----------------------------------------------------
C          CREATE XADJ AND ADJNCY
C          CORRESPONDING TO AX = B
C       ----------------------------------------------------
              .
              .
              .
           CALL  GENRCM(N,XADJ,ADJNCY,PERM,MASK,XLS)
           CALL  INVRSE(N,PERM,INVP)
           CALL  FNENV(N,XADJ,ADJNCY,PERM,INVP,XENV,ENVSZE,BANDW)
C
C       ----------------------------------------------------
C          PUT NUMERICAL VALUES IN DIAG, ENV AND RHS
C       ----------------------------------------------------
              .
              .
              .
           CALL  ESFCT(N,XENV,ENV,DIAG,IERR)
           CALL  ELSLV(N,XENV,ENV,DIAG,RHS,IERR)
           CALL  EUSLV(N,XENV,ENV,DIAG,RHS,IERR)
C
C       ----------------------------------------------------
C          PERMUTED SOLUTION IS NOW IN THE ARRAY RHS
C          RESTORE IT TO THE ORIGINAL ORDERING
C       ----------------------------------------------------
C
           CALL  PERMRV(N,RHS,PERM)
```

---

ordering PERM has been determined, is used to compute the inverse INVP of the ordering (or permutation) found. The vector INVP is required in setting up data structures for the solution scheme, and in putting numerical values into them.

After the numerical subroutines for factorization and triangular solutions have been executed, the solution $\tilde{x}$ obtained is that for the *permuted* system

$$(PAP^T)\tilde{x} = Pb \ .$$

The subroutine PERMRV is used to permute the vector $\tilde{x}$ back to the original given order.

After the data structure for the triangular factor has been successfully set up, the user must input the actual numerical values for the matrix A and the right hand side b. To insert values into the data structure, the user must understand the storage scheme in detail. In the next section, a sample subroutine is provided for matrix input. For different storage methods, these matrix input subroutines are obviously different. With the sample provided in the next section, the user should be able to write those for the other methods. It should be pointed out that they are all provided in SPARSPAK (see Appendix B).

Figure A.1.2 Skeleton driver for the tree partitioning method.

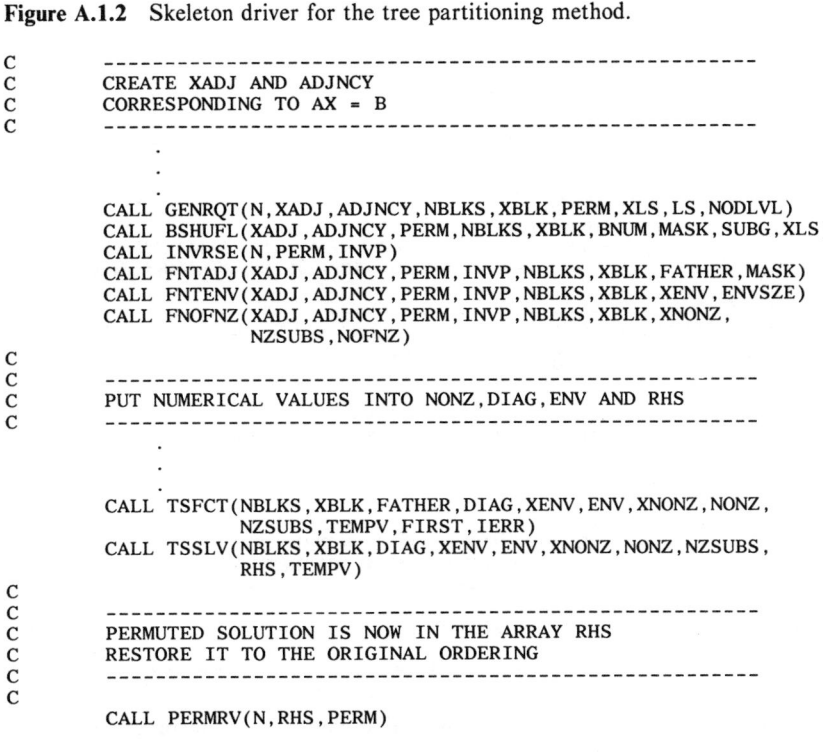

```
C      --------------------------------------------------------
C      CREATE XADJ AND ADJNCY
C      CORRESPONDING TO AX = B
C      --------------------------------------------------------
              .
              .
              .
       CALL  GENRQT(N,XADJ,ADJNCY,NBLKS,XBLK,PERM,XLS,LS,NODLVL)
       CALL  BSHUFL(XADJ,ADJNCY,PERM,NBLKS,XBLK,BNUM,MASK,SUBG,XLS)
       CALL  INVRSE(N,PERM,INVP)
       CALL  FNTADJ(XADJ,ADJNCY,PERM,INVP,NBLKS,XBLK,FATHER,MASK)
       CALL  FNTENV(XADJ,ADJNCY,PERM,INVP,NBLKS,XBLK,XENV,ENVSZE)
       CALL  FNOFNZ(XADJ,ADJNCY,PERM,INVP,NBLKS,XBLK,XNONZ,
                    NZSUBS,NOFNZ)
C
C      --------------------------------------------------------
C      PUT NUMERICAL VALUES INTO NONZ,DIAG,ENV AND RHS
C      --------------------------------------------------------
              .
              .
              .
       CALL  TSFCT(NBLKS,XBLK,FATHER,DIAG,XENV,ENV,XNONZ,NONZ,
                   NZSUBS,TEMPV,FIRST,IERR)
       CALL  TSSLV(NBLKS,XBLK,DIAG,XENV,ENV,XNONZ,NONZ,NZSUBS,
                   RHS,TEMPV)
C
C      --------------------------------------------------------
C      PERMUTED SOLUTION IS NOW IN THE ARRAY RHS
C      RESTORE IT TO THE ORIGINAL ORDERING
C      --------------------------------------------------------
       CALL  PERMRV(N,RHS,PERM)
```

A.2 A Sample Numerical Value Input Subroutine

Before the numerical subroutines for a sparse method are called, it is necessary to put the numerical values into the data structure. Here, we provide a sample subroutine for the tree-partitioning method, whereby the numerical values of an entry a_{ij} can be placed into the structure.

Recall from Chapter 6 that there are three vectors in the storage scheme containing numerical values. The vector DIAG contains the diagonal elements of the matrix. The entries within the envelope of the diagonal blocks are stored in ENV, while the vector NONZ keeps all the nonzero off-diagonal entries. For a given nonzero entry a_{ij}, the subroutine ADAIJ updates one of the three storage vectors DIAG, ENV, or NONZ, depending on where the value a_{ij} resides in the matrix.

The calling statement to the matrix input subroutine is

```
CALL ADAIJ (I,J,VALUE,INVP,DIAG,XENV,ENV,XNONZ,NZSUBS,IERR)
```

where I and J are the subscripts of the original matrix A (that is, unpermuted) and VALUE is the numerical value. This subroutine adds VALUE to the appropriate current value of a_{ij} in storage. This is used instead of an assignment so as to handle situations when the values of

---

Figure A.1.3   Skeleton driver for the nested dissection method.

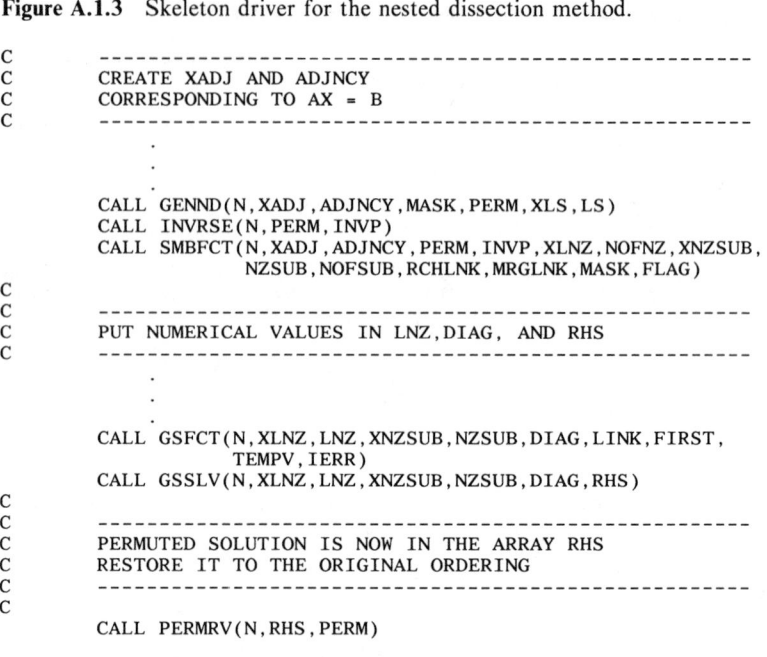

```
C        --------------------------------------------------------
C        CREATE XADJ AND ADJNCY
C        CORRESPONDING TO AX = B
C        --------------------------------------------------------
             .
             .
             .
         CALL  GENND(N,XADJ,ADJNCY,MASK,PERM,XLS,LS)
         CALL  INVRSE(N,PERM,INVP)
         CALL  SMBFCT(N,XADJ,ADJNCY,PERM,INVP,XLNZ,NOFNZ,XNZSUB,
                     NZSUB,NOFSUB,RCHLNK,MRGLNK,MASK,FLAG)
C
C        --------------------------------------------------------
C        PUT NUMERICAL VALUES IN LNZ,DIAG, AND RHS
C        --------------------------------------------------------
             .
             .
             .
         CALL  GSFCT(N,XLNZ,LNZ,XNZSUB,NZSUB,DIAG,LINK,FIRST,
                     TEMPV,IERR)
         CALL  GSSLV(N,XLNZ,LNZ,XNZSUB,NZSUB,DIAG,RHS)
C
C        --------------------------------------------------------
C        PERMUTED SOLUTION IS NOW IN THE ARRAY RHS
C        RESTORE IT TO THE ORIGINAL ORDERING
C        --------------------------------------------------------
C
         CALL  PERMRV(N,RHS,PERM)
```

---

a_{ij} are obtained in an incremental fashion (such as in certain finite element applications).

The subroutine checks to see if the nonzero component lies on the diagonal or within the envelope of the diagonal blocks. If so, the value is added to the appropriate location in DIAG or ENV. Otherwise, the subscript structure (XNONZ, NZSUBS) for off-diagonal block nonzeros is searched and VALUE is then added to the appropriate entry of the vector NONZ.

Since ADAIJ only adds new values to those currently in storage, the space used for *L* must be initialized to zero before numerical values of *A* are supplied. Therefore, the input of values of *A* for the tree-partitioning method would be done as follows:

- Initialize the vectors DIAG, ENV and NONZ to zeros.
- {Repeated calls to ADAIJ}.

The input of values for the right hand vector *b* can be performed in a similar way.

```
C************************************************************
C************************************************************
C********     ADAIJ  .....  ADD ENTRY INTO MATRIX    ********
C************************************************************
C************************************************************
C
C        PURPOSE - THIS ROUTINE ADDS A NUMBER INTO THE (I,J)-TH
C                  POSITION OF A MATRIX STORED USING THE
C                  IMPLICIT BLOCK STORAGE SCHEME.
C
C        INPUT PARAMETERS -
C            (ISUB,JSUB) - SUBSCRIPTS OF THE NUMBER TO BE ADDED
C                     ASSUMPTIONS- ISUB .GE. JSUB.
C            DIAG - ARRAY CONTAINING THE DIAGONAL ELEMENTS
C                     OF THE COEFFICIENT MATRIX.
C            VALUE - VALUE OF THE NUMBER TO BE ADDED.
C            INVP - INVP(I) IS THE NEW POSITION OF THE
C                     VARIABLE WHOSE ORIGINAL NUMBER IS I.
C            (XENV, ENV) - ARRAY PAIR CONTAINING THE ENVELOPE
C                     STRUCTURE OF THE DIAGONAL BLOCKS.
C            (XNONZ, NONZ, NZSUBS) - LEVEL STRUCTURE CONTAINING
C                     THE OFF-BLOCK DIAGONAL PARTS OF THE ROWS OF
C                     THE LOWER TRIANGLE OF THE ORIGINAL MATRIX.
C
C        OUTPUT PARAMETERS -
C            IERR - ERROR CODE....
C                   0 - NO ERRORS DETECTED
C                   5 - NO SPACE IN DATA STRUCTURE FOR NUMBER
C                       WITH SUBSCRIPTS (I,J), I>J.
C
C************************************************************
C
        SUBROUTINE ADAIJ  ( ISUB, JSUB, VALUE, INVP, DIAG,
     1                      XENV, ENV, XNONZ, NONZ, NZSUBS,
     2                      IERR )
C
C************************************************************
C
        REAL DIAG(1), ENV(1), NONZ(1), VALUE
        INTEGER   INVP(1), NZSUBS(1)
        INTEGER   XENV(1), XNONZ(1), KSTOP, KSTRT,
     1            I, IERR, ISUB, ITEMP, J, JSUB, K
C
C************************************************************
        I = INVP(ISUB)
        J = INVP(JSUB)
        IF ( I .EQ. J )  GO TO 400
           IF ( I .GT. J )  GO TO 100
              ITEMP = I
              I     = J
              J     = ITEMP
C
C           THE COMPONENT LIES WITHIN THE DIAGONAL ENVELOPE.
C           -------------------------------------------------
  100      K = XENV(I+1) - I + J
           IF ( K .LT. XENV(I) )  GO TO 200
              ENV(K) = ENV(K) + VALUE
              RETURN
C
C           THE COMPONENT LIES OUTSIDE DIAGONAL BLOCKS.
C           -------------------------------------------------
  200      KSTRT = XNONZ(I)
           KSTOP = XNONZ(I+1) - 1
           IF ( KSTOP .LT. KSTRT )  GO TO 500
C
           DO 300 K = KSTRT, KSTOP
              IF ( NZSUBS(K) .NE. J )  GO TO 300
```

```
                       NONZ(K) = NONZ(K) + VALUE
                       RETURN
        300            CONTINUE
                       GO TO 500
C              ----------------------------------------------------
C              THE COMPONENT LIES ON THE DIAGONAL OF THE MATRIX.
C              ----------------------------------------------------
        400            DIAG(I) = DIAG(I) + VALUE
                       RETURN
C              ------------------------------------
C              SET ERROR FLAG
C              ------------------------------------
        500            IERR = 5
                       RETURN
C
            END
```

A.3 Overlaying Storage in FORTRAN

Consider the skeleton driver in Figure A.1.1 for the envelope method. The ordering subroutine GENRCM generates an ordering PERM based on the adjacency structure (XADJ, ADJNCY). It also uses two working vectors MASK and XLS.

After the input of numerical values into the data structure for the envelope, note that the working vectors MASK and XLS are no longer needed. Moreover, even the adjacency structure (XADJ, ADJNCY) will no longer be used. To conserve storage, these vectors can be overlayed and re-used by the solution subroutines. Similar remarks apply to the other sparse methods.

In this section, we show how overlaying can be done in FORTRAN. The general technique involves the use of a large working storage array in the driver program. Storage management can be handled by this driver through the use of pointers into the main storage vector.

As an illustration, suppose that there are two subroutines SUB1 and SUB2:

```
SUBROUTINE SUB1 (X,Y,Z)
SUBROUTINE SUB2 (X,Y,U,V) .
```

The subroutine SUB1 requires two integer arrays X and Y of sizes 100 and 500 respectively, and a working integer array Z of size 400. On the other hand, SUB2 requires four vectors: the X and Y output vectors from SUB1 and two additional arrays U and V of sizes 40 and 200 respectively.

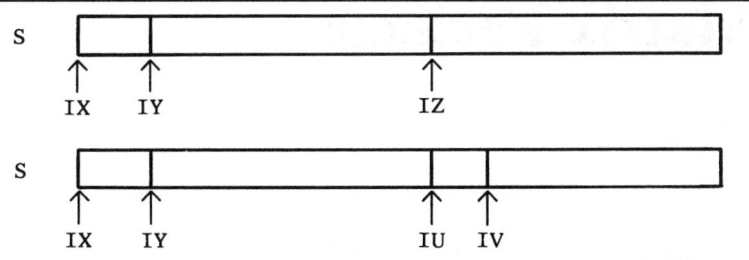

The following skeleton driver makes use of a main storage vector S(1000) and calls the subroutines SUB1 and SUB2 in succession. It manages the storage using pointers into the array S.

```
INTEGER S(1000)
   ⋮
IX = 1
IY = IX + 100
IZ = IY + 500
CALL SUB1 (S(IX),S(IY),S(IZ))
   ⋮
IU = IY + 500
IV = IU + 40
CALL SUB2 (S(IX),S(IY),S(IU),S(IV))
   ⋮
```

In this way, the storage used by the working vector Z can be overlayed by U and V.

The same overlay technique can be used in invoking the sequence of subroutines for a sparse solution method. The package SPARSPAK (Appendix B) uses essentially this same technique in a system of *user interface subroutines* which relieve the user of all the storage management tasks associated with using the subroutines in this book.

B/ SPARSPAK: A Sparse Matrix Package

B.1 Motivation

The skeleton programs in Appendix A illustrate several important characteristics of sparse matrix programs and subroutines. First, the unconventional data structures employed to store sparse matrices result in subroutines which have distressingly long parameter lists, most of which have little or no meaning to the user unless he or she understands and remembers the details of the data structure being employed. Second, the computation consists of several distinct phases, with numerous opportunities to overlay (re-use) storage. In order to use the subroutines effectively, the user must determine which arrays used in one module must be preserved as input to the next, and which ones are no longer required and can therefore be re-used. Third, in all cases, the amount of storage required for the solution phase is unknown until at least part of the computation has been performed. Usually we do not know the maximum storage requirement until the allocation subroutine (e.g., FNENV) has been executed. In some cases, the storage requirement for the successful execution of the allocation subroutine *itself* is not predictable (e.g., SMBFCT). Thus, often the computation must be suspended part way through because of insufficient storage, and if the user wishes to avoid repeating the successfully completed part, then he or she must be aware of all the information required to restart the computation.

These observations, along with our experience in using sparse matrix software, have prompted us to design and implement a *user interface* for the subroutines described in this book. This interface is simply a layer of subroutines between the user, who presumably has a sparse system of equations to solve, and subroutines which implement the various methods described in this book, as depicted in Figure B.1.1. The interface, along with the subroutines it serves, forms a package which has been given the name SPARSPAK (George 1979a). In addition to the subroutines from Chapters 4–8 and the interface subroutines, SPARSPAK also contains a number of utility subroutines for printing error messages, pictures of the structure of sparse matrices, etc.

Figure B.1.1   Schematic of the components of SPARSPAK.

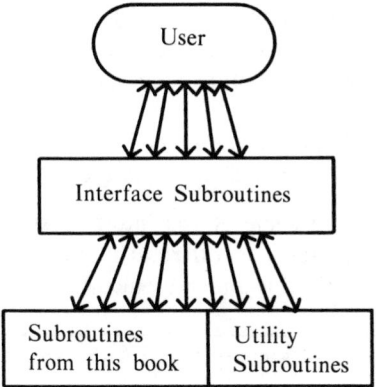

The interface provides a number of services. First, it relieves the user of all responsibility for the allocation of array storage. All storage is allocated by the interface from a user-supplied one-dimensional array, using a technique similar to that described in Section A.3. The interface also imposes sequencing control so that interface subroutines are called in the correct order. In addition, it provides a convenient means by which computation can be suspended and later restarted. Finally, it has comprehensive error diagnostics.

Our objective in subsequent sections is to give a brief survey of the various features of SPARSPAK, rather than to provide a detailed user guide. A comprehensive user guide and installation instructions are provided with the package. For information, the interested reader should write the authors.

B.2 Basic Structure of SPARSPAK

For all the methods described in Chapters 4 through 8, the user and SPARSPAK interact to solve the problem $Ax = b$ through the following basic steps.

Step 1   (Structure Input) The user supplies the nonzero structure of A to the package by calling the appropriate interface subroutines.

Step 2   (Order and Allocate) The execution by the user program of a single subroutine call instructs the package to find an ordering and set up the data structure for L.

Step 3   (Matrix Input) The user supplies the numerical values for A by calling appropriate interface subroutines.

Step 4   (Factor A) A single subroutine call tells SPARSPAK to factor A into LL^T.

Step 5   (Right Hand Side Input) The user supplies the numerical values for b by calling appropriate interface subroutines. (This step can be done before Step 4, and/or intermixed with Step 3.)

Step 6   (Solution) A single subroutine call instructs SPARSPAK to compute x, using L from Step 4 and the b supplied in Step 5.

A list of the names of some of the interface subroutines, along with their argument lists and general roles is given in Figure B.2.1. Details are provided later in this and subsequent sections.

Figure B.2.1   List of names of some of the SPARSPAK interface subroutines.

```
SPRSPK                             } Initialization

IJBEGN
INIJ(I, J, S)
INROW(I, NR, IR, S)                } Structure input
INIJIJ(NIR, II, JJ, S)                (Step 1)
INCLQ(NCLQ, CLQ, S)
IJEND(S)

ORDRxi(S)                          } Ordering and Allocation
                                       (Step 2. See Figure B.3.1
                                       for meanings of x and i.)

INAIJi(I, J, VALUE, S)
INROWi(I, NIR, IR, VALUES, S)      } Matrix Input
INMATi(NIJ, II, JJ, VALUES, S)         (Step 3)
INBI(I, VALUE, S)
INBIBI(NI, II, VALUES, S)          } Right hand side input
INRHS(RHS, S)                          (Step 5)

SOLVEi(S)                          } Factorization and Solution
                                       (Steps 4 and 6)
```

B.3   User Mainline Program and an Example

SPARSPAK allocates all its storage from a single one dimensional real array which for purposes of discussion we will denote by s. In addition, the user must provide its size MAXS, which is transmitted to the package via a common block /SPKUSR/, which has four variables:

COMMON /SPKUSR/ MSGLVL, IERR, MAXS, NEQNS

Here MSGLVL is the message level indicator which is used to control the amount of information printed by the package. The second variable IERR is an error code, which the user can examine in the mainline program for possible errors detected by the package. The variable NEQNS is the number of equations, set by the package.

The following program illustrates how one might use the envelope method of Chapter 4 to solve a system of equations, using SPARSPAK. The problem solved is a 10 by 10 symmetric tridiagonal system $Ax = b$ where the diagonal elements of A are all 4, the superdiagonal and subdiagonal elements are all -1, and the entries in the right hand side vector b are all ones.

The digit i and letter x in some of the interface subroutine names specify which method is to be used to solve the problem. We should note here that SPARSPAK handles both symmetric and unsymmetric A, but assumes that the *structure* of A is symmetric, and that no pivoting is required for numerical stability. (See Exercise 4.5.1.) The methods available are as indicated in Figure B.3.1.

Figure B.3.1   Choices of methods available in SPARSPAK.

ORDRxi		Description		Ref.
x	i			
A	1	Reverse Cuthill-McKee ordering;	symmetric A	Ch. 4
A	2	Reverse Cuthill-McKee ordering;	unsymmetric A	Ch. 4
A	3	One-way Dissection ordering;	symmetric A	Ch. 7
A	4	One-way Dissection ordering;	unsymmetric A	Ch. 7
B	3	Refined quotient tree ordering;	symmetric A	Ch. 6
B	4	Refined quotient tree ordering;	unsymmetric A	Ch. 6
A	5	Nested Dissection ordering;	symmetric A	Ch. 8
A	6	Nested Dissection ordering;	unsymmetric A	Ch. 8
B	5	Minimum Degree ordering;	symmetric A	Ch. 5
B	6	Minimum Degree ordering;	unsymmetric A	Ch. 5

```
C       SAMPLE PROGRAM ILLUSTRATING THE USE OF SPARSPAK
C       ------------------------------------------------------------
C
        COMMON /SPKUSR/ MSGLVL, IERR, MAXS, NEQNS
        REAL S(250)
C
        CALL SPRSPK
        MAXS = 250
C       ------------------------------------------------------------
C       INPUT THE MATRIX STRUCTURE. THE DIAGONAL IS ALWAYS
C       ASSUMED TO BE NONZERO, AND SINCE THE MATRIX IS SYM-
C       METRIC, ONLY THE  SUBDIAGONAL POSITIONS ARE INPUT.
C       ------------------------------------------------------------
        CALL IJBEGN
        DO 100 I = 2, 10
```

```
          CALL INIJ( I, I-1, S )
  100   CONTINUE
          CALL IJEND( S )
C         ----------------------------------------------------
C         FIND THE ORDERING AND ALLOCATE STORAGE ....
C         ----------------------------------------------------
          CALL ORDRA1( S )
C         ----------------------------------------------------
C         INPUT THE NUMERICAL VALUES. (LOWER TRIANGLE ONLY.)
C         ----------------------------------------------------
          DO 200 I = 1, 10
            IF ( I .GT.1 ) CALL INAIJ1( I, I-1, -1.0, S )
            CALL INAIJ1( I, I, 4.0, S )
            CALL  INBI( I, 1.0,  S )
  200   CONTINUE
C         ----------------------------------------------------
C         SOLVE THE SYSTEM. SINCE BOTH THE MATRIX AND RIGHT HAND
C         SIDE HAVE BEEN INPUT, BOTH THE FACTORIZATION AND THE
C         TRIANGULAR SOLUTION OCCUR.
C         ----------------------------------------------------
          CALL SOLVE1( S )
C         ----------------------------------------------------
C         PRINT THE SOLUTION, FOUND IN THE FIRST 10   POSITIONS OF
C         THE WORKING STORAGE ARRAY S.
C         ----------------------------------------------------
          WRITE(6, 11) (S(I), I = 1, 10)
   11   FORMAT(/ 10H SOLUTION  ,/, (5F10.6))
C         ----------------------------------------------------
C         PRINT SOME STATISTICS GATHERED BY THE PACKAGE.
C         ----------------------------------------------------
          CALL PSTATS
          STOP
          END
```

The subroutine SPRSPK must be called before any part of the package is used. Its role is to initialize some system parameters (e.g., the logical unit number for the printer), to set default values for options (e.g., the message level indicator), and to perform some installation dependent functions (e.g., initializing the timing subroutine). It needs only to be called once in the user program. Note that the only variable in the common block /SPKUSR/ that must be explicitly assigned a value by the user is MAXS.

SPARSPAK contains an interface subroutine called PSTATS which the user can call to obtain storage requirements, execution times, operation counts etc. for the solution of the problem.

It is assumed that the subroutines which comprise the SPARSPAK package have been compiled into a *library*, and that the user can reference them from a Fortran program just as the standard Fortran library subroutines, such as SIN, COS, etc., are referenced. Normally, a user will use only a small fraction of the subroutines provided in SPARSPAK.

B.4   Brief Description of the Main Interface Subroutines

B.4.1   Modules for Input of the Matrix Structure

SPARSPAK must know the matrix structure before it can determine an appropriate ordering for the system. SPARSPAK contains a group of subroutines which provide a variety of ways through which the user can inform the package where the nonzero entries are; that is, those subscripts (i, j) for which $a_{ij} \neq 0$. Before any of these input subroutines is called, the user must execute an initialization subroutine called IJBEGN, which tells the package that a matrix problem with a new structure is to be solved.

a)   Input of a nonzero location

To tell SPARSPAK that the matrix component a_{ij} is nonzero, the user simply executes the statement

```
CALL INIJ(I, J, S)
```

where I and J are the subscripts of the nonzero, and S is the working storage array declared by the user for use by the package.

b)   Input of the structure of a row, or part of a row.

When the structure of a row or part of a row is available, it may be more convenient to use the subroutine INROW. The statement to use is

```
CALL INROW(I, NIR, IR, S)
```

where I denotes the subscript of the row under consideration, IR is an array containing the column subscripts of some or all of the nonzeros in the I-th row, NIR is the number of subscripts in IR, and S is the user-declared working storage array. The subscripts in the array IR can be in arbitrary order, and the rows can be input in any order.

c)   Input of a submatrix structure

SPARSPAK allows the user to input the structure of a submatrix. The calling statement is

```
CALL INIJIJ(NIJ, II, JJ, S) ,
```

where NIJ is the number of input subscript pairs and II, JJ are the arrays containing the subscripts.

d)   Input of a full submatrix structure

The structure of an entire matrix is completely specified if all the full submatrices are given. In applications where they are readily available, the subroutine INCLQ is useful. Its calling sequence is

```
CALL INCLQ(NCLQ, CLQ, S) ,
```

where NCLQ is the size of the submatrix and CLQ is an array containing the subscripts of the submatrix.

Thus, to inform the package that the submatrix corresponding to subscripts 1, 3, 5 and 6 is full, we execute

```
CLQ(1) = 1
CLQ(2) = 3
CLQ(3) = 5
CLQ(4) = 6
CALL INCLQ(4, CLQ, S) .
```

The type of structure input subroutine to use depends on how the user obtains the matrix structure. Anyway, one can select those that best suit the application. The package allows *mixed use* of the subroutines in inputting a matrix structure. SPARSPAK automatically removes duplications so the user does not have to worry about inputting duplicate subscript pairs.

When all pairs have been input, using one or a combination of the input subroutines, the user is required to tell the package explicitly so by calling the subroutine IJEND. The calling statement is

```
CALL IJEND(S)
```

and its purpose is to transform the data from the format used during the recording phase to the standard (XADJ, ADJNCY) format used by all the subroutines in the book. The user does not have to be concerned with this input representation or the transformation process.

B.4.2   Modules for Ordering and Storage Allocation

With an internal representation of the nonzero structure of the matrix A, SPARSPAK is now ready to reorder the matrix problem. The user initiates this by calling an ordering subroutine, whose name has the form ORDRxi. Here i is a numerical digit between 1 and 6 that signifies the storage method, and the character x denotes the ordering strategy as summarized in Figure B.3.1. The subroutine ORDRxi determines the ordering and then sets up the data structure for the reordered matrix problem. The package is now ready for numerical inputs.

B.4.3   Modules for Inputting Numerical Values of A and b

The modules in this group are similar to those for inputting the matrix structure. They provide a means of transmitting the actual numerical values of the matrix problem to the package. Since the data structures for different storage methods are different, the package must have a different matrix input subroutine for each method. SPARSPAK uses the same set of subroutine names for all the methods (except for the last digit which distinguishes the method), and the parameter lists for all the methods are the same.

There are three ways of passing the numerical values to the package. In all of them, subscripts passed to the package always refer to those of the *original* given problem. The user need not be concerned about the various permutations to the problem which may have occurred during the ordering step.

a) Input of a single nonzero component

The subroutine INAIJ*i* is provided for this purpose and its calling sequence is

```
CALL INAIJi(I, J, VALUE, S)
```

where I and J are the subscripts, and VALUE is the numerical value. The subroutine INAIJ*i* adds the quantity VALUE to the appropriate current value in storage, rather than making an assignment. This is helpful in situations (e.g., in some finite element applications) where the numerical values are obtained in an incremental fashion. For example, the execution of

```
⋮
INAIJ2(3, 4, 9.5, S)
INAIJ2(3, 4, -4.0, S)
⋮
```

effectively assigns 5.5 to the matrix component a_{34}.

b) Input of a row of nonzeros

The subroutine INROW*i* can be used to input the numerical values of a row or part of a row in the matrix. Its calling sequence is similar to that of INROW, described in Section B.4.1.

```
CALL INROWi(I, NIR, IR, VALUES, S) .
```

Here the additional variable VALUES is an array containing the numerical values of the row. Again, the numerical values are added to the current values in storage.

c) Input of a submatrix

The subroutine for the input of a submatrix is called INMAT*i*. Its parameter list corresponds to that of INIJIJ with the additional parameter VALUES that stores the numerical quantities:

```
CALL INMATi(NIJ, II, JJ, VALUES, S) .
```

Again, the VALUES are added to those held by the package.

Mixed use of the subroutines INAIJ*i*, INROW*i*, and INMAT*i* is permitted. Thus, the user is free to use whatever subroutines are most convenient.

The same convenience is provided in the input of numerical values for the right hand side vector. The package includes the subroutine INBI which inputs an entry to the right hand vector.

```
CALL INBI(I, VALUE, S)
```

Here I is the subscript and VALUE is the numerical value.

The subroutine INBIBI can be used to input a subvector, and its calling sequence is

    CALL INBIBI(NI, II, VALUES, S)

where II and VALUES are vectors containing the subscripts and numerical values respectively. In both subroutines, incremental calculations of the numerical values are performed.

In some situations where the entire right hand vector is available, the user can use the subroutine INRHS which transmits the whole vector to the package. It has the form

    CALL INRHS(RHS, S)

where RHS is the vector containing the numerical values.

In all three subroutines, the numbers provided are added to those currently held by the package, and the use of the subroutines can be intermixed. The storage used for the right hand side by the package is initialized to zero the first time any of them is executed.

B.4.4   Modules for Factorization and Solution

The numerical computation of the solution vector is initiated by the Fortran statement

    CALL SOLVEi(S)

where S is the working storage array for the package. Again, the last digit *i* is used to distinguish between solvers for different storage methods.

Internally, the subroutine SOLVE*i* consists of *both* the factorization and forward/backward solution steps. If the factorization has been performed in a previous call to SOLVE*i*, the package will automatically skip the factorization step, and perform the solution step directly. The solution vector is returned in the first NEQNS locations of the storage vector S. If SOLVE*i* is called before any right hand side values are input, only the factorization will be performed. The solution returned will be all zeros.

B.5   Save and Restart Facilities

SPARSPAK provides two subroutines called SAVE and RESTRT which allow the user to stop the calculation at some point, save the results on an external sequential file, and then restart the calculation at exactly that point some time later. To save the results of the computation done thus far, the user executes the statement

```
CALL SAVE(K, S)
```

where K is the Fortran logical unit on which the results are to be written, along with other information needed to restart the computation. If execution is then terminated, the state of the computation can be re-established by executing the statement

```
CALL RESTRT(K, S) .
```

When an error is detected, so that the computation cannot proceed, a positive code is assigned to IERR. The user can simply check the value of IERR to see if the execution of the module has been successful. This error flag can be used in conjunction with the save/restart feature to retain the results of successfully completed parts of the computation, as shown by the program fragment below.

```
        ⋮
        CALL ORDRA1(S)
        IF (IERR.EQ.0) GO TO 100
        CALL SAVE(3, S)
        STOP
100     CONTINUE
        ⋮
```

Another potential use of the SAVE and RESTRT modules is to make the working storage array S available to the user in the middle of a sparse matrix computation. After SAVE has been executed, the working storage array S can be used by some other computation.

B.6   Solving Many Problems Having the Same Structure or the Same Coefficient Matrix A

In certain applications, many problems which have the same sparsity structure, but different numerical values, must be solved. This situation can be accommodated perfectly well by the package. The control sequence is depicted by the flowchart in Figure B.6.1. When the numerical input subroutines (INAIJi, INBI, etc.) are first called after SOLVEi has been called, this is detected by the package, and the computer storage used for A and b is initialized to zero.

Note that if such problems must be solved over an extended time period (i.e., in different runs), the user can execute SAVE after executing ORDRxi and thus avoid input of the structure of A and the execution of ORDRxi in subsequent equation solutions.

In other applications, numerous problems which differ only in their right hand sides must be solved. In this case, we only want to factor A once, and use the factors repeatedly in the calculation of x for each different b. Again, the package can handle this in a straightforward manner, as illustrated by the flowcharts in Figure B.6.2.

Figure B.6.1   Flowchart for using SPARSPAK to solve numerous problems having the same structure.

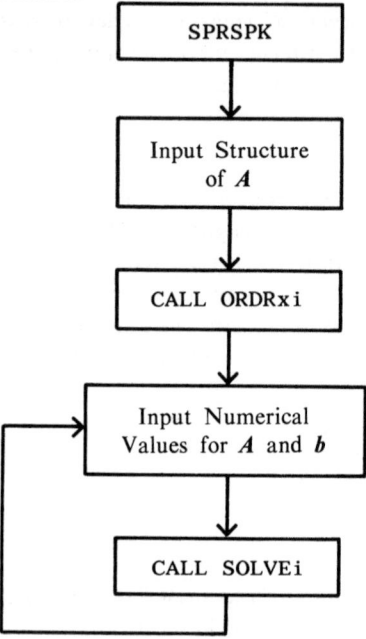

When SPARSPAK is used as indicated by flowchart (1) in Figure B.6.2, it detects that no right hand side has been provided during the first execution of SOLVEi, and only the factorization is performed. In subsequent calls to SOLVEi, the package detects that the factorization has already been performed, and that part of the SOLVEi module is by-passed. In flow-chart (2) of Figure B.6.2, both factorization and solution is performed during the first call to SOLVEi, with only the solve part performed in subsequent executions of SOLVEi.

Note that SAVE can be used after SOLVEi has been executed, if the user wants to save the factorization for use in some future calculation.

Figure B.6.2 Flowcharts for using SPARSPAK to solve numerous problems having the same coefficient matrix but different right hand sides.

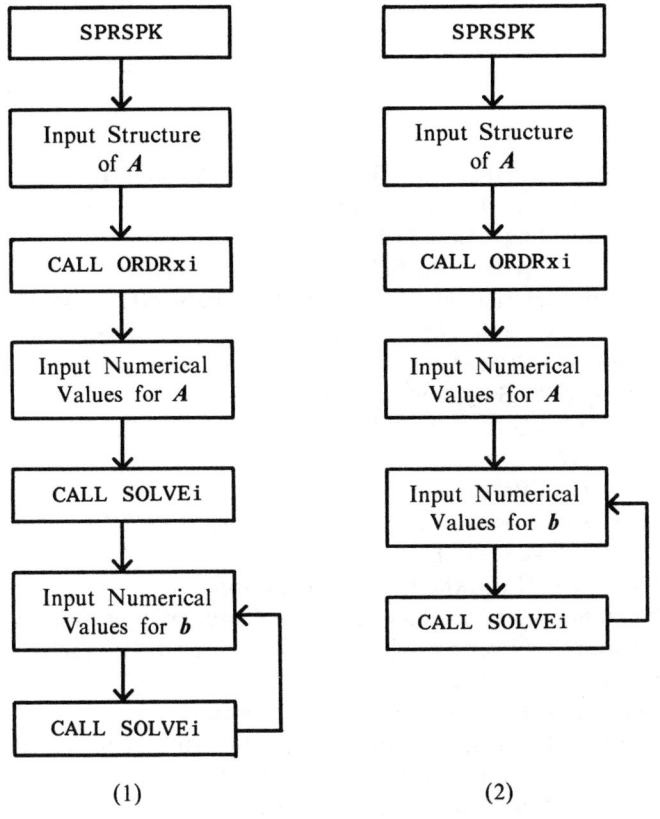

(1)   (2)

B.7 Output From the Package

As noted earlier, the user supplies a one-dimensional real array S, from which all array storage is allocated. In particular, the interface allocates the first NEQNS storage locations in S for the solution vector of the linear system. After all the interface modules for a particular method have been successfully executed, the user can retrieve the solution from these NEQNS locations.

In addition to the solution x, the package may print other information about the computation, depending upon the value of MSGLVL, whether or not errors occur, and whether or not the module PSTATS is called.

References

[Aho 1974] **A. V. Aho, J. E. Hopcroft**, and **J. D. Ullman**,
The Design and Analysis of Computer Algorithms, Addison-Wesley, Reading, Mass. (1974).

[Arany 1972] **I. Arany, W. F. Smyth**, and **L. Szoda**,
"An improved method for reducing the bandwidth of sparse symmetric matrices," *Information Processing 71:* Proc. of IFIP Congress, North-Holland, Amsterdam (1972).

[Baty 1967] **J. P. Baty**, and **K. L. Stewart**,
"Dissection of Structures," *J. Struct. Div.* ASCE 5 (1967), pp. 217–232.

[Berge 1962] **C. Berge**,
The Theory of Graphs and Its Applications, John Wiley & Sons Inc., New York (1962).

[Birkhoff 1973] **G. Birkhoff**, and **Alan George**,
"Elimination by nested dissection," in *Complexity of Sequential and Parallel Numerical Algorithms*, edited by J. F. Traub, Academic Press (1973).

[Bunch 1974] **J. R. Bunch**, and **J. E. Hopcroft**,
"Triangular factorization and inversion by fast matrix multiplication," *Math. Comp. 28* (1974), pp. 231–236.

[Bunch 1976] **J. R. Bunch**, and **D. J. Rose**, ed.,
Sparse Matrix Computations, Academic Press (1976).

[Calahan 1975] **D. A. Calahan**,
"Complexity of vectorized solution of two dimensional finite element grids," *Tech. Rept. 91*, Systems Engrg. Lab., University of Michigan (1975).

[Calahan 1976] **D. A. Calahan, W. N. Joy**, and **D. A. Orbits**,
"Preliminary report on results of matrix benchmarks on a vector processor," *Tech. Rept. 94*, Systems Engrg. Lab., University of Michigan (1976).

[Chan 1979] **W. M. Chan**, and **Alan George**,
"A linear time implementation of the Reverse Cuthill McKee algorithm," *B.I.T. 20* (1980), pp. 8–14.

[Chang 1969] **A. Chang**,
"Application of sparse matrix methods in electrical power system

analysis," in *Sparse Matrix Proceedings*, edited by R. A. Willoughby, IBM Research Rept. RA1 #11707, Yorktown Heights, N. Y. (1969).

[Crane 1976] **H. L. Crane Jr., N. E. Gibbs, W.G. Poole Jr.,** and **P. K. Stockmeyer,** "Algorithm 508: Matrix bandwidth and profile reduction," *ACM Trans. on Math. Software 2* (1976), pp. 375–377.

[Cuthill 1969] **E. Cuthill,** and **J. McKee,** "Reducing the bandwidth of sparse symmetric matrices," *Proc. 24th Nat. Conf. Assoc. Comput. Mach.*, ACM Publ. (1969), pp. 157–172.

[Cuthill 1972] **Elizabeth Cuthill,** "Several strategies for reducing the bandwidth of matrices" in *Sparse Matrices and their Applications*, edited by D. J. Rose and R. A. Willoughby, Plenum Press, New York (1972), pp. 157–166.

[Duff 1974] **I. S. Duff,** and **J. K. Reid,** "A comparison of sparsity orderings for obtaining a pivotal sequence in Gaussian elimination," *J. Inst. Math. Applics. 14* (1974), pp. 281–291.

[Duff 1976a] **I. S. Duff, A. M. Erisman,** and **J. K. Reid,** "On George's nested dissection method," *SIAM J. Numer. Anal. 13* (1976), pp. 686–695.

[Duff 1976b] **I. S. Duff,** and **J. K. Reid,** "A comparison of some methods for the solution of sparse overdetermined systems of linear equations," *J. Inst. Math. Applics. 17* (1976), pp. 267–280.

[Duff 1977] **I. S. Duff,** "A survey of sparse matrix research," *Proc. IEEE 65* (1977), pp. 500–535.

[Duff 1979] **I. S. Duff,** and **G. W. Stewart,** ed., *Sparse Matrix Proceedings 1978*, SIAM Publications, Philadelphia (1979).

[Eisenstat 1974] **S. C. Eisenstat,** and **A. H. Sherman,** "Subroutines for envelope solution of sparse linear systems," *Research Rept. 35*, Dept. of Computer Science, Yale University (1974).

[Eisenstat 1976] **S. C. Eisenstat, M. H. Schultz,** and **A. H. Sherman,** "Applications of an element model for Gaussian elimination," in *Sparse Matrix Computations*, edited by J. R. Bunch and D. J. Rose, Academic Press, New York (1976), pp. 85–96.

[Eisenstat 1981] **S. C. Eisenstat, M. C. Gursky, M. H. Schultz,** and **A. H. Sherman,** "Yale Sparse Matrix Package I. The Symmetric Codes," *ACM Trans. on Math. Software* (to appear).

[Erisman 1972] **A. M. Erisman,** and **G. E. Spies,** "Exploiting problem characteristics in the sparse matrix approach to frequency domain analysis," *IEEE Trans. on Circuit Theory*, CT–19 (1972), pp. 260–269.

[Felippa 1970] **C. Felippa**, and **R. W. Clough**,
"The finite element method in solid mechanics," in *Numerical Solution of Field Problems in Continuum Mechanics*, edited by G. Birkhoff and R. S. Varga, SIAM-AMS Proc., Amer. Math. Soc. (1970), pp. 210–252.

[Felippa 1975] **C. A. Felippa**,
"Solution of linear equations with skyline-stored symmetric matrix," *Computers and Structures 5* (1975), pp. 13–29.

[Forsythe 1967] **George E. Forsythe**, and **Cleve B. Moler**,
Computer Solution of Linear Algebraic Systems, Prentice-Hall Inc., Englewood Cliffs, N. J. (1967).

[George 1971] **Alan George**,
"Computer implementation of the finite element method," *Tech. Rept.* STAN–CS–208, Stanford University (1971).

[George 1973] **Alan George**,
"Nested dissection of a regular finite element mesh," *SIAM J. Numer Anal. 10* (1973), pp. 345–363.

[George 1974] **Alan George**,
"On block elimination for sparse linear systems," *SIAM J. Numer. Anal. 11* (1974), pp. 585–603.

[George 1975] **Alan George**, and **Joseph W-H Liu**,
"A note on fill for sparse matrices," *SIAM J. Numer. Anal. 12* (1975), pp. 452–455.

[George 1977] **Alan George**,
"Numerical experiments using dissection methods to solve n by n grid problems," *SIAM J. Numer. Anal. 14* (1977), pp. 161–179.

[George 1978a] **Alan George**, and **Joseph W-H Liu**,
"User guide for SPARSPAK: Waterloo Sparse Linear Equations Package," *Rept. CS–78–30*, Dept. of Computer Science, University of Waterloo (1978).

[George 1978b] **Alan George**,
"An automatic one-way dissection algorithm for irregular finite element problems," *Proc. 1977 Dundee Conf. on Numerical Analysis*, Lecture Notes No. 630, Springer Verlag (1978), pp. 76–89.

[George 1978c] **Alan George**, and **Joseph W-H Liu**,
"Algorithms for matrix partitioning and the numerical solution of finite element systems," *SIAM J. Numer. Anal. 15* (1978), pp. 297–327.

[George 1978d] **Alan George**, and **Joseph W-H Liu**,
"An automatic nested dissection algorithm for irregular finite element problems," *SIAM J. Numer. Anal. 15* (1978), pp. 1053–1069.

[George 1978e] **Alan George, W. G. Poole Jr.,** and **Robert G. Voigt,**
"Incomplete nested dissection for solving n by n grid problems,"
SIAM J. Numer. Anal. 15 (1978), pp. 662–673.

[George 1978f] **Alan George,** and **David R. McIntyre,**
"On the application of the minimum degree algorithm to finite element systems," *SIAM J. Numer. Anal. 15* (1978), pp. 90–112.

[George 1978g] **Alan George, W. G. Poole Jr.,** and **R. G. Voigt,**
"Analysis of dissection algorithms for vector computers," *J. Comp. and Maths. with Applics. 4* (1978), pp. 287–304.

[George 1979a] **Alan George,** and **Joseph W-H Liu,**
"The design of a user interface for a sparse matrix package,"
ACM Trans. on Math. Software 5 (1979), pp. 139–162.

[George 1979b] **Alan George,** and **Joseph W-H Liu,**
"An implementation of a pseudo-peripheral node finder," *ACM Trans. on Math. Software 5* (1979), pp. 286–295.

[George 1980] **Alan George,** and **Joseph W-H Liu,**
"A minimal storage implementation of the minimum degree algorithm," *SIAM J. Numer. Anal. 17* (1980), pp. 282–299.

[Gibbs 1976a] **N. E. Gibbs, W. G. Poole Jr.,** and **P. K. Stockmeyer,**
"A comparison of several bandwidth and profile reduction algorithms," *ACM Trans. on Math. Software 2* (1976), pp. 322–330.

[Gibbs 1976b] **N. E. Gibbs, W. G. Poole Jr.,** and **P. K. Stockmeyer,**
"An algorithm for reducing the bandwidth and profile of a sparse matrix," *SIAM J. Numer. Anal. 13* (1976), pp. 236–250.

[Gibbs 1976c] **N. E. Gibbs,**
"Algorithm 509: A hybrid profile reduction algorithm," *ACM Trans. on Math. Software 2* (1976), pp. 378–387.

[Gustavson 1972] **F. G. Gustavson,**
"Some basic techniques for solving sparse systems of equations," in *Sparse Matrices and their Applications,* edited by D. J. Rose and R. A. Willoughby, Plenum Press, New York (1972), pp. 41–52.

[Hachtel 1972] **G. D. Hachtel,**
"Vector and matrix variability type in sparse matrix algorithms," in *Sparse Matrices and their Applications,* edited by D. J. Rose and R. A. Willoughby, Plenum Press, New York (1972), pp. 53–66.

[Hansen 1978] **Richard J. Hansen,** and **John A. Wisniewski,**
"A single set of software that implements various storage methods for sparse matrices," *Rept. SAND78–0785,* Sandia Laboratories, Albuquerque, N. M. (1978).

[Hoffman 1973] **A. J. Hoffman, M. S. Martin,** and **D. J. Rose,**
"Complexity bounds for regular finite difference and finite element grids," *SIAM J. Numer. Anal. 10* (1973), pp. 364–369.

[Irons 1970] **Bruce M. Irons,**
"A frontal solution program for finite element analysis," *Internat.*

J. Numer. Meth. in Engrg. 2 (1970), pp. 5–32.

[Jennings 1966] **A. Jennings**
"A compact storage scheme for the solution of symmetric linear simultaneous equations," *Comput. J. 9* (1966), pp. 281–285.

[King 1970] **I. P. King**,
"An automatic reordering scheme for simultaneous equations derived from network problems," *Internat. J. Numer. Meth. Engrg. 2* (1970), pp. 523–533.

[Lambiotte 1975] **J. J. Lambiotte**,
"The solution of linear systems of equations on a vector computer," *Ph. D. Diss.*, Dept. Appl. Math. and Comp. Sci., Univ. of Virginia (1975).

[Levy 1971] **R. Levy**,
"Resequencing of the structural stiffness matrix to improve computational efficiency," *JPL Quart. Tech. Rev. 1* (1971), pp. 61–70.

[Lipton 1977] **Richard J. Lipton**, and **Robert E. Tarjan**,
"A separator theorem for planar graphs," *Proc. Conf. on Theoretical Computer Science*, Univ. of Waterloo (Aug. 15–17, 1977), pp. 1–10.

[Lipton 1979] **R. J. Lipton, D. J. Rose**, and **R. E. Tarjan**,
"Generalized nested dissection," *SIAM J. Numer. Anal. 16* (1979), pp. 346–358.

[Liu 1975] **Joseph W-H Liu**, and **Andrew H. Sherman**,
"Comparative analysis of the Cuthill-McKee and the reverse Cuthill-McKee ordering algorithms for sparse matrices," *SIAM J. Numer. Anal. 13* (1975), pp. 198–213.

[Liu 1976] **Joseph W-H Liu**,
"On reducing the profile of sparse symmetric matrices," *Rept. CS–76–07*, Dept. of Computer Science, Univ. of Waterloo (Feb. 1976).

[Martin 1971] **R. S. Martin**, and **J. H. Wilkinson**,
"Symmetric decomposition of positive definite band matrices," in *Handbook for Automatic Computation Vol. II*, edited by J. H. Wilkinson and C. Reinsch, Springer Verlag (1971).

[Markowitz 1957] **H. M. Markowitz**,
"The elimination form of the inverse and its application to linear programming," *Management Science 3* (1957), pp. 255–269.

[Melosh 1969] **R. J. Melosh**, and **R. M. Bamford**,
"Efficient solution of load deflection equations," *J. Struct. Div. ASCE*, Proc. Paper No. 6510 (1969), pp. 661–676.

[Meyer 1973] **C. Meyer**,
"Solution of linear equations — State-of-the-art," *J. Struct. Div. ASCE*, Proc. Paper No. 9861 (1973), pp. 1507–1526.

[Ng 1979] **Esmond Ng**, "A two-level one-way dissection scheme for finite element problems," *Proc. Ninth Manitoba Conference on*

Numerical Math. and Computing, Sept. 26–29, 1979.

[Parter 1961] **S. V. Parter**,
"The use of linear graphs in Gauss elimination," *SIAM Rev. 3*
(1961), pp. 119–130.

[Pooch 1973] **U. W. Pooch**, and **A Neider**,
"A survey of indexing techniques for sparse matrices," *ACM
Computing Surveys 5* (1973), pp. 109–133.

[Rose 1972a] **D. J. Rose**,
"A graph-theoretic study of the numerical solution of sparse
positive definite systems of linear equations," in *Graph Theory and
Computing*, edited by R. C. Read, Academic Press, New York
(1972).

[Rose 1972b] **D. J. Rose**, and **R. A. Willoughby**, ed.,
Sparse Matrices and their Applications, Plenum Press (1972).

[Rose 1974] **D. J. Rose**, and **G. F. Whitten**,
"Automatic nested dissection," *Proc. ACM Nat. Conf.* (1974),
pp. 82–88.

[Rose 1975] **D. J. Rose**, and **R. E. Tarjan**,
"Algorithmic aspects of vertex elimination," *Proc. 7th Annual
Symposium on the Theory of Computing* (1975), pp. 245–254.

[Rose 1976a] **D. J. Rose, R. E. Tarjan**, and **G. S. Lueker**,
"Algorithmic aspects of vertex elimination on graphs," *SIAM J.
on Computing 5* (1976), pp. 266–283.

[Rose 1976b] **D. J. Rose**, and **G. F. Whitten**,
"A recursive analysis of dissection strategies," in *Sparse Matrix
Computation*, edited by J. R. Bunch and D. J. Rose, Academic
Press (1976), pp. 59–84.

[Ryder 1974] **B. G. Ryder**,
"The PFORT Verifier," *Software Practice and Experience 4*
(1974), pp. 359–377.

[Sherman 1975] **A. H. Sherman**,
"On the efficient solution of sparse systems of linear and non-
linear equations," *Rept. No. 46*, Dept. of Computer Science, Yale
University (1975).

[Shier 1976] **D. R. Shier**,
"Inverting sparse matrices by tree partitioning," *J. Res. of the
Nat. Bur. of Standards, V80b, No. 2* (1976).

[Smyth 1974] **W. F. Smyth**, and **W. M. L. Benzi**,
"An algorithm for finding the diameter of a graph," *IFIP
Congress 74*, North-Holland Publ. Co. (1974), pp. 500–503.

[Stewart 1973] **G. W. Stewart**,
Introduction to Matrix Computations, Academic Press, New York
(1973).

[Strang 1973] **Gilbert W. Strang**, and **George J. Fix**,
An Analysis of the Finite Element Method, Prentice-Hall Inc.,

Englewood Cliffs, N. J. (1973).

[Strassen 1969] **V. Strassen,**
"Gaussian elimination is not optimal," *Numer. Math. 13* (1969),
pp. 354–356.

[Tinney 1969] **W. F. Tinney,**
"Comments on using sparsity techniques for power system
problems," in *Sparse Matrix Proceedings,* IBM Research Rept.
RAI 3–12–69 (1969).

[Varga 1962] **R. S. Varga,**
Matrix Iterative Analysis, Prentice-Hall, Englewood Cliffs, N. J.
(1962).

[Von Fuchs 1972] **G. von Fuchs, J. R. Roy,** and **E. Schrem,**
"Hypermatrix solution of large sets of symmetric positive definite
linear equations," *Comp. Meth. in Appl.   Mech. and Engrg. 1*
(1972), pp. 197–216.

[Wilkinson 1961] **J. H. Wilkinson,**
"Error analysis of direct methods of matrix inversion," *J. Assoc.
Comput. Mach. 8* (1961), pp. 281–330.

[Wilson 1974] **E. L. Wilson, K. L. Bathe,** and **W. P. Doherty,**
"Direct solution of large systems of linear equations," *Computers
and Structures 4* (1974), pp. 363–372.

[Young 1971] **D. M. Young,**
Iterative Solution of Large Linear Systems, Academic Press, New
York (1971).

[Zienkiewicz 1977] **O. C. Zienkiewicz,** *The Finite Element Method,*
McGraw Hill, London (1977).

Index